instructor's manual for

process dynamics, modeling, and control

BABATUNDE A. OGUNNAIKE
Chemical Engineering Department
University of Delaware

W. HARMON RAY
Department of Chemical Engineering
University of Wisconsin

New York Oxford
OXFORD UNIVERSITY PRESS
1997

Oxford University Press

Oxford New York
Athens Auckland Bangkok Bogota Bombay Buenos Aires
Calcutta Cape Town Dar es Salaam Delhi Florence Hong Kong
Istanbul Karachi Kuala Lumpur Madras Madrid Melbourne
Mexico City Nairobi Paris Singapore Taipei Tokyo Toronto

and associated companies in
Berlin Ibadan

Published by Oxford University Press, Inc.,
198 Madison Avenue, New York, New York 10016
http://www.oup-usa.org

ISBN: 978-0-19-511937-4

process dynamics, modeling, and control

BABATUNDE A. OGUNNAIKE & W. HARMON RAY

solutions

CHAPTER	Total # of Problems Solved	Left Unsolved	Pages
1	5/5	–	4
2	8/8	–	3
3	7/7	–	7
4	8/8	–	9
5	10/10	–	15
6	17/18	6.14	35
7	8/8	–	18
8	13/13	–	24
9	12/12	–	23
10	9/9	–	16
11	9/9	–	14
12	6/7	12.6	11
13	9/10	13.6	15
14	13/14	14.14	26
15	8/8	–	21
16	12/12	–	35
17	8/8	–	21
18	4/7	18.3, 18.4, 18.5	13
19	9/12	19.7, 19.11, 19.12	14
20	10/10	–	15
21	10/10	–	20
22	9/14	22.5, 22.6, 22.10, 22.12, 22.13	20
23	4/4	–	7
24	6/6	–	9
25	11/15	25.6, 25.7, 25.8, 25.15	22
26	6/12	26.5, 26.6, 26.7, 26.8, 26.10, 26.12	21

Total Number of problems in Textbook: 256
Total Number solved in Instr. Manual: 230

Total Number of handwritten pages: 438

CHAPTER 1

1.1 (a) Output variables : x_D, overhead composition
∴ x_B, bottoms composition

Control variables : L, reflux flow rate
Q, steam flow rate

Disturbance variables : F, Feed rate
T, Feed temperature
P, Steam supply pressure

(b) Feedforward Control.

So long as the effects of feed temperature variations on x_B are well characterized, and well understood, this feedforward control strategy will be effective enough.

<u>Problems</u>: The strategy does not call for a direct measurement of x_B, the variable to be controlled. If the effects of feed temperature variations on x_B are not well characterized, x_B may not attain its desired value under this strategy.

(c) Cascade Control. A flow controller (FC) is employed to ensure that, in the face of steam supply pressure variations, the steam flow rate demanded by the bottoms composition controller is delivered through the valve. (see Figure)

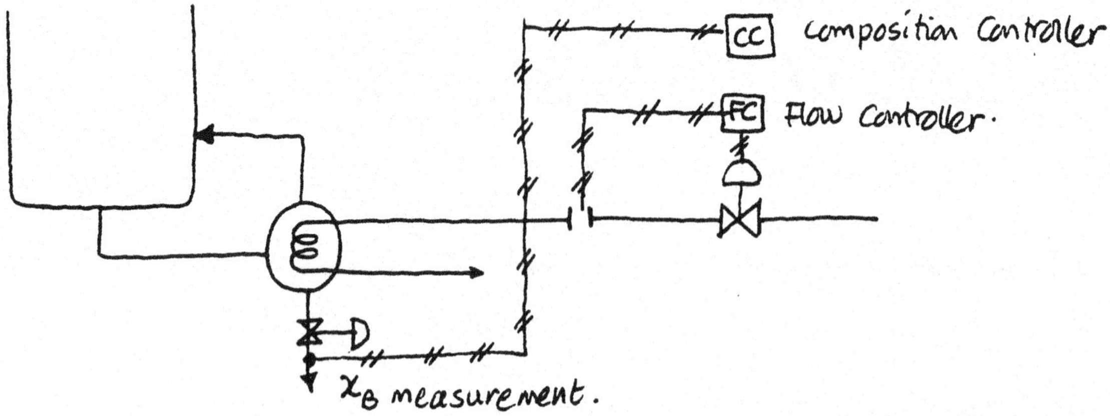

Figure S1.1. Cascade Control Configuration for Bottoms composition control

1.2 (a) From the given steady state energy balance, set $T = T^*$ and solve for Q_F to obtain

$$Q_F = \frac{FC_p}{\lambda_F}\left(T^* - T_i\right). \qquad (1.1)$$

This is the required control law for determining the steady-state value of Q_F required to heat the feed from T_i to T^*.

(b) Potential Problems:

i) Process parameters, λ_F and C_p may not be known precisely. Consequently, the Q_F prescribed by (1.1) will be inaccurate and T will not equal the desired T^* at steady-state;

ii) The process measurements F, (fuel rate) and T_i (feed temperature) may be subject to errors;

iii) This control law only holds for steady-state

conditions. Even if λ_F and C_p are perfectly known, and both F and T_i can be measured perfectly, T will only equal T^* at steady state. There will be a transient period during which $T \neq T^*$

1.3 The appropriate mathematical model under these conditions is:

$$\frac{dy}{dt} = \frac{\xi}{A_c} - \frac{Ky}{A_c} \qquad (1.2)$$

since $u = Ky$. The appropriate parameter values are:

$\xi = (0.075 - 0.05) = 0.025 \ m^3/s$, the disturbance;
$A_c = 1.5 \ m^2$; $K = 0.05$.

The solution to (1.2) with these parameters is

$$y(t) = \tfrac{1}{2}\left[1 - \exp(-t/30)\right] \qquad (1.3)$$

As $t \to \infty$ $y(t) \to 0.5m$. Since this represents the <u>change</u> in the liquid level, then we deduce that after steady-state has been achieved, the liquid level in the tank settles to:

$$h = 3 + 0.5 = 3.5 \, m$$

Since the liquid level is desired to remain at 3m, there is an offset of $3.5 - 3 = 0.5m$.

1.4 Under these new conditions, the mathematical model becomes:

$$\frac{dy}{dt} = \frac{\xi}{A_c} - \frac{K}{A_c}\left\{y + \int_0^t y(\tau)\,d\tau\right\} \qquad (1.4)$$

Differentiating this equation once, recalling that ξ, A_c, K are constants, we obtain:

$$\frac{d^2y}{dt^2} = \frac{-K}{A_c}\frac{dy}{dt} - \frac{K}{A_c}y(t) \qquad (1.5)$$

At steady state all derivatives vanish and (1.5) reduces to

$$y(t) = 0 \qquad (1.6)$$

implying <u>no change</u> in the liquid level at <u>steady-state</u> and hence <u>no</u> steady-state offset.

1.5 Under these conditions, the original mathematical model becomes

$$A_c\frac{dh}{dt} = F_i - (F_i + \delta)$$

or

$$A_c\frac{dh}{dt} = -\delta$$

whose solution is:

$$\boxed{h(t) = h(0) - \left(\frac{\delta}{A_c}\right)t} \qquad (1.7)$$

Thus $h(t)$, the liquid level, will decrease linearly with time at a rate determined by (δ/A_c). In physical terms the interpretation is as follows:

As a result of <u>overestimating</u> the inlet flow rate by δ (due to flow measurement errors) the feedforward scheme is misled into overcompensation; Material is withdrawn faster than it is actually being introduced into the tank; outflow becomes consistently larger than inflow, and the liquid level therefore <u>drops</u> linearly with time.

Chapter 2

2.1 (a) 47075 (b) 52563 (c) 71643

2.2 (a) $D = 7 \times 8^2 + 4 \times 8 + 7 = 448 + 32 + 7 = 487$

(b) $D = 4 \times 64 + 4 \times 8 = 288$

(c) $D = 1 \times 64 = 64$

(d) $D = 5 \times 64 + 5 \times 8 + 6 = 320 + 40 + 6 = 366$

2.3 Total # characters = 100 lines × 25 char/line = 2500 char.
Need one byte (= ½ word) per character;

∴ 1250 words memory for 16 bit machine

2.4 Devices where speed and timing are critical should have the highest priority. Thus the real time clock, disk drives and magtapes are at priority 7. A/D converters which receive data should also receive high priority because data acquisition timing is crucial. However, it can be slower than disks and magtapes. Slow I/O devices such as line printers and terminals are only at level 4 because their access to the CPU is not so crucial in real time.

2.5

For 12 bit, -10 v to +10 v converter, formula is

V(volts) = I/204.7 $\pm$ 0.00244 v / V = I/204.8 for V < 0

$\therefore$

(a) $V = 570/204.7 = 2.78456$ v, $\varepsilon_{relative} = \frac{.00244}{2.78456} = .00088$ (.088%)

(b) $V = 2000/204.7 = 9.77040$ v, $\varepsilon_{rel} = \frac{.00244}{9.77040} = 0.00025$ (.025%)

(c) $V = -960/204.8 = -4.6875$ v, $\varepsilon_{rel} = \frac{0.00244}{4.6875} = 0.00052$ (.052%)

(d) $V = 25/204.7 = 0.12213$ v, $\varepsilon_{rel} = \frac{.00244}{.12213} = .01998$ (1.998%)

2.6

For 10 bit converter, -5 v to + 5 v span:

Total number of integers is $2^{10} = 1024$, but to include

0 we have $V = \frac{5I}{511}$ because we have + 5 v $\Rightarrow$ + 511; -5v $\Rightarrow$ -512

Absolute error is $\frac{1}{2}\left(\frac{5\text{ v}}{511}\right) = .00489$ v

Thus relative error = $\frac{0.00489}{\text{measured voltage}}$

Thus

(a) $V = \frac{5(450)}{511} = 4.40313$, $\varepsilon_{rel} = \frac{.00489}{4.40313} = 0.00111 = 0.111\%$

(b) $V = \frac{5(-200)}{512} = -1.95313$, $\varepsilon_{rel} = \frac{0.00489}{1.95313} = 0.00250 = .250\%$

(c) $V = \frac{5(100)}{511} = 0.97847$, $\varepsilon_{rel} = \frac{0.00489}{0.97847} = 0.00500 = 0.500\%$

(d) $V = \frac{5(25)}{511} = 0.24462$, $\varepsilon_{rel} = \frac{0.00489}{0.24462} = 0.01999 = 1.999\%$

2.7 For converter noted

$I = 204.7\,(V)$

(a) $I = -204.8 \cong \underline{-205}$

(b) $I = 204.7(2.7) = 552.6 \cong \underline{553}$

(c) $I = 204.7(8.7) = 1780.89 \cong \underline{1781}$

(d) $I = -204.8(6) = -1228.2 \cong \underline{-1228}$.

2.8 (a) 2400 baud ≡ 240 cps

FORTRAN program = 500 lines x 30 char/line = 15000 char.

Time = 15000 char/240 cps = 62.5 seconds.

(b) For 9600 baud line ≡ 960 cps

Time = 15000 char/960 cps

= 15.63 seconds.

CHAPTER 3

3.1 (i) By definition

$$\mathcal{L}\{\cos \omega t\} = \int_0^\infty e^{-st} \cos \omega t \, dt \qquad (3.1)$$

Integrating by parts, with

$$u = e^{-st} \Rightarrow du = -se^{-st}$$

$$dv = \cos \omega t \Rightarrow v = \frac{1}{\omega}\sin \omega t$$

we have

$$\int_0^\infty e^{-st}\cos \omega t \, dt = \frac{1}{\omega}e^{-st}(\sin \omega t)\Big|_0^\infty + \int_0^\infty se^{-st}\frac{\sin \omega t}{\omega}\, dt \qquad (3.2)$$

with

$$u = e^{-st} \Rightarrow du = -se^{-st}$$

$$dv = \sin \omega t \Rightarrow v = -\frac{1}{\omega}\cos \omega t$$

we have

$$\int_0^\infty se^{-st}\frac{\sin \omega t}{\omega} = -\frac{e^{-st}\cos \omega t}{\omega}\Big|_0^\infty - \int_0^\infty \frac{s^2}{\omega^2}e^{-st}\cos \omega t \, dt$$

which reduces (3.2) to:

$$\left(1 + \frac{s^2}{\omega^2}\right)\int_0^\infty e^{-st}\cos \omega t \, dt = \frac{1}{\omega}e^{-st}(\sin \omega t)\Big|_0^\infty - \left[\frac{1}{\omega}e^{-st}\cos \omega t\right]\Big|_0^\infty$$

$$= \frac{1}{\omega}$$

Thus

$$\mathcal{L}\{\cos \omega t\} = \frac{1}{\omega}\cdot\frac{1}{1 + s^2/\omega^2} \qquad (3.3)$$

or

$$\boxed{\mathcal{L}\{\cos \omega t\} = \frac{s}{s^2 + \omega^2}}$$

as required.

(ii) $\mathcal{L}\{\sin \omega t\} = \int_0^\infty e^{-st} \sin \omega t \, dt$

Integrating by parts gives:

$$\int_0^\infty e^{-st} \sin \omega t \, dt = -\frac{1}{\omega} e^{-st} \cos \omega t \Big|_0^\infty - \frac{s}{\omega} \int_0^\infty e^{-st} \cos \omega t \, dt$$

$$= -\frac{1}{\omega}(0 - 1) - \frac{s}{\omega}\left(\frac{s}{s^2+\omega^2}\right)$$

Simplifying further gives (Result from part (i))

$$\boxed{\mathcal{L}\{\sin \omega t\} = \frac{\omega}{s^2 + \omega^2}}$$ as required.

3.2 To establish that

$$\mathcal{L}\{c_1 f_1(t) + c_2 f_2(t)\} = c_1 \mathcal{L}\{f_1(t)\} + c_2 \mathcal{L}\{f_2(t)\} \quad (3.4)$$

Observe that, by definition,

$$\text{LHS} = \int_0^\infty [c_1 f_1(t) + c_2 f_2(t)] e^{-st} \, dt$$

$$= \int_0^\infty c_1 f_1(t) e^{-st} \, dt + \int_0^\infty c_2 f_2(t) e^{-st} \, dt$$

$$= c_1 \int_0^\infty f_1(t) e^{-st} \, dt + c_2 \int_0^\infty f_2(t) e^{-st} \, dt$$

$$= c_1 \mathcal{L}\{f_1(t)\} + c_2 \mathcal{L}\{f_2(t)\} = \text{RHS}, \text{ as reqd.}$$

3.3 (i) $\mathcal{L}\{\cos \omega t\} = \frac{1}{2}\mathcal{L}\{(e^{j\omega t} + e^{-j\omega t})\}$

$$= \frac{1}{2}\left(\frac{1}{s - j\omega}\right) + \frac{1}{2}\left(\frac{1}{s + j\omega}\right) \quad (3.5)$$

since $\mathcal{L}\{e^{-at}\} = \frac{1}{s+a}$

Simplifying (3·5) further gives:

$$\mathcal{L}\{\cos \omega t\} = \frac{1}{2} \frac{2s}{s^2+\omega^2}$$

$$= \frac{s}{s^2+\omega^2} \quad , \text{ as required.}$$

ii) Similarly

$$\mathcal{L}\{\sin \omega t\} = \frac{1}{2j} \mathcal{L}\{e^{j\omega t} - e^{-j\omega t}\}$$

$$= \frac{1}{2j}\left(\frac{1}{s-j\omega}\right) - \frac{1}{2j}\left(\frac{1}{s+j\omega}\right)$$

$$= \frac{1}{2j} \frac{2j\omega}{s^2+\omega^2}$$

$$= \frac{\omega}{s^2+\omega^2} \quad , \text{ as required.}$$

3·4 Given $f(t) = t^n$ (4·1)
to find $\bar{f}(s) = \mathcal{L}\{f(t)\}$

proceed by differentiating $f(t)$ n times:

$$f(t) = t^n \Rightarrow f(0) = 0$$

and

$$f'(t) = \frac{df}{dt} = nt^{n-1} \; ; \Rightarrow f'(0) = 0$$

$$f''(t) = \frac{d^2f}{dt^2} = n(n-1)t^{n-2} ; \Rightarrow f''(0) = 0$$

$$\vdots$$

$$f^{(k)}(t) = \frac{d^kf}{dt^k} = n(n-1)\cdots(n-k+1)t^{n-k} ; \Rightarrow f^{(k)}(0) = 0$$

$$\vdots$$

finally

$$f^{(n)}(t) = \frac{d^nf}{dt^n} = n(n-1)(n-2)\cdots 2\cdot 1$$

$$= \Gamma(n+1) \quad \text{a constant} \qquad (4·2)$$

Now, from (4.2), since $\Gamma(n+1)$ is a constant for any given integer n, taking Laplace transforms, we obtain:

$$\mathcal{L}\left\{\frac{d^n f}{dt^n}\right\} = \frac{\Gamma(n+1)}{s} \qquad (4.3)$$

But from the results on the transform of derivatives we know that

$$\mathcal{L}\left\{\frac{d^n f}{dt^n}\right\} = s^n \bar{f}(s) \qquad (4.4)$$

since $f(0) = 0$, and all other initial conditions on the n-1 derivatives of f are identically zero. Equating (4.3) and (4.4) gives:

$$s^n \bar{f}(s) = \frac{\Gamma(n+1)}{s}$$

or

$$\boxed{\bar{f}(s) = \frac{\Gamma(n+1)}{s^{n+1}}} \qquad (4.5)$$

as required.

3.5 Given $y(s) = \dfrac{K(\xi s+1)}{s(\tau s+1)}$

(i) $$\lim_{t\to 0} y(t) = \lim_{s\to\infty} s y(s)$$

$$= \lim_{s\to\infty}\left\{\frac{K(\xi s+1)}{(\tau s+1)}\right\} = \lim_{s\to\infty}\left\{\frac{K(\xi + 1/s)}{(\tau + 1/s)}\right\}$$

$$\therefore \boxed{\lim_{t\to 0} y(t) = \frac{K\xi}{\tau}} \qquad (4.6)$$

(ii) $$\lim_{t\to\infty} y(t) = \lim_{s\to 0} s y(s) = \lim_{s\to 0}\left\{\frac{K(\xi s+1)}{(\tau s+1)}\right\}$$

so that $$\boxed{\lim_{t\to\infty} y(t) = K} \qquad (4.7)$$

The initial value (result (ii)) will be greater than the final value (result (i)) provided:

$$\xi > \tau$$

and $K > 0$.

3.6 Given $y(s) = \dfrac{3(s+2)(s-2)}{5s^4+6s^3+2s^2+s}$

$$\lim_{t\to\infty} y(t) = \lim_{s\to 0} s\,y(s)$$

$$= \lim_{s\to 0} \frac{3(s+2)(s-2)}{5s^3+6s^2+2s+3} = -4$$

3.7 There are several ways by which this problem may be solved. Here are two.

Method 1. By direct integration using the definition of the Laplace transform.

Since
$$u(t) = \begin{cases} 0 & t<0 \\ \sigma t & 0 \le t < b \\ A & t \ge b \end{cases}$$

then by definition,

$$u(s) = \int_0^\infty e^{-st} u(t)\,dt$$

$$= \int_0^b e^{-st}\sigma t\,dt + \int_b^\infty e^{-st} A\,dt \qquad (5.1)$$

Proceed to obtain each integral by integration by parts:

$$\sigma\int_0^\infty e^{-st}\,t\,dt = \sigma\left\{-\frac{1}{s}e^{-st}.t\Big|_0^b + \int_0^b \frac{1}{s}e^{-st}\,dt\right\}$$

$$= \sigma\left\{\frac{-b}{s}e^{-bs} - \frac{1}{s^2}e^{-st}\Big|_0^b\right\}$$

$$= \sigma\left\{-\frac{b}{s}e^{-bs} - \frac{1}{s^2}e^{-bs} + \frac{1}{s^2}\right\}$$

and since $\sigma = A/b$ (see the Diagram, Fig. P3.1),

$$\sigma\int_0^\infty e^{-st}t\,dt = \frac{A}{bs^2} - \frac{A}{bs^2}e^{-bs} - \frac{A}{s}e^{-bs} \qquad (5.2)$$

Also,

$$\int_0^\infty e^{-st}A\,dt = A\left\{-\frac{1}{s}e^{-st}\Big|_b^\infty\right\}$$

$$= \frac{A}{s}e^{-bs} \qquad (5.3)$$

Thus (5.1) becomes

$$u(s) = \frac{A}{s}e^{-bs} + \left\{\frac{A}{bs^2} - \frac{A}{bs^2}e^{-bs} - \frac{A}{s}e^{-bs}\right\}$$

or

$$\boxed{u(s) = \frac{A}{b}\cdot\frac{1}{s^2}\left(1 - e^{-bs}\right)} \qquad (5.4)$$

Method 2 Based on representing u(t) as a combination of two ramp functions.

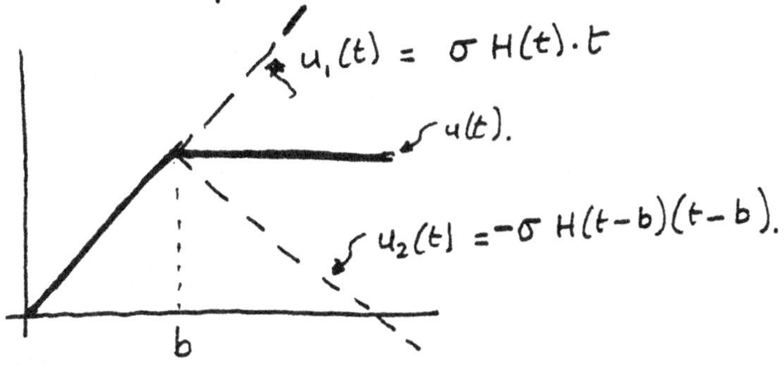

Recognising that u(t) can be represented as $u(t) = u_1(t) + u_2(t)$ where

$$u_1(t) = \sigma H(t)\, t$$
$$u_2(t) = -\sigma H(t-b)(t-b)$$

with $H(t)$ as the unit Heaviside function, we have

$$u(t) = \sigma \left\{ H(t)t - H(t-b)(t-b) \right\} \qquad (5.5)$$

From where, by taking Laplace transforms, we obtain

$$u(s) = \sigma \left[\frac{1}{s^2} - \frac{e^{-bs}}{s^2} \right]$$

and since $\sigma = A/b$, we obtain, finally

$$\boxed{u(s) = \frac{A}{b} \frac{1}{s^2} \left(1 - e^{-bs}\right)}$$

CHAPTER 4

4.1 Material balance:

$$\frac{d}{dt}(\rho V) = F_1\rho - F_2\rho - F_3\rho \qquad (4.1)$$

- $\frac{d}{dt}(\rho V)$: Rate of accumulation
- $F_1\rho$: Rate of Inflow
- $F_2\rho$: Rate of Outflow by sales
- $F_3\rho$: Rate of outflow by leakage

$$V = Ah,$$
$$F_3 = \beta h$$

so that (4.1) simplifies to

$$\frac{dh}{dt} = \frac{1}{A}\left(F_1 - F_2 - \beta h\right) \qquad (4.2)$$

assuming ρ = constant, as specified, and A = constant.

At steady state,

$$0 = \frac{1}{A}\left(F_{1_s} - F_{2_s} - \beta h_s\right) \qquad (4.3)$$

Considering the outflow F_2 (the sales flow) as the control variable (on the notion that the government can regulate what it sells to the public), so that the supply flow, F_1 is the disturbance variable (the government has no control over the supply flow) we may then define the following deviation variables:

$$y = h - h_s\ ;\quad u = F_2 - F_{2_s}\ ;\quad d = F_1 - F_{1_s}$$

Subtracting (4.3) from (4.2) then gives:

$$\boxed{A\frac{dy}{dt} = d - u - \beta y} \qquad (4.4)$$

Taking laplace transforms in (4.4) and rearranging gives the required transfer function model.

$$y(s) = \frac{-u(s)}{(As+\beta)} + \frac{d(s)}{(As+\beta)}$$

or

$$y(s) = \frac{-\frac{1}{\beta}u(s)}{\left(\frac{A}{\beta}s+1\right)} + \frac{\frac{1}{\beta}d(s)}{\left(\frac{A}{\beta}s+1\right)}. \qquad (4.5)$$

4.2 (a) Assuming that:

- ρ, C_p are constant over the operating range;
- $T_{water} = T_1$ (ideal mixing in the kettle)
- Instantaneous change in hot plate temperature, and that energy input from plate to kettle is directly proportional to $(T_2 - T_1)$
- Heat losses to the atmosphere are negligible,

An energy balance yields:

$$\frac{d}{dt}(\rho V C_p T_1) = c(T_2 - T_1) + F\rho C_p(T_0 - T^*) - F(\rho C_p T_1 - T^*)$$

Rate of Accumulation; Rate in through hotplate heat; Rate in from Feed; Datum Temp.; Rate out

Simplifies to:

$$\rho V C_p \frac{dT_1}{dt} = c(T_2 - T_1) + F\rho C_p T_0 - F\rho C_p T_1$$

as in (P4.1)

(b) The appropriate energy balance will require

$$E_{accumulated} = E_{in} - E_{out}$$

In this case,

$$E_{accumulated} = mC_{p_2}\frac{dT_2}{dt}$$

$$E_{in} = Q$$

$$E_{out} = A_c h(T_2 - T_a) + c(T_2 - T_1)$$

$A_c h(T_2 - T_a)$: loss to atmosphere; $c(T_2 - T_1)$: Transferred to water in kettle.

Thus, the energy balance becomes

$$\boxed{mC_{p_2}\frac{dT_2}{dt} = Q - A_c h(T_2 - T_a) - c(T_2 - T_1)} \quad (4.6)$$

4.3 (a) Input variable : T_2
Disturbance variable : T_0

(b) At steady state (P4.1) becomes

$$0 = c(T_2^* - T_1^*) + F\rho C_p T_0^* - F\rho C_p T_1^*$$

Subtracting from (P4.1), with $y = T - T_1^*$; $u = T_2 - T_2^*$, and $d = T_0 - T_1^*$, gives

$$\rho V C_p \frac{dy}{dt} = c(u - y) + F\rho C_p d - F\rho C_p y$$

Rearrange to obtain:

$$\frac{dy}{dt} = -\left(\frac{F}{V} + \frac{c}{\rho V C_p}\right)y + \frac{c}{\rho V C_p}u + \frac{F}{V}d \quad (4.7)$$

which is of the form given in (P4.2) with

$$a = -\left(\frac{F}{V} + \frac{c}{\rho V C_p}\right); \quad b = \frac{c}{\rho V C_p}; \quad \gamma = \frac{F}{V}$$

(c) Take Laplace transforms in (4.7) and rearrange to give:

$$y(s) = \underbrace{\left[\frac{c/\rho V C_p}{s + \left(\frac{F}{V} + \frac{c}{\rho V C_p}\right)}\right]}_{g(s)} u(s) + \underbrace{\left[\frac{F/V}{s + \left(\frac{F}{V} + \frac{c}{\rho V C_p}\right)}\right]}_{g_d(s)} d(s)$$

The required transfer functions are as indicated.

(d) For the current purposes, it is convenient to let:

$$k = \frac{\frac{c}{\rho V C_p}}{\frac{F}{V} + \frac{c}{\rho V C_p}} \quad ; \text{ i.e. } \quad k = \frac{c}{F\rho C_p + c} \qquad (4.8a)$$

$$k_d = \frac{F/V}{\frac{F}{V} + \frac{c}{\rho V C_p}} \quad ; \text{ i.e } \quad k_d = \frac{F\rho C_p}{F\rho C_p + c} \qquad (4.8b)$$

$$\tau = \frac{1}{\left(\frac{F}{V} + \frac{c}{\rho V C_p}\right)} \quad ; \text{ i.e. } \quad \tau = \frac{\rho V C_p}{F\rho C_p + c} \qquad (4.8c)$$

Then

$$g(s) = k/\tau s + 1 \quad \Rightarrow \quad g(t) = \frac{k}{\tau} e^{-t/\tau} \qquad (4.9a)$$

$$g_d(s) = \frac{k_d}{\tau s + 1} \quad \Rightarrow \quad g_d(t) = \frac{k_d}{\tau} e^{-t/\tau} \qquad (4.9b)$$

The impulse response model is therefore:

$$\boxed{y(t) = \int_0^t \frac{c}{\rho V C_p} \exp\left\{\frac{-(t-\sigma)}{\tau}\right\} u(\sigma)\, d\sigma + \int_0^t \frac{F}{V} \exp\left\{\frac{-(t-\sigma)}{\tau}\right\} d(\sigma)\, d\sigma}$$

(4.10)

4.4 (a) Take Laplace transforms in (P4.3), (P4.4) to obtain, upon rearrangement:

$$x_1(s) = \frac{u(s)}{(s+k_1)} \qquad (4.11)$$

and

$$x_2(s) = \frac{k_1 x_1(s)}{(s+k_2)} \qquad (4.12)$$

Introducing (4.11) into (4.12) for x_1 gives

$$x_2(s) = \frac{k_1 u(s)}{(s+k_1)(s+k_2)}$$

or, from (P4.5),

$$\boxed{y(s) = \frac{k_1 u(s)}{(s+k_1)(s+k_2)}} \qquad (4.13)$$

(b) In vector-matrix form, the model is:

$$\dot{\underset{\sim}{x}} = \frac{d}{dt}\begin{pmatrix} x_1 \\ x_2 \end{pmatrix} = \begin{bmatrix} -k_1 & 0 \\ k_1 & -k_2 \end{bmatrix}\begin{bmatrix} x_1 \\ x_2 \end{bmatrix} + \begin{bmatrix} 1 \\ 0 \end{bmatrix} u$$

$$y = \begin{bmatrix} 0 & 1 \end{bmatrix}\begin{bmatrix} x_1 \\ x_2 \end{bmatrix}$$

so that $\underset{\sim}{A} = \begin{bmatrix} -k_1 & 0 \\ k_1 & -k_2 \end{bmatrix}$; $\underset{\sim}{B} = \begin{bmatrix} 1 \\ 0 \end{bmatrix}$; $\underset{\sim}{c}^T = \begin{bmatrix} 0 & 1 \end{bmatrix}$.

4.5 (a) As in Example 4.2, let

$$x_1(s) = \frac{12.8}{16.7s+1}u(s) \qquad (4.14a)$$

$$x_2(s) = \frac{3.8}{14.9s+1}d(s) \qquad (4.14b)$$

so that $y(s) = x_1(s) + x_2(s)$ (4.14c)

Then, as in Example 4.1, observe that the <u>most obvious</u> differential equation which would result in (4.14a) upon Laplace transformation is:

$$16.7\frac{dx_1}{dt} + x_1 = 12.8\,u(t) \qquad (4.15a)$$

Similarly, for (4.14b), we have

$$14.9\frac{dx_2}{dt} + x_2 = 3.8\,d(t) \qquad (4.15b)$$

Thus, these two equations, along with

$$y(t) = x_1(t) + x_2(t) \qquad (4.15c)$$

constitute the required set of equations to be solved in order to generate a time domain response for the top portion of the distillation column.

(b) From the equation given in (P4.7),

$$g(s) = \frac{12.8}{16.7s+1} \Rightarrow g(j\omega) = \frac{12.8}{16.7(j\omega)+1}$$

$$g_d(s) = \frac{3.8}{14.9s+1} \Rightarrow g_d(j\omega) = \frac{3.8}{14.9(j\omega)+1}$$

These transfer function expressions are simplified to

$$g(j\omega) = Re(g) + j\,Im(g) \qquad (4.16a)$$

where $Re(g) = \dfrac{12.8}{1 + 278.9\omega^2}$; $Im(g) = \dfrac{-213.76\omega}{278.9\omega^2 + 1}$ (4.16b)

along with

$$g_d(j\omega) = Re(g_d) + j\,Im(g_d) \qquad (4.17a)$$

where $Re(g_d) = \dfrac{3.8}{1 + 222\omega^2}$; $Im(g_d) = \dfrac{-56.62\omega}{1 + 222\omega^2}$ (4.17b).

- Choose your 10 values of ω;
- Compute $Re(g)$, $Im(g)$; $Re(g_d)$, $Im(g_d)$ from (4.16b) and (4.17b) respectively for each value of ω;
- Generate 10 numerical values for $g(j\omega)$ and $g_d(j\omega)$.

4.6 From $y(k) = ay(k-1) + bu(k-4)$; $y(0) = 0$

and given that $u(k) = 0$ for $k \leq 0$

we immediately obtain the following:

$k = 1$; $y(1) = a\,y(0) + bu(-3) = 0$

$k = 2$; $y(2) = a\,y(1) + bu(-2) = 0$

$k = 3$; $y(3) = a\,y(2) + bu(-1) = 0$

$k = 4$; $y(4) = a\,y(3) + bu(0) = 0$

$k = 5$; $y(5) = ay(4) + bu(1) = bu(1)$

$k = 6$; $y(6) = a\,y(5) + bu(2) = abu(1) + bu(2)$

$k = 7$; $y(7) = ay(6) + bu(3)$

$= a^2bu(1) + abu(2) + bu(3)$

Similarly, for $k = 8$,

$$y(8) = a^3bu(1) + a^2bu(2) + abu(3) + bu(4)$$

or

$$y(8) = bu(4) + abu(3) + a^2bu(2) + a^3bu(1)$$

For any general k, therefore,

$$y(k) = a^0bu(k-4) + abu(k-5) + a^2bu(k-6) + \cdots$$

which may be written as:

$$y(k) = \sum_{i=4}^{k} a^{i-4}\, b\, u(k-i) \tag{4.18}$$

along with $y(k) = 0$ for $k < 4$

These may now be written jointly as:

$$y(k) = \sum_{i=1}^{k} g(i)\, u(k-i) \tag{4.19}$$

as required in Equation (P4.9), with $g(i)$ given by:

$$g(i) = \begin{cases} 0\,; & i < 4 \\ a^{i-4}b\,; & i \geq 4 \end{cases} \tag{4.20}$$

4.7 The impulse response model form we seek is:

$$y(t) = \int_0^t g(t-\sigma)\, u(\sigma)\, d\sigma$$

where $g(t)$ is to be obtained from the given $g(s)$. Since

$$g(s) = \frac{K}{(\tau_1 s + 1)(\tau_2 s + 1)} = \frac{A}{(\tau_1 s + 1)} + \frac{B}{(\tau_2 s + 1)}$$

by applying the methods of partial fraction expansion presented in Appendix C (Section C.3.4), we obtain

$$\left.\begin{aligned} A &= \frac{K\tau_1}{\tau_1-\tau_2} \\ B &= \frac{K\tau_2}{\tau_2-\tau_1} \end{aligned}\right\} \qquad (4.21)$$

This facilitates Laplace inversion, to obtain

$$g(t) = Ae^{-t/\tau_1} + Be^{-t/\tau_2} \qquad (4.22)$$

The impulse response model form is therefore:

$$y(t) = \int_0^t \left[\left(\frac{K\tau_1}{\tau_1-\tau_2}\right)e^{-(t-\sigma)/\tau_1} + \left(\frac{K\tau_2}{\tau_2-\tau_1}\right)e^{-(t-\sigma)/\tau_2}\right]u(\sigma)\,d\sigma.$$

4.8
(a) 1 pole, $s = -1/\tau$; no zero
(b) 1 pole, $s = -1/\tau$; 1 zero $s = -1/\xi$
(c) 2 poles, $s = -1/\tau_1, -1/\tau_2$; no zero
(d) 2 poles, $s = 1/3, -1/4$; no zero
(e) 2 poles, $s = -1/3, -1/4$; no zero
(f) 2 complex conjugate poles, $s = -1/2 \pm \sqrt{3}/2\, j$; no zero
(g) 2 poles, $s = -1/5, 1/3$; one zero, $s = -4$
(h) 4 poles, $s = -1/4, -1/2, -0.4, -2/3$
3 zeros, $s = -1/3, -1, +1/5$

CHAPTER 5

5.1 Assuming:

(i) Constant water density ρ, and specific heat c_p;
(ii) Rate of energy input via heater = $Q(t)$
(iii) Well mixed content in the tank;
(iv) No heat losses to the atmosphere

An energy balance yields:

$$\rho V C_p \frac{dT}{dt} = F\rho C_p T_i - F\rho C_p T + Q(t) \tag{5.1}$$

which rearranges to give:

$$\frac{V}{F}\frac{dT}{dt} = T_i - T + \frac{Q}{F\rho C_p} \tag{5.2}$$

By defining the following variables:

$$y = T - T^*\;;\quad u = Q - Q^*\;;\quad \tau = \frac{V}{F}\;;\quad K = \frac{1}{F\rho C_p}$$

with T^*, Q^* as steady state values of T and Q respectively, Eq. (5.2) becomes

$$\boxed{\tau\frac{dy}{dt} = -y + Ku} \tag{5.3}$$

which is immediately recognized as the standard differential equation model for a first-order system.

For the specific process and conditions in question we have

$$V = 100\,L\;;\quad F = 10\,L/min;\quad T_i = 30^\circ C;\; T^* = T(0) = 80^\circ C$$

and the heater failure corresponds to $Q = 0$.
The process response to this condition involves investigating the solution to (5.3) with

$$u = -Q^*$$

Q^* is obtained from (5.2) by setting the LHS to zero and rearranging, to give

$$Q^* = (F\rho C_p)\,50$$

the required response is therefore obtained from

$$10\frac{dy}{dt} = -y - 50$$

or

$$y(t) = -50\left(1 - e^{-t/10}\right) \qquad (5.4)$$

In terms of the original tank temperature, Eq.(5.4) is

$$T(t) = T^* - 50\left(1 - e^{-t/10}\right)$$

$$T(t) = 80 - 50\left(1 - e^{-t/10}\right)$$

From which we easily obtain

$$\boxed{T(5) = 60.33\,°C.}$$

5.2 (a) The variable to be controlled is T, therefore define $y = T - T^*$;

The manipulated variable is R, therefore

$$u = R - R^*$$

Similarly, $d = F - F^*$ since F is the disturbance.

The process model may therefore be presented as:

$$y(s) = \underbrace{\left(\frac{-5}{6s+1}\right)}_{g(s)} u(s) + \underbrace{\left(\frac{20}{30s+1}\right)}_{g_d(s)} d(s) \qquad (5.5)$$

(b) The specified process conditions imply

$$u(s) = 1/s \quad ; \quad d(s) = 0$$

so that

$$y(s) = \frac{-5}{s(6s+1)}$$

which, upon Laplace inversion results in:

$$y(t) = -5\left(1 - e^{-t/6}\right)$$

or

$$\boxed{T(t) = 250 - 5\left(1 - e^{-t/6}\right).} \qquad (5.6)$$

$$\lim_{t\to\infty} T(t) = 250 - 5 = 245\,°F$$

is the value to which the tray temperature ultimately settles after the unit step change in R.

(c) Under the current conditions,

$$y(s) = \frac{20}{30s+1} \cdot \frac{1}{s}$$

or

$$y(t) = 20\left(1 - e^{-t/30}\right).$$

or

$$\boxed{T(t) = 250 + 20\left(1 - e^{-t/30}\right).} \qquad (5.7)$$

this time,

$$\lim_{t \to \infty} T(t) = 250 + 20 = 270°F$$

is the value to which the tray temperature settles.

5.3 (a) The responses are shown below in Fig S5.1

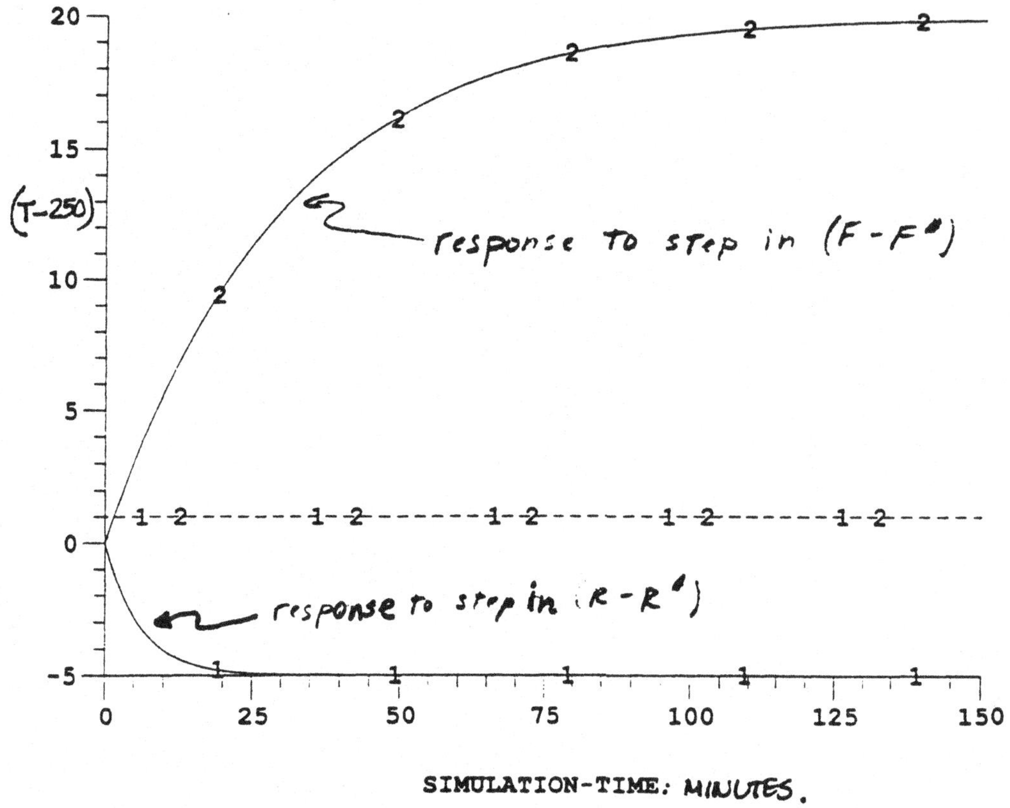

FIG. S5.1

The feed rate, F has the largest effect on T_j whereas the reflux flow rate, R has the quickest effect.

(c) The tray temperature response to a simultaneous implementation of unit step changes in both F and R is given by:

$$y(t) = -5(1 - e^{-t/6}) + 20(1 - e^{-t/30}).$$

or

$$T(t) = 250 - 5(1 - e^{-t/6}) + 20(1 - e^{-t/30}). \qquad (5.8)$$

A plot of this response is shown below in Fig. S5.2. The most important aspect of this response is the initial period of inverse response. The tray temperature drops first before eventually rising to its final value of 250 + 15 = 265°F. (see chapter 7).

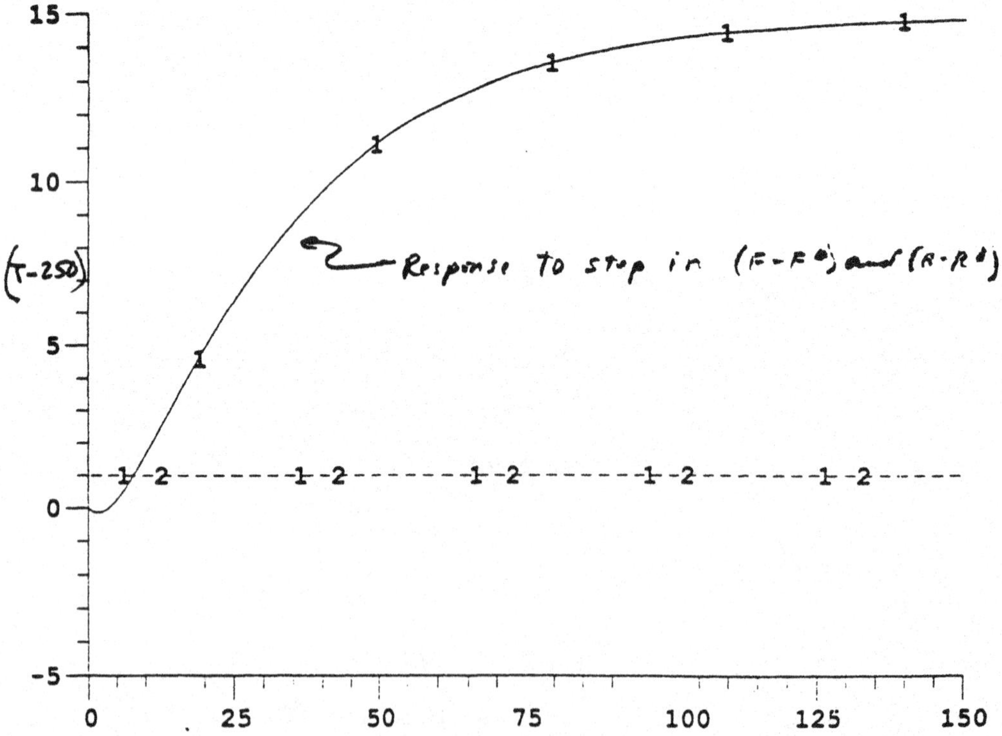

FIG. S5.2

5.4 (a) From Eq. (P5.1) we observe that at steady-state,

$$\Delta T = 20\,\Delta F \tag{5.9a}$$

and from Eq. (P5.2),

$$\Delta T = -5\Delta R \tag{5.9b}$$

also at steady state. Thus, the indicated change $\Delta F = 2$ will result in $\Delta T = 40°F$ at steady state. To counteract this will require a change in reflux rate that will cause ΔT to be $-40°F$.

From (5.9b), observe that the required ΔR is given by

$$\Delta R = 8 \text{ Mlb/hr.}$$

Recommended change: an <u>increase</u> of 8 Mlb/hr in Reflux flow rate.

(b)

$$\Delta R = \frac{4(6s+1)}{(30s+1)}\,\Delta F$$

Transfer function: lead/lag form with $K = 4$; $\tau = 30$; $\xi = 6$; ρ and $\rho = 1/5$.

The reflux rate response to $\Delta F = 2$ is now given by

$$\Delta R = 8\left[1 - 0.8e^{-t/30}\right] \tag{5.10}$$

using Eq. (5.66) in the text.

This response is shown in Figure S5.3; the ultimate change in reflux flow rate is indicated as $\Delta R(\infty) = 8$, precisely as obtained in part (a). Also $\lim_{t\to\infty} \Delta R$ from (5.10) also indicates the same value.

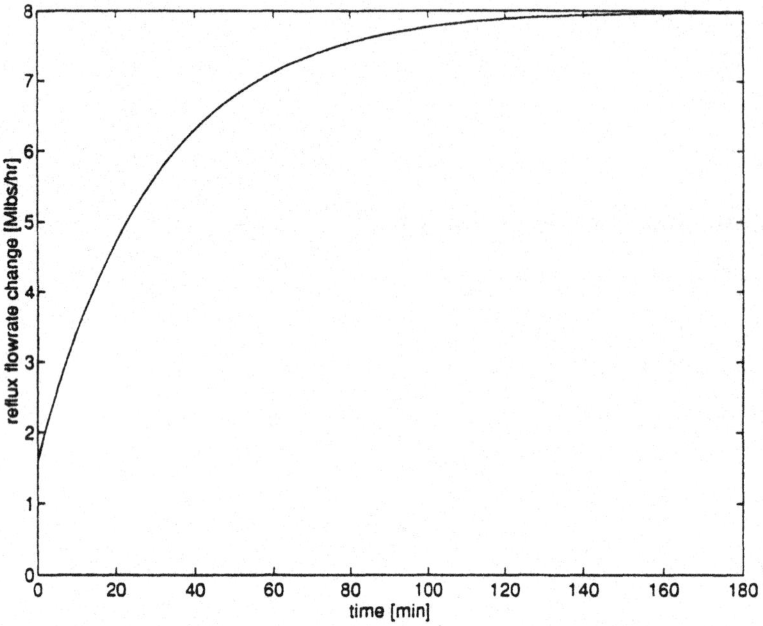

FIG S5.3

5.5 (a) Given $g(s) = \frac{4}{5s+1}$

the required theoretical unit impulse response is:

$$y^*(t) = \frac{4}{5} e^{-t/5} \quad (5.11)$$

(b) For

$$u_1(t) = \begin{cases} 1 & 0 < t \leq 1 \\ 0 & t \geq 1 \end{cases}$$

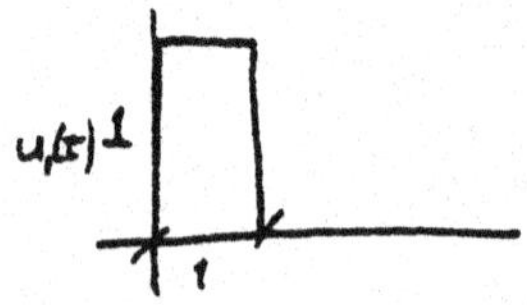

$$u_1(s) = (1 - e^{-s})$$

and the required response is given by:

$$y_1(t) = \begin{cases} 4(1-e^{-t/5}) & ; 0<t\leq 1 \\ 4\{e^{-(t-1)/5} - e^{-t/5}\} & ; t>1 \end{cases}$$

(c) For the given $u_2(t)$ (height 2, width 0.5), we have that the required response is given by:

$$y_2(t) = \begin{cases} 8(1-e^{-t/5}), & 0<t\leq 0.5 \\ 8\{e^{-(t-0.5)/5} - e^{-t/5}\} & t>0.5 \end{cases}$$

(d) The required plot is shown below in Fig. S5.4, where it is clear that y_2 approximates y^* better than y_1 does.

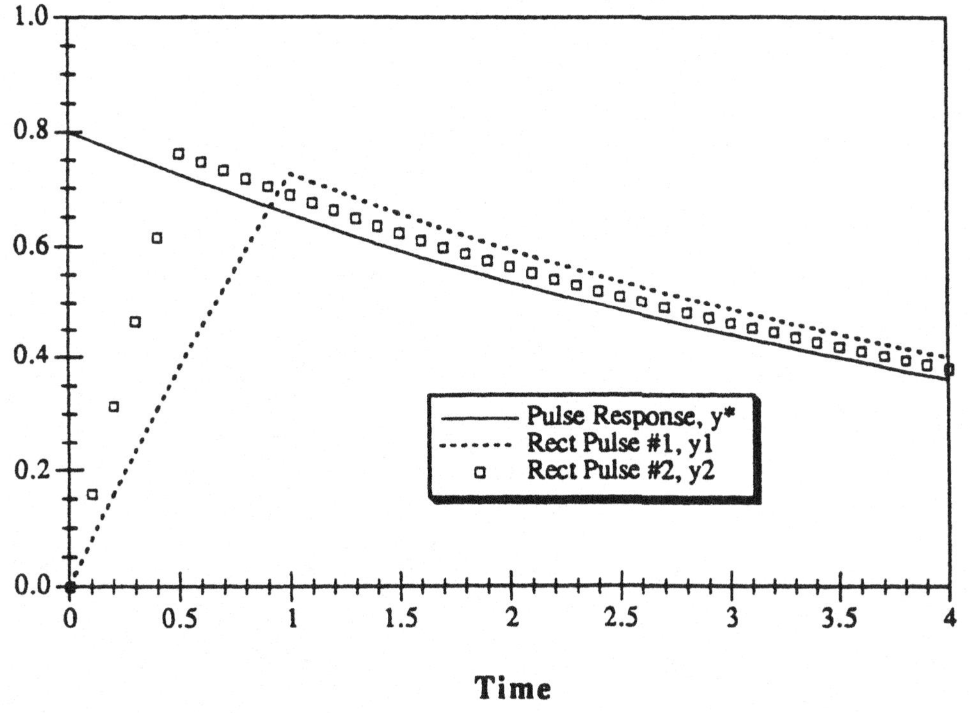

FIG. S5.4

5.6 In general, given $g(s) = \frac{K}{\tau s+1}$, $g(t) = \frac{K}{\tau} e^{-t/\tau}$

For this problem therefore,

$$g(t) = \frac{0.32}{15.5} e^{-t/15.5} \qquad (5.12)$$

It is more convenient in this case to use the alternative form of the impulse response model:

$$y(t) = \int_0^t g(t-\sigma)\,u(\sigma)\,d\sigma \qquad (5.13)$$

We now apply (5.13) for the staircase input function.

(i) For $0 < t \leq 2$,

$$y(t) = \int_0^{t(\leq 2)} \frac{K}{\tau} e^{-\left(\frac{t-\sigma}{\tau}\right)} 1.\,d\sigma = K\left[1 - e^{-t/\tau}\right]$$

(ii) $0 < t \leq 4$,

$$y(t) = \int_0^2 \frac{K}{\tau} e^{-\frac{(t-\sigma)}{\tau}} .1\,d\sigma + \frac{K}{\tau}\int_2^{t(\leq 4)} e^{-\frac{(t-\sigma)}{\tau}} .2\,d\sigma$$

$$= K\left[e^{-\frac{(t-\sigma)}{\tau}}\right]_0^2 + 2K\left[1 - e^{-\left(\frac{t-2}{\tau}\right)}\right]$$

$$\boxed{y(t) = 2K\left[1 - \frac{1}{2}\left(e^{-t/\tau} + e^{-\left(\frac{t-2}{\tau}\right)}\right)\right]}$$

(iii) $0 < t \leq 6$

$$y(t) = \int_0^2 \frac{K}{\tau} e^{-\frac{(t-\sigma)}{\tau}} .1\,d\sigma + \int_2^4 \frac{K}{\tau} e^{-\frac{(t-\sigma)}{\tau}} .2.\,d\sigma + \int_4^{t(\leq 6)} \frac{K}{\tau} e^{-\frac{(t-\sigma)}{\tau}} .3.\,d\sigma$$

$$= K\left[e^{-\frac{(t-2)}{\tau}} - e^{-t/\tau}\right] + 2K\left[e^{-\frac{(t-4)}{\tau}} - e^{-\frac{(t-2)}{\tau}}\right] + 3K\left[1 - e^{-(t-4)/\tau}\right]$$

$$\boxed{y(t) = 3K\left[1 - \tfrac{1}{3}\left(e^{-t/\tau} + e^{-(t-2)/\tau} + e^{-(t-4)/\tau}\right)\right]}$$

(iv) $t \geq 6$, via the same procedure obtain

$$\boxed{y(t) = 4K\left[1 - \tfrac{1}{4}\left(e^{-t/\tau} + e^{-(t-2)/\tau} + e^{-(t-4)/\tau} + e^{-(t-6)/\tau}\right)\right]}$$

Thus for this specific case with $K = 0.32$ and $\tau = 15.5$, the response to the staircase function is given by:

$$y(t) = \begin{cases} 0.32\left[1 - e^{-t/15.5}\right]; & 0 < t \le 2 \\ 0.64\left[1 - \frac{1}{2}\left(e^{-t/15.5} + e^{-(t-2)/15.5}\right)\right]; & 2 < t \le 4 \\ 0.96\left[1 - \frac{1}{3}\left(e^{-t/15.5} + e^{-(t-2)/15.5} + e^{-(t-4)/15.5}\right)\right]; & 4 < t \le 6 \\ 1.28\left[1 - \frac{1}{4}\left(e^{-t/15.5} + e^{-(t-2)/15.5} + e^{-(t-4)/15.5} + e^{-(t-6)/15.5}\right)\right]; & t > 6 \end{cases}$$

5.7

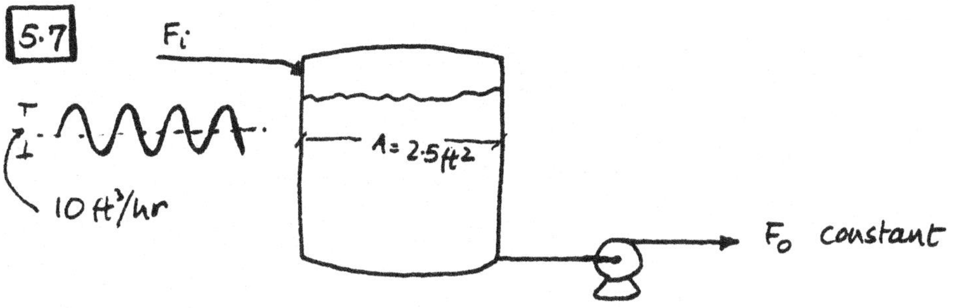

(a)

For this process, the transfer function model is:

$$y(s) = \frac{1/2.5}{s} u(s)$$

$$\boxed{y(s) = \frac{0.4}{s} u(s)} \qquad (5.18)$$

Given that $u(t) = A \sin \omega t = 10 \sin \omega t$, we have

$$u(s) = \frac{10\omega}{s^2 + \omega^2}$$

and ∴

$$y(s) = 4\omega / s(s^2 + \omega^2). \qquad (5.19)$$

From Table C.1, entry 19, we obtain

$$\boxed{y(t) = \frac{4}{\omega}\left[1 - \cos\omega t\right]} \qquad (5.20)$$

For $\omega = 0.2$, the specific response will be

$$y(t) = 20\left[1 - \cos\omega t\right]$$

A plot of this along with $u(t) = 10 \sin 0.2t$ is shown in FIG S5.5 below.

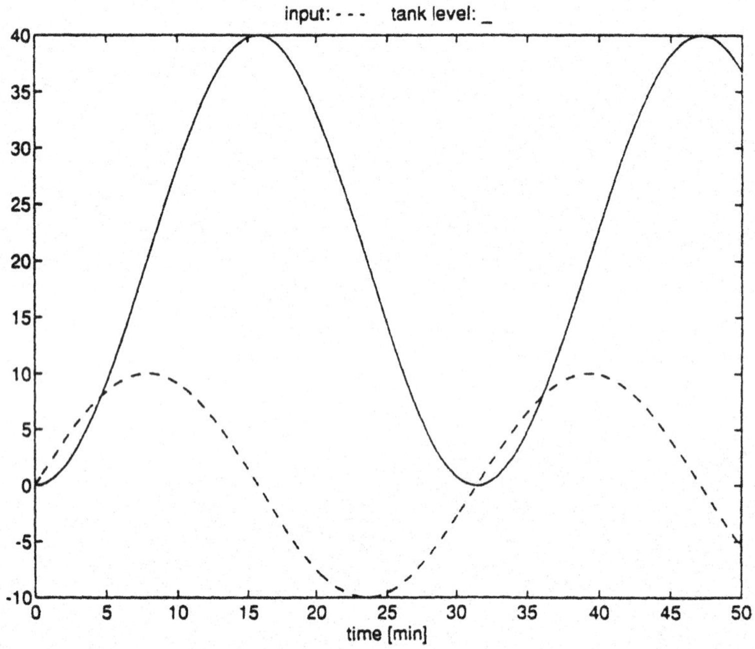

FIG S5.5

(b) Observe from (5.20) that since $\cos\omega t$ takes a max value of +1 and a min value of −1, $y(t)$ in turn takes a <u>max</u> value of

$$y_{max} = \frac{4}{\omega}\left[1 + 1\right] = \frac{8}{\omega} \text{ ft.}$$

and a min value of zero. Since y is already in

terms of deviation from nominal tank level, the required maximum deviation is thus $8/\omega$ ft. (Note that for Fig. S5.5, $\omega = 0.2$, and $8/0.2 = 40$ ft. This is precisely what is shown in the Figure.)

(c) Nominal level = 5 ft; Total tank height = 10 ft, thus max allowable deviation either way is 5 ft.
The critical frequency ω_c is determined from

$$\delta_{max} = \frac{8}{\omega_c} = 5$$

$$\Rightarrow \quad \omega_c = 8/5 = 1.6 \text{ rad/hr}$$

Thus to avoid overflowing the tank (it is impossible to drain the tank) the loading frequency must be greater than 1.6 rad/hr.
Note that in Fig S5.5 $\omega = 0.2 < \omega_c$ and the tank will overflow.

5.8 (a) An energy balance around the iron yields

$$mC_p \frac{dT}{dt} = Q - hA(T - T_a) \tag{5.21}$$

(b) At steady state (5.21) becomes

$$0 = Q_s - hA(T_s - T_a) \tag{5.22}$$

and given $y = T - T_s$, $u = Q - Q_s$, subtracting (5.22) from (5.21) gives

$$mC_p \frac{dy}{dt} = u - hAy$$

which rearranges to give

$$\boxed{\frac{mC_p}{hA}\frac{dy}{dt} = -y + \frac{1}{hA}u.} \tag{5.23}$$

Upon taking Laplace transforms and rearranging, we now obtain

$$y(s) = \frac{1/hA}{\left(\frac{mC_p}{hA}s + 1\right)} u(s)$$

and the indicated transfer function has the first-order form with

$$K = 1/hA \ ; \quad \tau = mC_p/hA$$

(c) The response of the first-order system to a step input of magnitude 10, is given by

$$y(t) = 10K(1 - e^{-t/\tau})$$

For the iron in question, this implies a temperature response represented by

$$T(t) = 24 + 50(1 - e^{-t/\tau}) \qquad (5.24)$$

If $T(6) = 55$, then from (5.24)

$$55 = 24 + 50(1 - e^{-6/\tau})$$

or

$$\left(1 - \frac{31}{50}\right) = e^{-6/\tau}$$

which is easily solved for τ, yielding

$$\tau = -6/\ln(0.38)$$

$$\tau = 6.20 \text{ minutes.}$$

5.9 (a) The consolidated transfer function is given by

$$g(s) = \frac{(K_1\tau_2 + K_2\tau_1)s + (K_1 + K_2)}{(\tau_1 s + 1)(\tau_2 s + 1)}$$

or

$$g(s) = \frac{(K_1 + K_2)\left[\left(\frac{K_1\tau_2 + K_2\tau_1}{K_1 + K_2}\right)s + 1\right]}{(\tau_1 s + 1)(\tau_2 s + 1)} \tag{5.25}$$

which is of the form

$$g(s) = \frac{K(\xi s + 1)}{(\tau_1 s + 1)(\tau_2 s + 1)}$$

with

$$K = (K_1 + K_2) \tag{5.26}$$

$$\xi = \frac{K_1\tau_2 + K_2\tau_1}{K_1 + K_2} \tag{5.27}$$

Thus, the characteristics of the consolidated transfer function are as follows:

(i) A steady state gain K given in Eq. (5.26) a sum of the steady-state gains of the contributing transfer functions;

(ii) Two poles located at $s = -1/\tau_1$, $s = -1/\tau_2$; identical to the poles of the contributing transfer functions;

(iii) A single zero located at $s = -1/\xi$, where ξ, the lead time constant, is given in (5.27) — a weighted average of τ_1 and τ_2 (correspondingly weighted by K_1/K_1+K_2 and K_2/K_1+K_2 respectively).

(b) For the composite system zero to be positive requires that ξ be negative, i.e. that

$$\xi = (K_1\tau_2 + K_2\tau_1)/(K_1 + K_2) < 0 \tag{5.28}$$

5.10 (a) By taking Laplace transforms of the differential equation, with B = 0, we obtain

$$W_R(s) + W_B(s) = \frac{1}{s}(L - V)$$

so that

$$\frac{W_B(s)}{Q(s)} = \frac{1}{s}\,\frac{V(s)}{Q(s)}\left[\frac{L(s)}{V(s)} - 1\right] - \frac{W_R(s)}{Q(s)}$$

and upon introducing the given transfer function relationships, we obtain

$$\frac{W_B(s)}{Q(s)} = \frac{1}{s}\left(\frac{0.01}{10s+1}\right)\left[\frac{1}{1000s+1} - 1\right] + \frac{8}{10s+1}$$

Further simplification now gives

$$\boxed{\frac{W_B(s)}{Q(s)} = \frac{-2(1 - 4000s)}{(1000s+1)(10s+1)}} \qquad (5.29)$$

as the required transfer function.

(b) The W_B response to a step change Q^* in Q may be obtained from (5.29). By partial fraction expansion and further simplification, we obtain

$$w_B(t) = -2Q^*\left[1 - \frac{5000}{990}e^{-t/1000} + \frac{4010}{990}e^{-t/10}\right]$$

as the required expression.

Chapter 6

6.1 (a) Laplace transforms of Eq. (6.8) in the text gives

$$\left[\tau_1\tau_2 s^2 + (\tau_1+\tau_2)s + 1\right] y(s) = K_1 K_2 u(s)$$

which rearranges easily to give

$$y(s) = \frac{K_1 K_2}{(\tau_1 s+1)(\tau_2 s+1)} u(s)$$

from where the implied $g(s)$ is seen to be identical to that given in Eq. (6.10) in the text.

(b) Differentiating the unit step response in Eq. (6.12) of the text gives

$$\frac{dy_2}{dt} = K\left[\frac{1}{(\tau_1-\tau_2)} e^{-t/\tau_1} + \frac{1}{(\tau_2-\tau_1)} e^{-t/\tau_2}\right]$$

and for $t=0$, therefore

$$\left.\frac{dy_2}{dt}\right|_{t=0} = K\left[\frac{1}{\tau_1-\tau_2} + \frac{1}{\tau_2-\tau_1}\right] = 0 \quad \text{as required.}$$

When $\tau_1=\tau_2=\tau$, the appropriate step response is

$$y_2(t) = K\left[1 - e^{-t/\tau} - \frac{t}{\tau} e^{-t/\tau}\right]$$

so that

$$\left.\frac{dy_2}{dt}\right|_{t=0} = K\left[\frac{1}{\tau}e^{-t} + \left(-\frac{1}{\tau}e^{-t/\tau} + \frac{t}{\tau^2}e^{-t/\tau}\right)\right]\Bigg|_{t=0}$$

$$= K\left[\frac{t}{\tau^2}e^{-t/\tau}\right]\Bigg|_{t=0} = 0 \text{ as required.}$$

6.2 The unit step response is obtained from

$$y(s) = g(s)\cdot\frac{1}{s} \qquad (6.1a)$$

where

$$g(s) = \frac{K\prod_{i=1}^{q}(\xi_i s+1)}{\prod_{i=1}^{p}(\tau_i s+1)} \qquad (6.1b)$$

Let the slope of this step response be $z(t)$, i.e.

$$z(t) = \frac{dy}{dt};$$

we may then invoke the initial value theorem of Laplace transforms to determine the value of the step response's slope at the origin. i.e.

$$\lim_{t\to 0}\{z(t)\} = \lim_{s\to\infty}\{s\,z(s\}$$

$$= \lim_{s\to\infty}\{s^2 y(s)\} \qquad (6.2)$$

since $\mathcal{L}\{z(t)\} = sy(s)$.

Introducing (6.1) into (6.2) now gives

$$\lim_{t\to 0}\left\{\frac{dy}{dt}\right\} = \lim_{s\to\infty}\left\{s^2\frac{K\prod_{i=1}^{q}(\xi_i s+1)}{\prod_{i=1}^{p}(\tau_i s+1)}\cdot\frac{1}{s}\right\} \qquad (6.3)$$

Let us rearrange (6.3) to the following form:

$$\lim_{t\to 0}\left\{\frac{dy}{dt}\right\} = \lim_{s\to\infty}\left\{\frac{K\prod_{i=1}^{q}(\xi_i s+1)}{\prod_{i=1}^{q}(\tau_i s+1)}\cdot\frac{s}{\prod_{i=q+1}^{p}(\tau_i s+1)}\right\} \qquad (6.4)$$

where we have simply broken up the p products in the denominator into one set of q products and the remainder. And now, we may observe that if $p > q+1$, i.e. $(p-q) > 1$, then there will be at least 2 terms left over in the denominator of the rightmost quotient in Eq (6.4) above. We may now rearrange (6.4) further to give:

$$\lim_{t\to 0}\left\{\frac{dy}{dt}\right\} = \lim_{s\to\infty}\left\{\frac{K\prod_{i=1}^{q}(\xi_i+1/s)}{\prod_{i=1}^{q}(\tau_i+1/s)}\cdot\frac{s}{\prod_{i=q+1}^{p}(\tau_i s+1)}\right\}$$

or

$$\lim_{t\to 0}\left\{\frac{dy}{dt}\right\} = \left(\frac{K\prod_{i=1}^{q}\xi_i}{\prod_{i=1}^{q}\tau_i}\right)\lim_{s\to\infty}\left\{\frac{s}{\prod_{i=q+1}^{p}(\tau_i s+1)}\right\} \qquad (6.5)$$

And now, with 2 or more products in the rightmost quotient in Eq. (6.5), we see that the required limit is zero, and therefore

$$\lim_{t\to 0}\left\{\frac{dy}{dt}\right\} = 0 \qquad (6.6)$$

whenever $(p-q) > 1$.

On the other hand, if $p-q=1$, then $p=q+1$ and Eq 6.5 becomes:

$$\lim_{t\to 0}\left\{\frac{dy}{dt}\right\} = \left(\frac{K\prod_{i=1}^{q}\xi_i}{\prod_{i=1}^{q}\tau_i}\right)\lim_{s\to\infty}\left\{\frac{s}{\tau_{q+1}s+1}\right\}$$

$$= \left(\frac{K\prod_{i=1}^{q}\xi_i}{\prod_{i=1}^{q}\tau_i}\right)\lim_{s\to\infty}\left\{\frac{1}{\tau_{q+1}+\frac{1}{s}}\right\}$$

so that the initial slope will be given by

$$\lim_{t\to 0}\left\{\frac{dy}{dt}\right\} = \left(\frac{K\prod_{i=1}^{q}\xi_i}{\prod_{i=1}^{q+1}\tau_i}\right) \qquad (6.7)$$

which is nonzero.

Observe from (6.7) that

(i) As the values of the ξ_i's increase, so does the value of the initial slope, implying a steeper initial response; whereas,

(ii) As the values of the τ_i's increase, the result is a <u>decrease</u> in the value of the initial slope, implying an initial response which is less steep.

6.3 (a) The required impulse response is obtained from

$$y(s) = \frac{K_1K_2}{(\tau_1 s+1)(\tau_2 s+1)}$$

which upon inversion yields

$$y(t) = \frac{K}{\tau_1 - \tau_2}\left(e^{-t/\tau_1} - e^{-t/\tau_2}\right) \qquad (6.8)$$

with $K = K_1 K_2$.

(b) $$\frac{d}{dt}\left[K\left(1 - \frac{\tau_1}{\tau_1 - \tau_2} e^{-t/\tau_1} - \frac{\tau_2}{\tau_2 - \tau_1} e^{-t/\tau_2}\right)\right]$$

$$= \frac{K}{\tau_1 - \tau_2}\left(e^{-t/\tau_1} - e^{-t/\tau_2}\right) \qquad (6.9)$$

identical to Eq. (6.8) above.

(c) For a step input, $u(s) = 1/s$, and for the impulse, $u(s) = 1$. Therefore,

$$y_{step}(s) = g(s) \cdot \frac{1}{s}$$

and $$y_{impulse}(s) = g(s)$$

so that

$$s y_{step}(s) = y_{impulse}(s)$$

from where inverse Laplace transformation gives

$$\frac{d}{dt}\left(y_{step}\right) = y_{impulse}$$

as required.

6.4 The consolidated transfer function $g(s)$ is given by

$$g(s) = \frac{K_1(\tau_2 s + 1) + K_2(\tau_1 s + 1)}{(\tau_1 s + 1)(\tau_2 s + 1)}$$

(a) This transfer function rearranges easily to

$$g(s) = \frac{(K_1+K_2)\left[\left(\frac{K_1\tau_2 + K_2\tau_1}{K_1+K_2}\right)s + 1\right]}{(\tau_1 s+1)(\tau_2 s+1)} \quad (6.10)$$

of the form

$$g(s) = \frac{K[\xi s + 1]}{(\tau_1 s+1)(\tau_2 s+1)}$$

with:

$$K = K_1 + K_2 \quad (6.11a)$$

$$\xi = \frac{K_1\tau_2 + K_2\tau_1}{K_1 + K_2} \quad (6.11b)$$

The properties of this transfer function to be established are:

1. Gain, K is as given in (6.11a);
2. The single zero is located at $s = -1/\xi$ where ξ is as given in (6.11b) explicitly. Observe that this expression involves all the 4 contributing system parameters.
3. If K_1 and K_2 are both positive, then, (6.11b) can be rearranged as

$$\xi = \alpha\tau_2 + (1-\alpha)\tau_1 \quad (6.12)$$

where

$$\alpha = \frac{K_1}{K_1+K_2} \quad \text{and} \quad 0 < \alpha < 1$$

And now because (6.12) indicates ξ is a weighted average of τ_1 and τ_2, then if $\tau_1 < \tau_2$,

$$\tau_1 < \xi < \tau_2$$

conversely, if $\tau_2 < \tau_1$ then $\tau_2 < \xi < \tau_1$, as required.

And now, since the condition for overshoot is that $\xi > \tau_2$ and τ_1, and since under the current conditions ξ is "sandwiched" between the two time constants, we conclude that a parallel arrangement of two first-order systems, with $K_1, K_2 > 0$, will never exhibit overshoot.

(b) If $K_1 > 0$ and $K_2 < 0$, but $|K_1| > |K_2|$, then $(K_1 + K_2) > 0$, and, for ξ given in (6.11b) to be negative requires

$$K_1\tau_2 + K_2\tau_1 < 0$$

or

$$\boxed{\frac{K_1}{\tau_1} < -\frac{K_2}{\tau_2}.} \qquad (6.13)$$

(see chapter 7 and keep in mind that K_2 is now negative.)

6.5 In terms of the given deviation variables, and introducing the specified parameters, the process model becomes,

$$\frac{dx_1}{dt} = \frac{-1}{\tau_1}x_1 + \frac{1}{\tau_1}x_2 + \frac{K}{\tau_1}u$$

$$\frac{dx_2}{dt} = \frac{K_2}{\tau_2}x_1 - \left(\frac{1+K_2}{\tau_2}\right)x_2$$

$$y = x_2$$

or:

$$\dot{\underset{\sim}{x}} = \begin{bmatrix} \frac{-1}{\tau_1} & \frac{1}{\tau_1} \\ \frac{K_2}{\tau_2} & -\frac{(1+K_2)}{\tau_2} \end{bmatrix} \underset{\sim}{x} + \begin{bmatrix} K_1\tau_1 \\ 0 \end{bmatrix} u$$

$$y = [0 \quad 1]\, \underset{\sim}{x}$$

(a) The indicated matrices are:

$$\underset{\sim}{A} = \begin{bmatrix} -1/\tau_1 & 1/\tau_1 \\ \frac{K_2}{\tau_2} & -\frac{(1+K_2)}{\tau_2} \end{bmatrix}; \quad \underset{\sim}{B} = \begin{bmatrix} K_1\tau_1 \\ 0 \end{bmatrix}; \quad \underset{\sim}{c}^T = [0 \quad 1]$$

(b) By comparing this current $\underset{\sim}{A}$ matrix with the one in the text (Eq. (6.29)), we observe that the interaction is primarily due to the presence of $1/\tau_1$ in the a_{12} term; but also note that instead of $-1/\tau_2$ alone as the a_{22} term, there is the additional $-K_2/\tau_2$ term.

6.6 (a) The expression in question is:

$$y(t) = K\left[1 - \left(\frac{\tau_1 - \xi_1}{\tau_1 - \tau_2}\right)e^{-t/\tau_1} - \left(\frac{\tau_2 - \xi_1}{\tau_2 - \tau_1}\right)e^{-t/\tau_2}\right]$$

For a maximum, $\left.\frac{dy}{dt}\right|_{t^*} = 0$ AND $\left.\frac{d^2y}{dt^2}\right|_{t^*} < 0$

Now,

$$\frac{dy}{dt} = K\left[\frac{1}{\tau_1}\left(\frac{\tau_1 - \xi_1}{\tau_1 - \tau_2}\right)e^{-t/\tau_1} + \frac{1}{\tau_2}\left(\frac{\tau_2 - \xi_1}{\tau_2 - \tau_1}\right)e^{-t/\tau_2}\right] \qquad (6.14)$$

setting to zero and rearranging gives:

$$\left(\frac{\tau_1 - \xi_1}{\tau_1}\right) e^{-t^*/\tau_1} = \left(\frac{\tau_2 - \xi_1}{\tau_2}\right) e^{-t^*/\tau_2} \qquad (6.15)$$

where t^* is the time at which the max. occurs. Differentiating (6.14) once more gives:

$$\frac{d^2y}{dt^2} = \frac{1}{\tau_1 - \tau_2}\left\{\frac{1}{\tau_2^2}(\tau_2 - \xi_1)e^{-t/\tau_2} - \frac{1}{\tau_1^2}(\tau_1 - \xi_1)e^{-t/\tau_1}\right\} \qquad (6.16)$$

Introducing (6.15) first for eliminating the term involving e^{-t/τ_1}, we have

$$\frac{d^2y}{dt^2} = \frac{1}{\tau_1 - \tau_2}\left\{\frac{1}{\tau_2}\left(\frac{\tau_2 - \xi_1}{\tau_2}\right)e^{-t/\tau_2} - \frac{1}{\tau_1}\left(\frac{\tau_2 - \xi_1}{\tau_2}\right)e^{-t/\tau_2}\right\}$$

$$= \frac{(\tau_2 - \xi_1)e^{-t/\tau_2}}{\tau_2(\tau_1 - \tau_2)}\left\{\frac{1}{\tau_2} - \frac{1}{\tau_1}\right\}$$

$$= \frac{(\tau_2 - \xi_1)e^{-t/\tau_2}}{\tau_2(\tau_1 - \tau_2)}\left\{\frac{(\tau_1 - \tau_2)}{\tau_1\tau_2}\right\}$$

$$= \frac{(\tau_2 - \xi_1)}{\tau_1\tau_2^2}e^{-t/\tau_2}$$

which will be negative if, and only if, $\xi_1 > \tau_2$.

Similarly by eliminating the term involving e^{-t/τ_2} in (6.16) and rearranging yields the corresponding result – that a max occurs only if $\xi_1 > \tau_1$.

(b) From (6.15) we obtain

$$\underbrace{\frac{\tau_2}{\tau_1}\left(\frac{\tau_1 - \xi_1}{\tau_2 - \xi_1}\right)}_{\theta^*} = e^{t^*[-1/\tau_2 + 1/\tau_1]}$$

easily solved to give

$$t^* = t_{max} = \frac{\ln \theta^*}{\frac{1}{\tau_1} - \frac{1}{\tau_2}}$$

or

$$\boxed{t_{max} = \left(\frac{\tau_1\tau_2}{\tau_2-\tau_1}\right)\ln \theta^*}$$

Given that $y(t_1) = K$, we have

$$K = K\left[1 - \left(\frac{\tau_1-\xi_1}{\tau_1-\tau_2}\right)e^{-t_1/\tau_1} - \left(\frac{\tau_2-\xi_1}{\tau_2-\tau_1}\right)e^{-t_1/\tau_2}\right]$$

which simplifies to give

$$\left(\frac{\tau_1-\xi_1}{\cancel{\tau_1-\tau_2}}\right)e^{-t_1/\tau_1} = \left(\frac{\tau_2-\xi_1}{\cancel{\tau_1-\tau_2}}\right)e^{-t_1/\tau_2}$$

solving for t_1 gives:

$$t_1 = \frac{\tau_1\tau_2}{\tau_2-\tau_1}\ln\theta_1$$

where $\theta_1 = \left(\frac{\tau_1-\xi_1}{\tau_2-\xi_1}\right)$.

6.7 The unit step responses are shown in Figs S6.1(a) and S6.1(b). For the system in (a), $J = 2+1-3=0$; and for the system in (b), $J = 2+1-4 = -1$.

The unusual values for J are due to the strong zeros giving rise to overshoots, and hence negative areas.

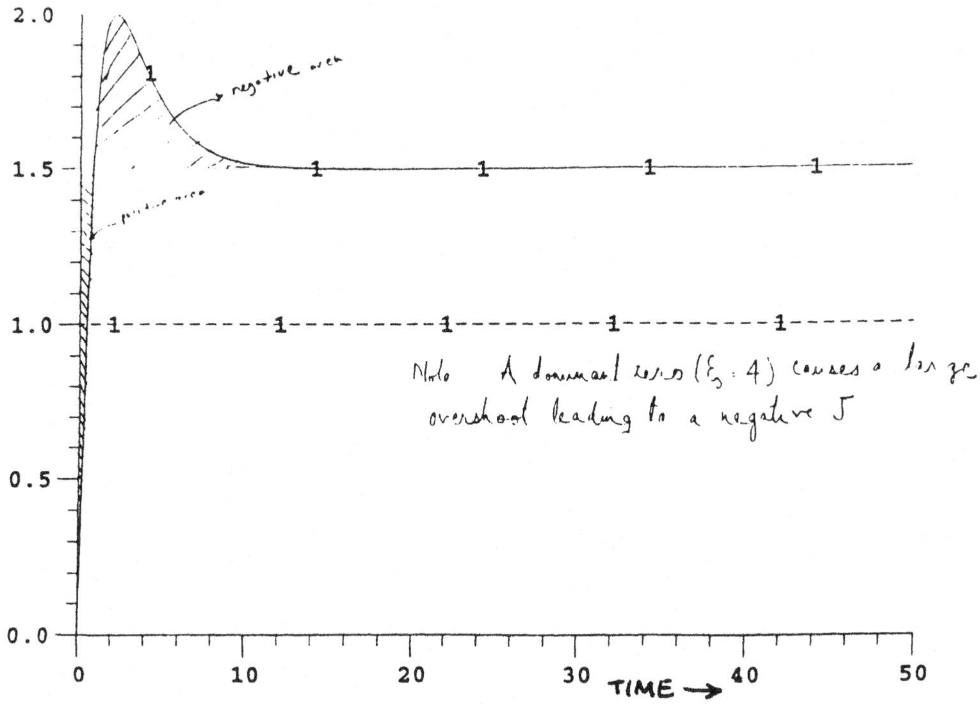

FIG S6.1(a)

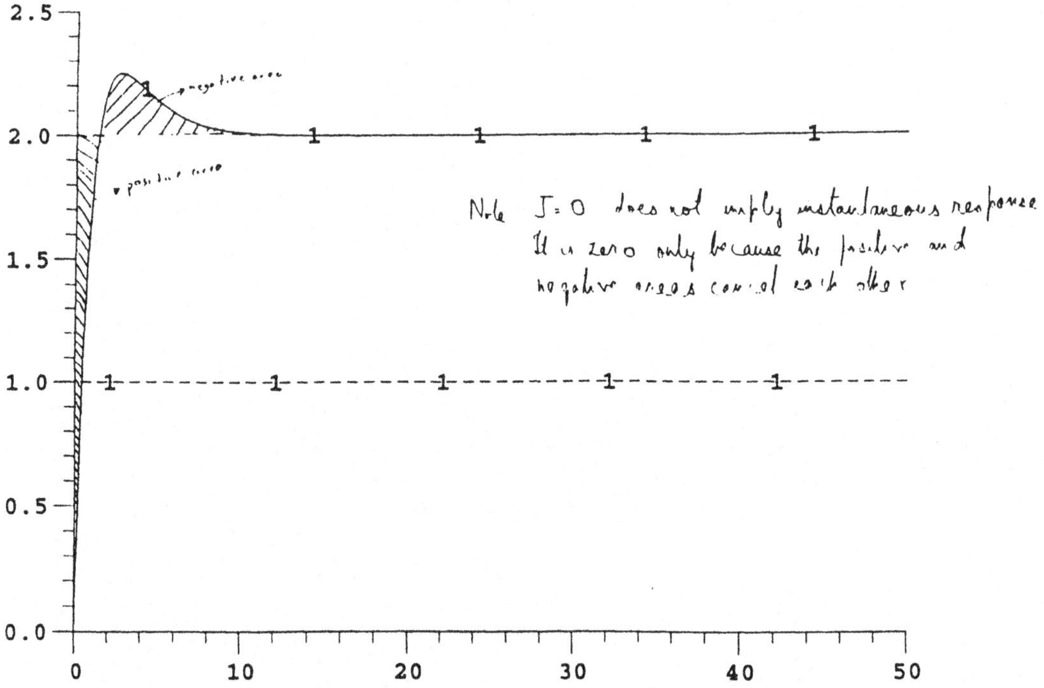

FIG. S6.1(b)

6.8 (a) The system transfer function has:

- 3 poles, located at: $s = -1/6$; $s = \dfrac{-3-\sqrt{5}}{2}$; $s = \dfrac{-3+\sqrt{5}}{2}$
- 2 zeros located at: $s = -\dfrac{1}{12} + \dfrac{\sqrt{23}j}{12}$; $s = -\dfrac{1}{12} - \dfrac{\sqrt{23}j}{12}$

(b) The system's unit step response is shown below in FIG S6.2a. The unusual response is due to the nature of the complex conjugate zeros.

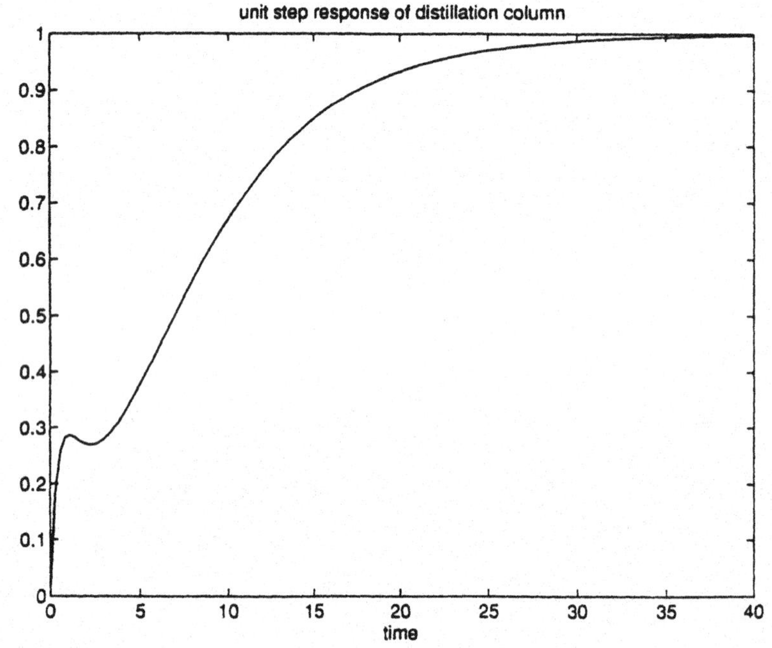

FIG. S6.2a

(c) $g(s)$ can be rewritten as

$$g(s) = \frac{As + B}{s^2 + 3s + 1} + \frac{C}{6s + 1}$$

with $A = 13/19$; $B = -17/19$; $C = 36/19$

Thus:

$$g(s) = \frac{1/19\,(13s - 17)}{s^2 + 3s + 1} + \frac{36/19}{6s + 1} \qquad (6.17)$$

From where we observe the following:

(i) The overall system is composed of two subsystems: a first order system with $\tau_1 = 6$, and $K_1 = 36/19$, and a second order system with $\tau = 1$, $\zeta = 1.5$ and $K_2 = -17/9$ but also with a single zero in the right half of the complex plane, at $s = +17/13$. The second order part is <u>overdamped</u>.

(ii) The first order part has the higher gain value (in absolute values), therefore it ultimately dominates; also, first order systems respond faster than second order systems. So the initial response is also dominated by the first order part.

(iii) The second order system has a <u>negative</u> gain so its contribution is to take the response in a direction which is opposite to that of the first order portion. However, as is shown in Chapter 7, the RHP zero causes the initial response to go in the opposite direction first, thereby reinforcing, first, the first order subsystem response before ultimately opposing it (by virtue of the negative 2nd order subsystem gain). (See Fig S6.2b.)

(iv) The negative gain of the 2nd order subsystem is much smaller in absolute value than the positive gain of the first order subsystem: thus the "opposition" does not last long, and the more substantial 1st order contribution dominates. The composition of all this is Fig S6.2a

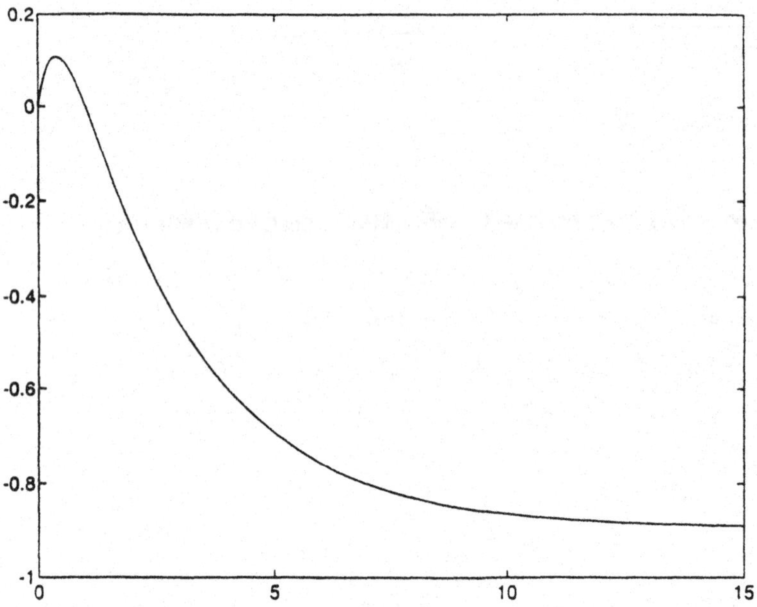

FIG S 6.2b

Unit step response of the second order subsystem with transfer function

$$g(s) = \frac{1/19\,(13s-17)}{s^2+3s+1}.$$

Compare with FIG S6.2a and recall Eq (6.17). This illustrates what is responsible for the "hump" in FIG S 6.2a for the overall system.

6.9 Laplace transforms of the given model equations:

$$sx_1(s) = -k_1 x_1(s) + u(s)$$
$$sx_2(s) = k_1 x_1(s) - k_2 x_2(s)\ ; \quad y(s) = x_2(s)$$

Easily rearranges to give

$$\boxed{y(s) = \frac{k_1}{(s+k_1)(s+k_2)} u(s)} \qquad (6.18)$$

(a) Given $k_1 = 5.63$, $k_2 = 12.62$, and $u(t)$ as in Fig P6.1, we have

$$y(s) = \frac{5.63}{(s+5.63)(s+12.62)}\left[\frac{10(1-e^{-5s})}{s}\right] \qquad (6.19)$$

By partial fraction expansion, this becomes:

$$y(s) = \left[\frac{0.7923}{s} + \frac{0.638}{(s+12.62)} - \frac{1.4306}{(s+5.63)}\right](1-e^{-5s})$$

with the following inverse Laplace transform:

$$y(t) = \begin{cases} 0.7923 + 0.638e^{-12.62t} - 1.4306e^{-5.63t}\ ; & 0 < t < 5 \\ 0.638\left(e^{-12.62t} - e^{-12.62(t-5)}\right) - 1.4306\left(e^{-5.63t} - e^{-5.63(t-5)}\right); & t \geq 5 \end{cases}$$

Max occurs at $t = 5$, and $y(5) = 0.7923$

(b) With $u(t) = 10e^{-k_2 t}$; $t \geq 0$, we have

$$u(s) = \frac{10}{(s+k_2)}$$

and the required response is obtained from

$$y(s) = \frac{10k_1}{(s+k_1)(s+k_2)^2} = \frac{56.3}{(s+5.63)(s+12.62)^2}$$

$$= \frac{A}{(s+5.63)} + \frac{B}{(s+12.62)} + \frac{C}{(s+12.62)^2}$$

Obtain $A = 1.152$; $B = -1.152$; $C = -8.054$

and therefore

$$y(t) = 1.152e^{-5.63t} - 1.152e^{-12.62t} - 8.054te^{-12.62t}$$

(c) With the given $u(t)$, we have

$$u(s) = \frac{20}{s^2+4}$$

and now

$$y(s) = \frac{112.6}{(s+5.63)(s+12.62)(s^2+4)} \cdot \frac{1}{}$$

$$= \frac{A}{(s+5.63)} + \frac{B}{(s+12.62)} + \frac{Cs+D}{s^2+4}.$$

Obtain $A = 0.4512$; $B = -0.0986$; $C = -0.3519$;
$D = 1.2974$

Hence

$$y(t) = 0.4512\,e^{-5.63t} - 0.0986\,e^{-12.62t} + 0.6473\sin 2t - 0.3519\cos 2t$$

6.10 The approximate linear set of equations

$$\dot{x}_1 = \alpha_{11}x_1 - \alpha_{12}x_2 + bu \qquad (6.20)$$

$$\dot{x}_2 = \alpha_{21}x_1 - \alpha_{22}x_2 \qquad (6.21)$$

is to be rearranged to eliminate x_1. This is achieved in two stages.

First, differentiate (6.21) wrt time to obtain,

$$\ddot{x}_2 = \alpha_{21}\dot{x}_1 - \alpha_{22}\dot{x}_2$$

and introduce (6.20) for $\dot{x}_1$, to give, upon rearrangement:

$$\ddot{x}_2 = -\alpha_{22}\dot{x}_2 - \alpha_{12}\alpha_{21}x_2 + \alpha_{11}\alpha_{21}x_1 + \alpha_{21}bu$$

Next, eliminate the surviving x_1 term using (6.21). The result is

$$\ddot{x}_2 = -\alpha_{22}\dot{x}_2 - \alpha_{12}\alpha_{21}x_2 + \alpha_{11}(\dot{x}_2 + \alpha_{22}x_2) + \alpha_{21}bu$$

Further rearrangement now gives

$$\ddot{x}_2 + (\alpha_{22} - \alpha_{11})\dot{x}_2 + (-\alpha_{11}\alpha_{22} + \alpha_{21}\alpha_{12})x_2 = \alpha_{21}bu \qquad (6.22)$$

Comparing this with (P6.8), we have

$$\beta_2 = 1; \quad \beta_1 = \alpha_{22} - \alpha_{11}; \quad \beta_0 = (-\alpha_{11}\alpha_{22} + \alpha_{21}\alpha_{12})$$

and $\phi = \alpha_{21}b$.

or, numerically, $\beta_2 = 1$, $\beta_1 = 1$, $\beta_0 = 1$; $\phi = 0.5$

These parameters imply a second order underdamped system since

$$\tau^2 = \beta_2 = 1 \Rightarrow \tau = 1$$

$$2\zeta\tau = \beta_1 = 1 \Rightarrow \zeta = 1/2 .$$

A plot of the actual unit step response is shown in FIG S6.3

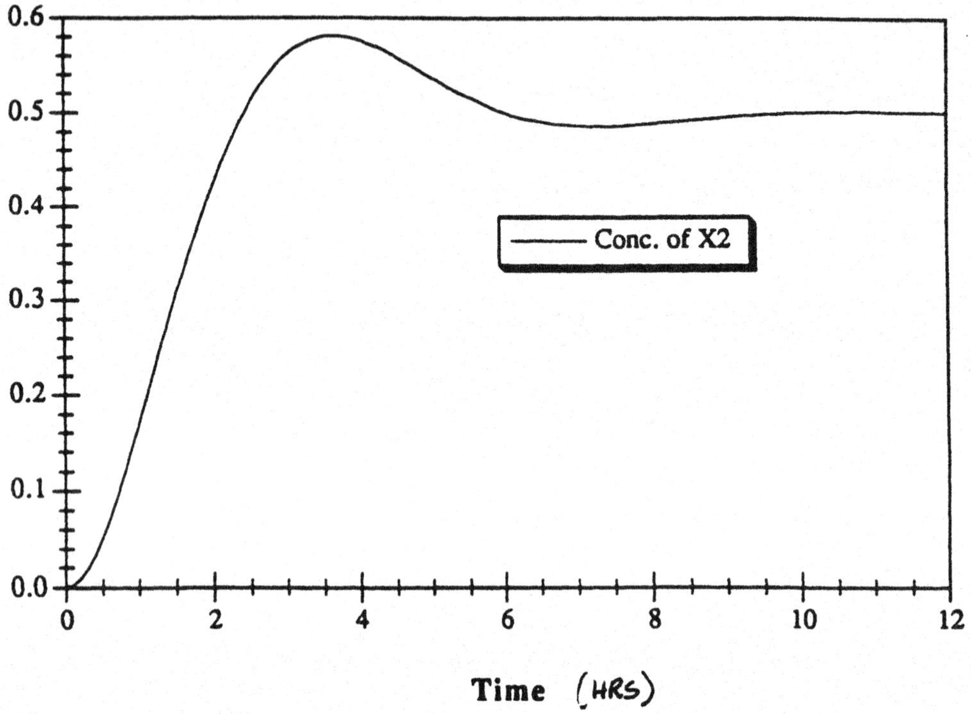

FIG S6.3

The predictions of the simplified model show some fluctuations; they are thus consistent with the general character observed via the pond's monitoring system.

6.11 Taking Laplace transforms of the approximate linear model equations, we obtain

$$sx_1(s) = \alpha_{11}x_1(s) - \alpha_{12}x_2(s) + bu(s) \qquad (6.23a)$$

$$sx_2(s) = \alpha_{21}x_1(s) - \alpha_{22}x_2(s) \qquad (6.23b)$$

From (6.23b) we obtain, first, that

$$x_2(s) = \left(\frac{\alpha_{21}}{s+\alpha_{22}}\right) x_1(s) \qquad (6.24)$$

which may now be introduced into (6.23a) for $x_2(s)$ to obtain, upon rearrangement,

$$x_1(s) = \left(\frac{b(s+\alpha_{22})}{(s-\alpha_{11})(s+\alpha_{22})+\alpha_{12}\alpha_{21}}\right) u(s) \qquad (6.25)$$

This gives the first transfer function, $g_1(s)$. Reintroducing (6.25) into (6.24) now gives

$$x_2(s) = \frac{\alpha_{21}b}{(s-\alpha_{11})(s+\alpha_{22})+\alpha_{12}\alpha_{21}} u(s) \qquad (6.26)$$

indicating the second transfer function, $g_2(s)$.

(a) From (6.25) and (6.26) we have

$$g_1(s) = \frac{b(s+\alpha_{22})}{(s-\alpha_{11})(s+\alpha_{22})+\alpha_{12}\alpha_{21}}$$

and

$$g_2(s) = \frac{\alpha_{21} b}{(s-\alpha_{11})(s+\alpha_{22}) + \alpha_{12}\alpha_{21}}$$

g_1 and g_2 have identical denominators, hence, identical poles; g_1 has steady state gain $K_1 = \alpha_{22} b$; g_2 has gain $K_2 = \alpha_{21} b$; g_1 has a single zero at $s = -\alpha_{22}$; g_2 has no zero.

Introducing numerical values given in Problem (6.10), we obtain

$$g_2(s) = \frac{0.5}{s^2+s+1} \tag{6.27}$$

upon introducing the same values into Eq (6.22) and taking Laplace transforms, we obtain upon rearrangement

$$x_2(s) = \frac{0.5}{s^2+s+1}\, u(s)$$

with the same transfer function as shown in (6.27) above.

(b) The required unit step response, shown in (solid line) FIG S6.4 is obtained from

$$x_1(s) = \frac{0.25(s+2)}{s^2+s+1}$$

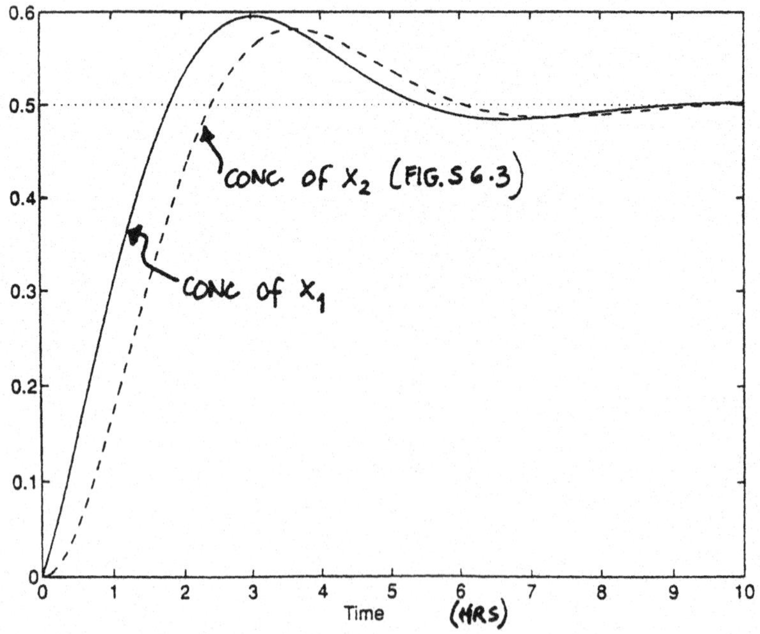

FIG. S 6.4

6.12 (a) The system response to the step change of manitude 0.1 is obtained from

$$y(s) = \frac{3.6}{(12s+1)(15s+1)(20s+1)} \cdot \frac{0.1}{s}$$

$$= \frac{A}{s} + \frac{B}{12s+1} + \frac{C}{15s+1} + \frac{D}{20s+1}$$

obtain $A = 0.36$; $B = -25.92$; $C = 81$; $D = -72$

and by Laplace inversion, obtain the required response as

$$y(t) = 0.36 - 2.16e^{-t/12} + 5.4e^{-t/15} - 3.6e^{-t/20}$$

a plot of which is shown below in FIG S6.5

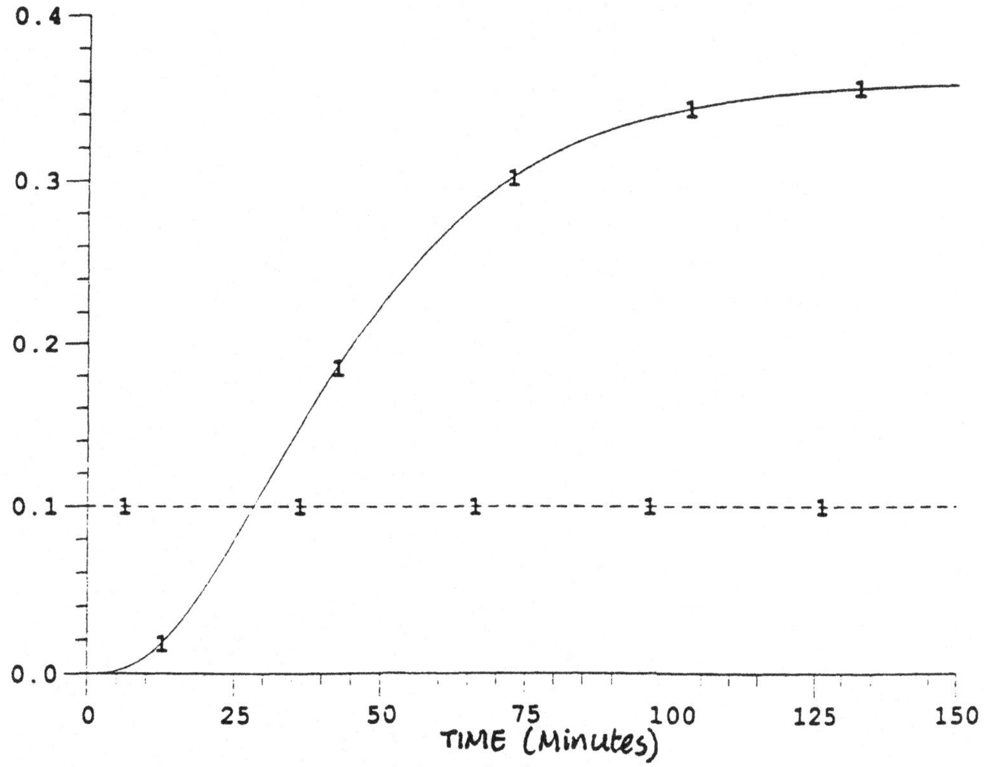

FIG S 6.5

(b) Let $g^+(s)$ be the transfer function of the surviving two-reactor ensemble after having eliminated the 3rd reactor whose transfer function has been given as

$$g_3(s) = \frac{1.5}{20s+1}$$

since
$$g^+(s)\,g_3(s) = \frac{3.6}{(12s+1)(15s+1)(20s+1)}$$

we now obtain immediately that

$$g^+(s) = \frac{2.4}{(12s+1)(15s+1)}. \tag{6.29}$$

The response of this system to a 0.1 step input is easily obtained as: (see FIG. S6.6)

$$y(t) = 0.24 + 0.96\,e^{-t/12} - 1.2\,e^{-t/15}$$

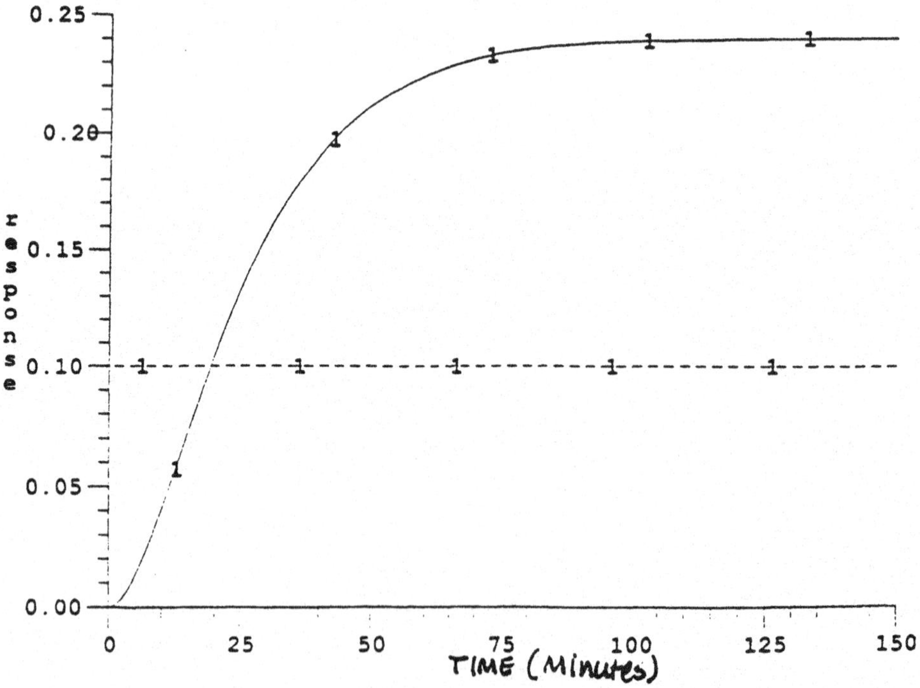

FIG. S6.6

6.13 (a) Model 1 ("One Tank" Model)

$$V\frac{dc}{dt} = Fc_i - Fc$$

or

$$\boxed{y(s) = \frac{1}{(V/F)s + 1}\, u(s)} \qquad (6.30)$$

interms of deviation variables $y = c - c_s$; $u = c_i - c_{is}$.

Model 2: ("Two Tank" Model)

$$\frac{V}{2}\frac{dc_1}{dt} = Fc_i - Fc_1$$

$$\frac{V}{2}\frac{dc_2}{dt} = Fc_1 - Fc_2$$

or

$$y_1 = \frac{1}{(V/2F)s + 1}\, u(s) \qquad (6.31a)$$

$$y_2 = \frac{1}{\left[(V/2F)s + 1\right]^2}\, u(s) \qquad (6.31b)$$

in terms of deviation variables $y_1 = c_1 - c_{1s}$; $y_2 = c_2 - c_{2s}$.

The given Lake superior parameters ⇒ $V/F = 184$ and therefore we have the following step responses ($u = 5$)

- "ONE TANK" model

$$y = \frac{5}{(184s + 1)} \cdot \frac{1}{5} =$$

so that

$$c(t) - c_s(t) = 5\left(1 - e^{-t/184}\right) \qquad (6.32)$$

- "TWO TANK" model

$$y_1 = \frac{5}{(92s+1)}\cdot\frac{1}{s} \quad ; \quad y_2 = \frac{5}{(92s+1)^2}\cdot\frac{1}{s}$$

so that

$$c_1(t) - c_{1s}(t) = 5\left(1 - e^{-t/92}\right) \qquad (6.33a)$$

$$c_2(t) - c_{2s}(t) = 5\left(1 - e^{-t/92} - \frac{t}{92}e^{-t/92}\right) \qquad (6.33b)$$

A plot of $c(t)$, $c_1(t)$ and $c_2(t)$ are shown in FIG S6.7.

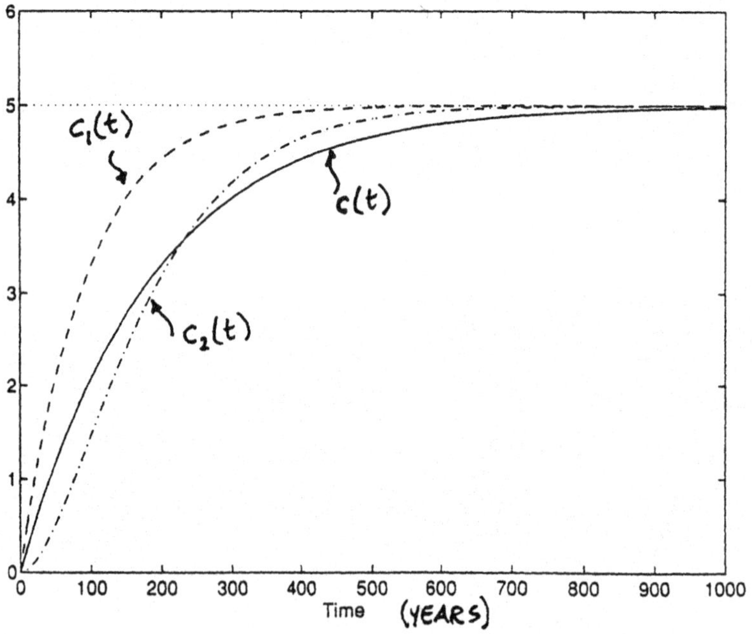

FIG S 6.7

(b) Model 3 ("Three Tank" Model)

$$\frac{V}{3}\frac{dc_1}{dt} = Fc_i - Fc_1$$

$$\frac{V}{3}\frac{dc_2}{dt} = Fc_1 - Fc_2 \quad ; \quad \frac{V}{3}\frac{dc_3}{dt} = Fc_2 - Fc_3$$

or

$$y_1 = \frac{1}{(V/3F)s + 1} u(s)$$

$$y_2 = \frac{1}{\left[\left(\frac{V}{3F}\right)s + 1\right]^2} u(s)$$

$$y_3 = \frac{1}{\left[\left(\frac{V}{3F}\right)s + 1\right]^3} u(s)$$

in terms of the usual deviation variables ($y_3 = c_3 - c_{3s}$). We now have the following step responses.

$$c_1 - c_{1s} = y_1(t) = 5\left(1 - e^{-t/61.3}\right)$$

$$c_2 - c_{2s} = y_2(t) = 5\left(1 - e^{-t/61.3} - \frac{t}{61.3} e^{-t/61.3}\right)$$

$$c_3 - c_{3s} = y_3(t) = 5\left(1 - e^{-t/61.3} - \frac{t}{61.3} e^{-t/61.3} - \frac{t^2}{2 \times (61.3)^2} e^{-t/61.3}\right)$$

A plot of these responses is shown in FIG S6.8.

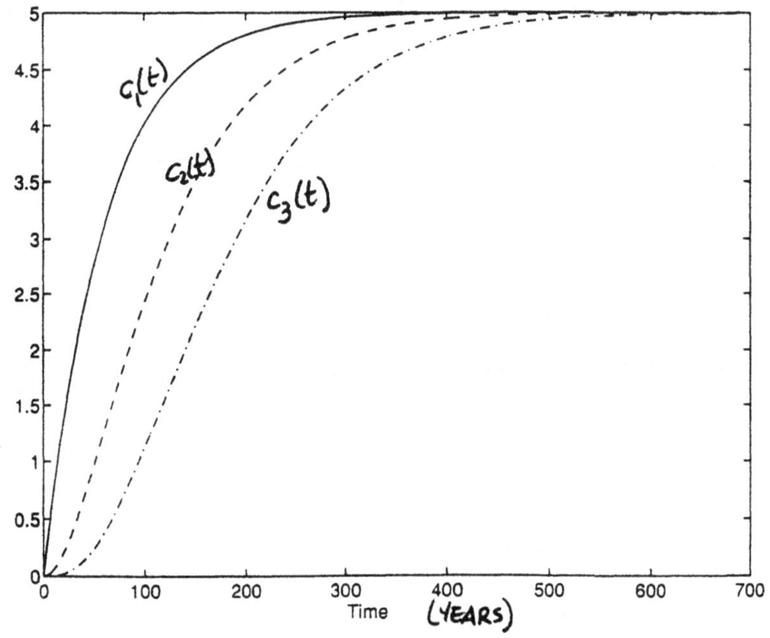

FIG S 6.8

A final comparison of the "outlet" concentration predicted by each of the three different models is shown in FIG S 6.9 below (C for "one tank" model; C_2 for "two-tank" model, and C_3 for "three tank" model).

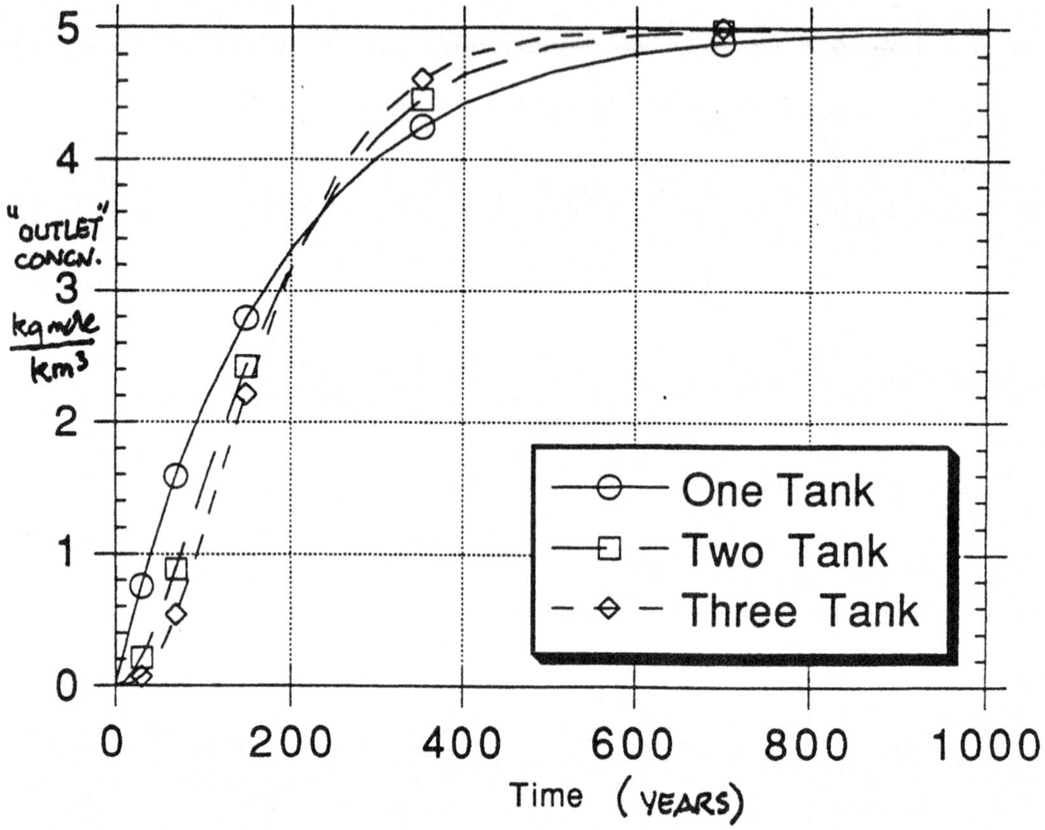

FIG S 6-9

6.14 Solution not provided: ideal for take-home exam.

6.15 (a) From the supplied differential equation model deduce that

$$\tau^2 = 2 \quad \Rightarrow \quad \tau = \sqrt{2}$$

$$2\zeta\tau = 1.414 \quad \Rightarrow \quad \zeta = 0.5$$

characteristic of an underdamped 2nd order system.

(b) For this particular case, $K = 5$, $A = 1$, $\tau = \sqrt{2}$, $\zeta = 0.5$ and from Eqs. (6.37) and (6.38) in the text, the required response is given by:

$$y(t) = 5\left[1 - 1.155\, e^{-0.353t} \sin\left(0.612t + \phi\right)\right] \quad (6.34)$$

with $\phi = \tan^{-1}(1.732) = 60°$ or $\pi/3$ radians.

A plot of this response is shown in FIG S6.10

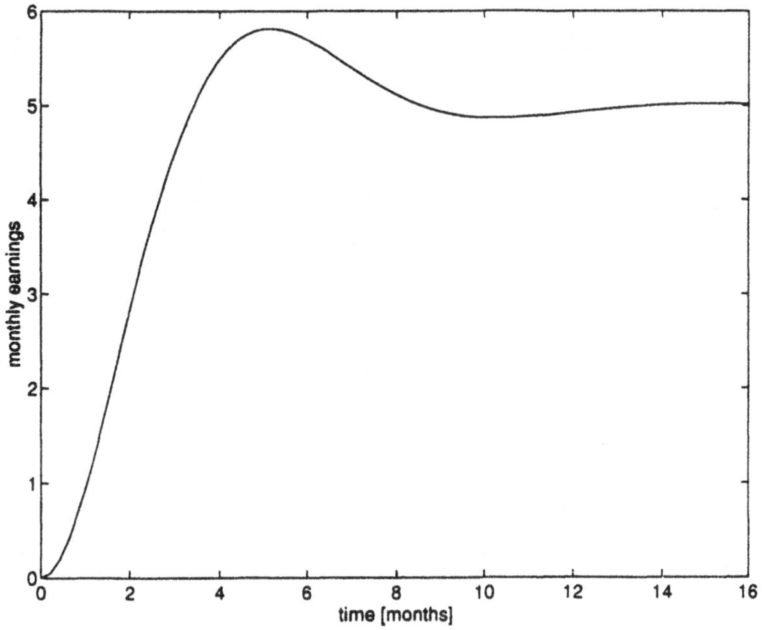

FIG S6.10

(c) From the expression in Eq (6.34) above (or from a knowledge of the fact that the "economic process" has a gain of $K=5$, we conclude that the ultimate value to which the earnings settle is \$5 million. (This is also evident by inspection of FIG. S6.10).

From Eq. (6.53) in the text, the expression for the max value y_{max} is given as

$$y_{max} = 5\left[1 + \exp\left(\frac{-0.5\pi}{0.866}\right)\right]$$
$$= 5(1 + 0.163) = 5.815$$

and from the text's Eq. (6.55) this max occurs at

$$t_{max} = \pi/\omega = \frac{\pi}{\beta/\tau} = \frac{\sqrt{2}\,\pi}{0.866}$$
$$= 5.130 \text{ months.}$$

Thus the maximum earnings made on this product is \$5.815 million; and this occurs 5.130 months after its introduction. Because monthly earnings are computed on a month-by-month basis, in "reality" therefore, the max will occur (to the nearest integer value) in the 5th month.

Note that these answers could also be obtained graphically (from FIG S6.10) but obviously not as accurately.

(d) For this problem, we must return to the original discrete econometric model in Eq (P6.13); specify the input as:

$$u(k) = \begin{cases} 0 & k < 0 \\ 1 & k = 0, 1, 2, \ldots 12 \end{cases}$$

compute $y(k)$ with the initial conditions $y(k) = 0$ for $k < 0$. It is easy to write a simple program to carry out these computations.

The result is shown in Table S6.1. (Note that the max indicated here (5.7826) is close to the value 5.815 obtained with the continuous model; it occurs at $k=5$.

TABLE S6.1.

MONTH k	EARNINGS $y(k)$
1	0.95
2	2.7425
3	4.4459
4	5.4444
5	5.7826 ← MAX
6	5.6822
7	5.4011
8	5.1269
9	4.9494
10	4.8797
11	4.8865
12	4.9280
TOTAL	55.2192

The total earnings realized on this product over its 12-month lifetime is \$55.2192 million.

6.18 By defining

$$\gamma_1 = \frac{BA_sMR}{\dot{V}_A} + \frac{MR}{Q} + A_i$$

and $\gamma_2 = \dfrac{MR}{\dot{V}_A}$, both constants,

and upon using the given consolidated parameters ξ_1, ξ_2 and $\xi_1\xi_2$, the two modeling equations (P6.15) and (P6.16) become, respectively,

$$\xi_1 \xi_2 \ddot{\theta}_T + \left[\xi_1 + \xi_{12} + \xi_2\right] \dot{\theta}_T + \theta_T = BA_S F_{CO_2}^I(t) + \gamma_1 \qquad (6.35)$$

$$\xi_1 \xi_2 \ddot{\theta}_A + \left[\xi_1 + \xi_{12} + \xi_2\right] \dot{\theta}_A + \theta_A = F_{CO_2}^I(t) + \xi_2 \dot{F}_{CO_2}^I(t) + \gamma_2 \qquad (6.36)$$

In terms of the deviation variables

$$y_1 = \theta_T - \theta_T^* \; ; \; y_2 = \theta_A - \theta_A^* \; ; \; u = F_{CO_2}^I(t) - \left(F_{CO_2}^I\right)^*$$

Eqs (6.35) and (6.36) above become

$$\tau_1 \tau_2 \frac{d^2 y_1}{dt^2} + (\tau_1 + \tau_2) \frac{dy_1}{dt} + y_1 = BA_S u \qquad (6.37)$$

$$\tau_1 \tau_2 \frac{d^2 y_2}{dt^2} + (\tau_1 + \tau_2) \frac{dy_2}{dt} + y_2 = u + \xi_2 \frac{du}{dt} \qquad (6.38)$$

where

$$\tau_1 \tau_2 = \xi_1 \xi_2 \qquad (6.39a)$$

$$\tau_1 + \tau_2 = \xi_1 + \xi_{12} + \xi_2 \qquad (6.39b)$$

Taking Laplace Transforms of (6.37) and (6.38) yields

$$(\tau_1 s + 1)(\tau_2 s + 1)\, y_1(s) = K u(s) \quad ; \quad K = BA_S$$

$$(\tau_1 s + 1)(\tau_2 s + 1)\, y_2(s) = (\xi_2 s + 1)\, u(s)$$

which are easily rearranged to give

$$y_2(s) = \frac{(\xi_2 s + 1)}{(\tau_1 s + 1)(\tau_2 s + 1)}\, u(s)$$

$$y_1(s) = \frac{K}{(\tau_1 s + 1)(\tau_2 s + 1)}\, u(s)$$

and thus

$$y_1(s) = \frac{K}{\xi_2 s + 1} \; y_2(s)$$

or

$$y_2(s) = \frac{\xi_2 s + 1}{K} \; y_1(s)$$

as required.

6.17 (a) For overshoot to occur, the zero of the system should be related to the poles as follows:

$$\xi_2 > \tau_2 \qquad (6.40a)$$

$$\xi_2 > \tau_1 \qquad (6.40b)$$

From which we obtain, first by addition, that

$$2\xi_2 > \tau_1 + \tau_2 \qquad (6.41a)$$

and, by multiplication (since all parameters are strictly positive)

$$\xi_2^2 > \tau_1 \tau_2 \qquad (6.41b)$$

Whenever these conditions are satisfied, overshoot will occur.

Now, in terms of the physiological process, we introduce (6.39) into (6.41) and obtain

$$2\xi_2 > \xi_1 + \xi_{12} + \xi_2$$

or $\boxed{\xi_2 > \xi_1 + \xi_{12}}$ (6.42a)

on the one hand, and, on the other

$\xi_2^2 > \xi_1 \xi_2$ or $\boxed{\xi_2 > \xi_1}$ (6.42b)

and now since ξ_1 and ξ_{12} are nonegative physiological parameters, we have finally that:

The step response of the alveolar concentration of CO_2 will show an overshoot if

$$\boxed{\xi_2 > \xi_1 + \xi_{12}}$$

In terms of the actual physiological variables, this implies

$$\frac{K_T}{Q} > \frac{K_A}{\dot{V}_A} + \frac{K_T}{\dot{V}_A} BA_s \tag{6.43}$$

which rearranges to give

$$\dot{V}_A > Q\left[\frac{K_A}{K_T} + BA_S\right]$$

Thus is ventilation rate ($\dot{V}_A$) is high enough) and/or cardiac output (Q) is low enough, it is possible to obtain an overshoot in alveolar concentration of CO_2. Eq (6.43) may also be rearranged as

$$\frac{\dot{V}_A}{Q} > \frac{K_A}{K_T} + BA_S$$

providing a lower limit for the ratio of ventilation rate to the cardiac output, above which overshoot can occur in the alveolar concentration of CO_2.

(b) The given physiological parameters correspond to:

$$\xi_1 = \frac{K_A}{\dot{V}_A} = 0.6 \; ; \; \xi_2 = \frac{K_T}{Q} = 40/6 \; ; \; \xi_{12} = 25.84$$

Thus $\tau_1 \tau_2 = 4$; and $\tau_1 + \tau_2 = 33.106$. The required step responses to a change of 0.1 in the inspired fraction of CO_2 are shown in FIGS S6.11a and b.

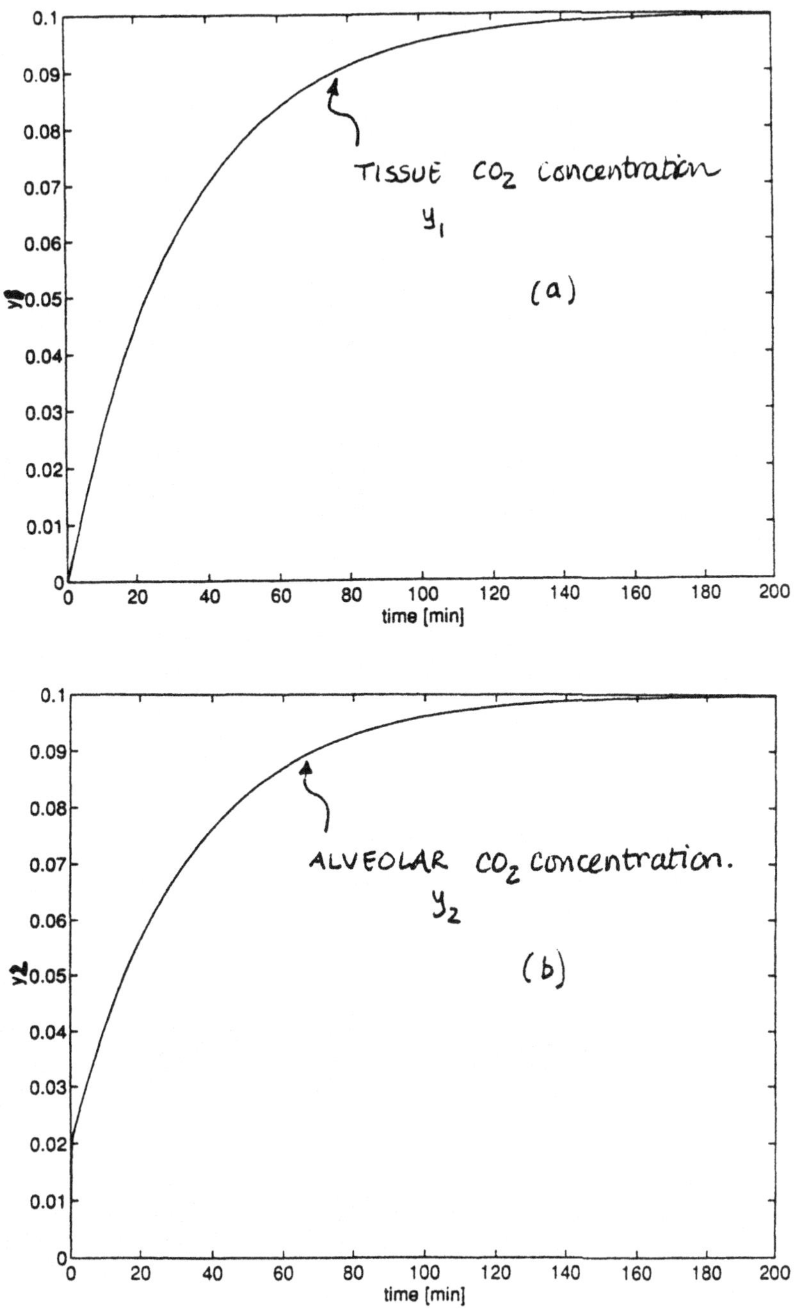

FIG S6.11

6.18 (a)
$$P_{AS}(s) = \left[\frac{-K_1}{\tau s+1} - \frac{K_2}{s(\tau s+1)}\right] Q_H(s)$$

$$P_{AS}(s) = \frac{-(K_1 s + K_2)}{s(\tau s+1)} Q_H(s) \qquad (6.44)$$

so that the required $g(s)$ is given by

$$\boxed{g(s) = \frac{-(K_2 + K_1 s)}{s(\tau s + 1)}} \qquad (6.45)$$

A transfer function with

2 poles: $s = 0;\ s = -1/\tau$

1 zero: $s = -K_2/K_1$

To use this model for studying the rate of transfusion requires reversing the sign associated with the overall transfer function, i.e. Eq. (6.45) should be modified to

$$g_{trans}(s) = \frac{(K_2 + K_1 s)}{s(\tau s + 1)}$$

(b) The response of P_{AS} to a unit step change in Q_H is obtained by Laplace inversion of

$$P_{AS}(s) = \frac{-(K_2 + K_1 s)}{s^2(\tau s + 1)}$$

or, by partial fractions,

$$P_{AS}(s) = \frac{K_2}{s^2} + \frac{(K_1 - K_2\tau)}{s} + \frac{\tau(K_2\tau - K_1)}{\tau s + 1}$$

yielding

$$P_{AS}(t) = -\left\{K_2 t + (K_1 - K_2\tau)\left[1 - e^{-t/\tau}\right]\right\} \qquad (6.46)$$

Observe that this indicates a ramp as the ultimate response; and since the slope of this ramp is $-K_2$ the indication is that the long term effect of hemorrhaging is a linear drop in systemic arterial load pressure. In reality, when P_{AS} falls below a critical value, the patient "passes out" and death ultimately follows if the hemorrhaging persists.

(c) The rectangular pulse responses obtained using the given data are shown in FIG S6.12. From (6.46), it is easy to show that for a rectangular pulse of height A and width b, the ultimate value of $P_{AS}(t)$ is given by

$$P_{AS}(\infty) = -AbK_2. = \begin{cases} -0.8 & \text{for } C_{AS} = 250 \\ -0.48 & \text{for } C_{AS} = 750 \end{cases}$$

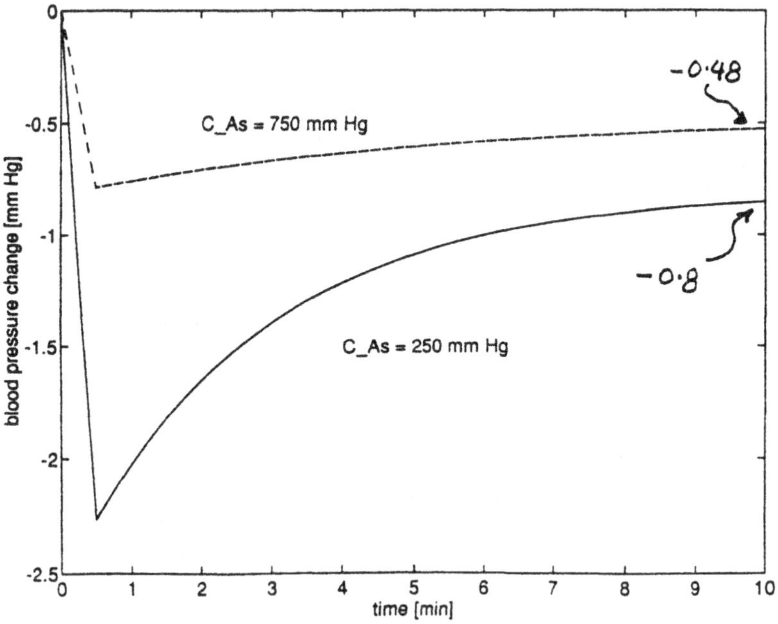

FIG S6.12

CHAPTER 7

7.1 (a) The transfer function in question is

$$g(s) = \frac{-3s+1}{(2s+1)(5s+1)}$$

1. The unit impulse response is obtained from

$$y(s) = \frac{-3s+1}{(2s+1)(5s+1)}$$
$$= \frac{-5/3}{(2s+1)} + \frac{8/3}{5s+1}$$

so that

$$y(t) = -\frac{5}{6}e^{-t/2} + \frac{8}{15}e^{-t/5} \qquad (7.1)$$

2. For the pulse input of magnitude A and duration b,

$$u(s) = A\left(1-e^{-bs}\right)$$

and the required response is obtained from

$$y(s) = \frac{-3s+1}{(2s+1)(5s+1)} A\left(1-e^{-bs}\right)$$

from where:

$$y(t) = \begin{cases} A\left[1+\frac{5}{3}e^{-t/2} - \frac{8}{3}e^{-t/5}\right] & ; \; 0<t<b \\ A\left[\frac{5}{3}\left(e^{-t/2} - e^{-(t-b)/2}\right) - \frac{8}{3}\left(e^{-t/5} - e^{-(t-b)/5}\right)\right] & t \geq b. \end{cases}$$

(b) The required responses are shown in FIG S7.1 where it is clear that the response to the taller and narrower pulse (pulse input 2) is closer to the unit impulse response.

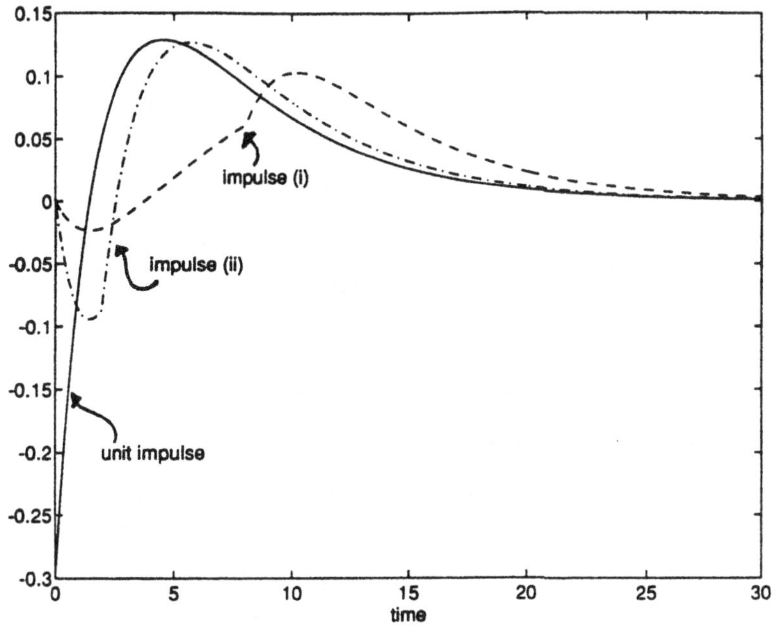

FIG S7.1

(c) For the system with the given transfer function

$$g(s) = \frac{3s+1}{(2s+1)(s+1)} \quad \left(\text{note strong zero!}\right)$$

the unit impulse response is obtained from

$$y(s) = \frac{3s+1}{(2s+1)(s+1)}$$

$$= \frac{-1}{2s+1} + \frac{2}{s+1}$$

so that

$$y(t) = 2e^{-t} - \tfrac{1}{2}e^{-t/2} \tag{7.2}$$

A comparison of the expressions in Eqs (7.1) and (7.2) above indicates:

(i) The initial value ($t = 0$) for the system with the RHP zero is -0.3; the initial value for the system with the <u>strong</u> LHP zero is $+1.5$.

(ii) For the system with the RHP zero, the longer time constant ($\tau = 5$) is associated with the smaller, positive number $8/15$; the shorter time constant ($\tau = 2$) is associated with the larger, negative number, $-5/6$. On the other hand, for the system with the strong LHP zero, the smaller time constant ($\tau = 1$) is associated with the larger positive number 2; the longer time constant ($\tau = 2$) is associated with the smaller negative number $-1/2$.

A plot of these two responses is shown below in FIG S7.2

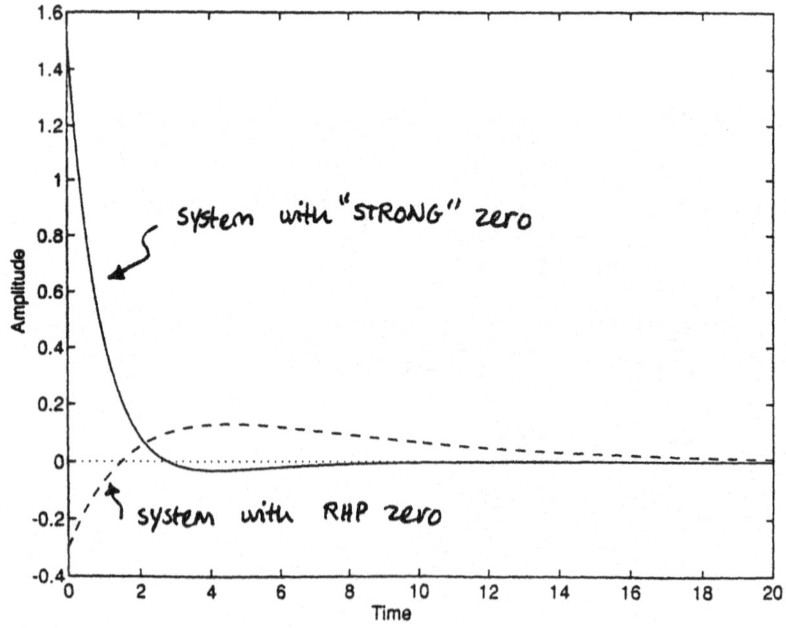

FIG S7.2

7.2 From the step response given in Eq 7.15 in the text, we have, using the definition for J, that

$$J = \int_0^\infty \left(\frac{\tau_1+\eta}{\tau_1-\tau_2}\right)e^{-t/\tau_1} + \left(\frac{\tau_2+\eta}{\tau_2-\tau_1}\right)e^{-t/\tau_2}\,dt$$

$$= K\left[-\frac{\tau_1(\tau_1+\eta)}{\tau_1-\tau_2}e^{-t/\tau_1}\Big|_0^\infty - \frac{\tau_2(\tau_2+\eta)}{\tau_2-\tau_1}e^{-t/\tau_2}\Big|_0^\infty\right]$$

$$= K\left[\frac{\tau_1(\tau_1+\eta)}{(\tau_1-\tau_2)} - \tau_2\frac{(\tau_2+\eta)}{(\tau_1-\tau_2)}\right]$$

$$= K\left[\frac{\tau_1^2-\tau_2^2+\tau_1\eta-\tau_2\eta}{\tau_1-\tau_2}\right]$$

$$= K\left[\frac{(\tau_1+\tau_2)(\tau_1-\tau_2)+\eta(\tau_1-\tau_2)}{(\tau_1-\tau_2)}\right]$$

$$\therefore\ J_{(IR)} = K(\tau_1+\tau_2+\eta) \quad \text{as required.} \qquad (7.3)$$

For the corresponding system with a left-half plane zero, the expression given in Eq (6.89) in the text is

$$J_{(2,1)} = K(\tau_1+\tau_2-\xi_1)$$

where we observe that the lead time constant ξ_1 contributes to the reluctance factor by <u>reducing</u> it; for the system with the RHP zero, on the other hand, the expression in Eq (P7.2) or (7.3) above shows that the lead time constant's contribution <u>increases</u> the reluctance.

7.3 The unit step responses are shown in FIG S7.3(a), (b). The one shows an overshoot; the other shows an "undershoot".

FIG S7.3(a) and (b).

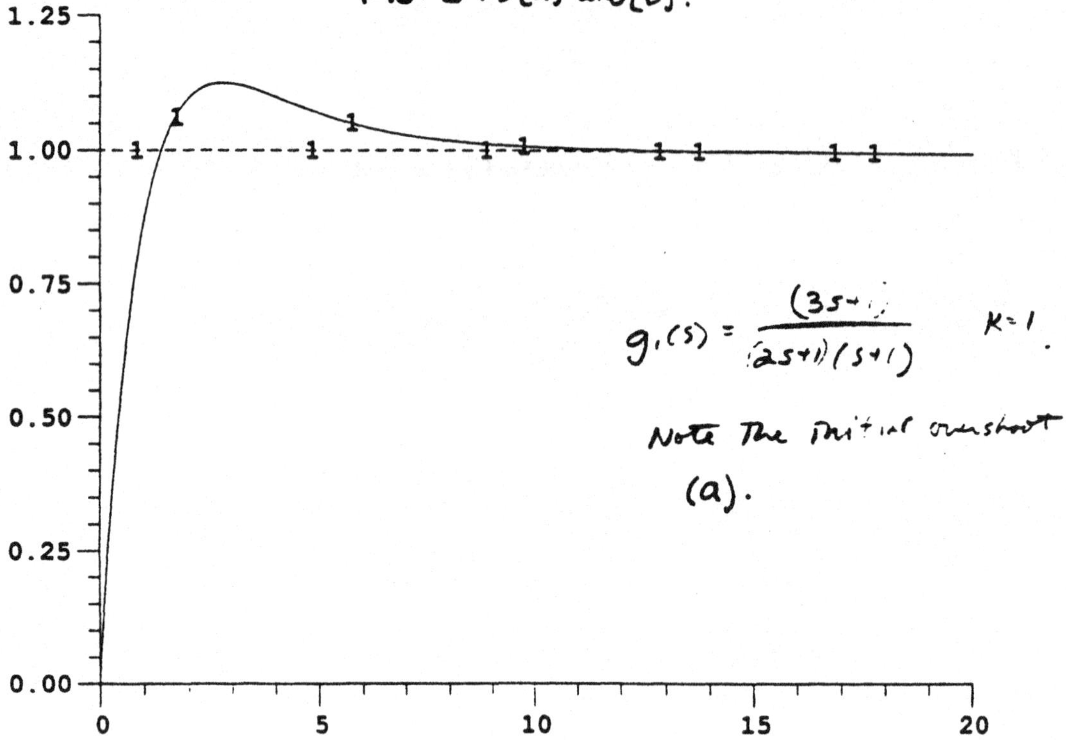

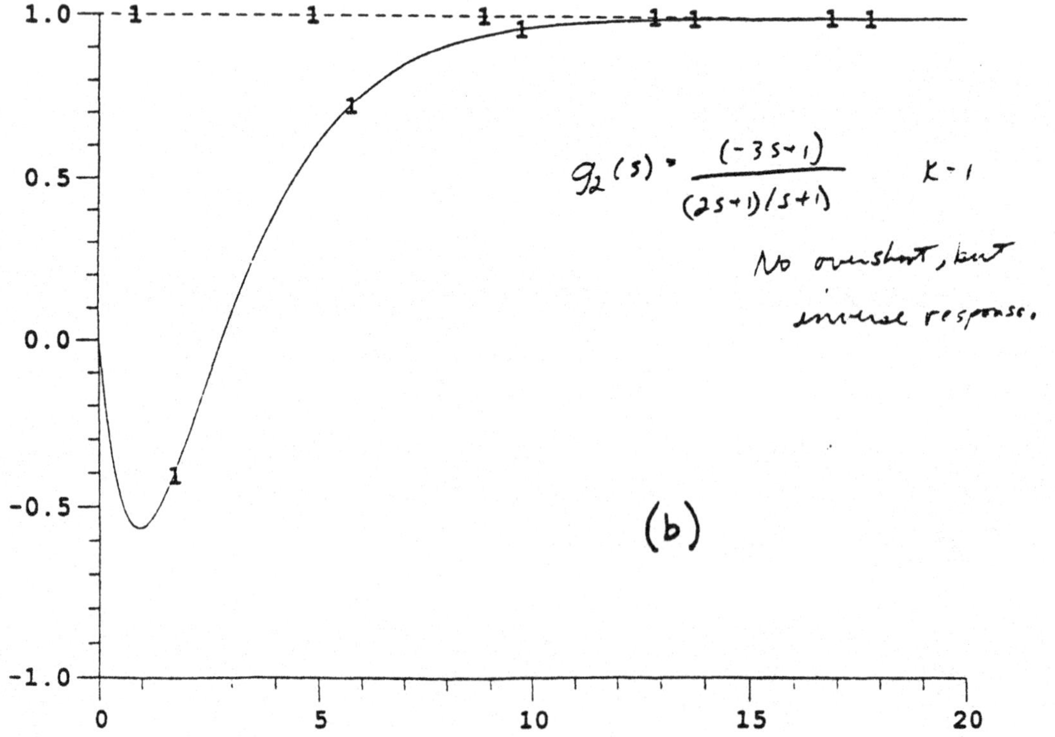

7.4 (a) The composite transfer function in this case is

$$g(s) = \frac{K_1(-\eta s+1)}{(\tau_1 s+1)(\tau_2 s+1)} + \frac{K_2}{(\lambda s+1)}$$

$$= \frac{K_1(-\eta s+1)(\lambda s+1) + K_2(\tau_1 s+1)(\tau_2 s+1)}{(\tau_1 s+1)(\tau_2 s+1)(\lambda s+1)}.$$

which simplifies to

$$g(s) = \frac{as^2 + bs + c}{(\tau_1 s+1)(\tau_2 s+1)(\lambda s+1)} \tag{7.4}$$

where
$$\left.\begin{aligned} a &= K_2\tau_1\tau_2 - K_1\lambda\eta \\ b &= K_1(\lambda-\eta) + K_2(\tau_1+\tau_2) \\ c &= K_1 + K_2 \end{aligned}\right\} \tag{7.5}$$

Thus $g(s)$ is seen to have 3 poles and 2 zeros.

(b) The zeros of $g(s)$ are obtained from the roots of the quadratic numerator polynomial in (7.4), i.e. from

$$as^2 + bs + c = 0 \tag{7.6}$$

The conditions required for these roots to have real parts which are negative are that $a, b, c > 0$. From (7.5) we observe

$c > 0$ because $K_1, K_2 > 0$;

For $b > 0 \Rightarrow$
$$K_1(\lambda-\eta) + K_2(\tau_1+\tau_2) > 0 \tag{7.7}$$

For $a > 0 \Rightarrow$
$$K_2\tau_1\tau_2 > K_1\lambda\eta \tag{7.8}$$

These conditions could be simplified somewhat as follows. By multiplying Eq (7.7) by η (which is positive) we obtain

$$K_1\lambda\eta - K_1\eta^2 + K_2\eta(\tau_1+\tau_2) > 0$$

or

$$K_1\lambda\eta > K_1\eta^2 - K_2\eta(\tau_1+\tau_2) \qquad (7.9)$$

And now using (7.8) we have, from (7.9) that

$$K_2\tau_1\tau_2 > K_1\lambda\eta > K_1\eta^2 - K_2\eta(\tau_1+\tau_2)$$

or simply

$$K_2\tau_1\tau_2 > K_1\eta^2 - K_2\eta(\tau_1+\tau_2)$$

which rearranges to give

$$\boxed{K_2 > \frac{K_1\eta^2}{\tau_1\tau_2 + \eta(\tau_1+\tau_2)}.} \qquad (7.10)$$

Eq (7.8) itself rearranges to

$$\boxed{\lambda < \frac{K_2\tau_1\tau_2}{K_1\eta}.} \qquad (7.11)$$

Thus, K_2 must satisfy the condition in (7.10); and once K_2 is given, (7.11) is the condition λ must satisfy so that the composite system will show no inverse response.

(c) The system in question is

$$g_1(s) = \frac{-3s+1}{(2s+1)(5s+1)}$$

for which $K_1 = 1$; $\eta = 3$; $\tau_1 = 2$; $\tau_2 = 5$.
For the given g_2, $K_2 = 1$ and $\lambda = 3$.

Now, the RHS of (7.10)

$$\frac{K_1 \eta^2}{\tau_1\tau_2 + \eta(\tau_1 + \tau_2)} = \frac{9}{10 + 21} = \frac{9}{31}$$

Observe now that $K_2 > 9/31$, thus the given K_2 satisfies the required condition. Next, the RHS of (7.11)

$$\frac{K_2 \tau_1 \tau_2}{K_1 \eta} = \frac{10}{3}$$

and now we observe that $\lambda = 3 < 10/3$ and therefore we conclude that the proposition is right.

It is easy to show that the consolidated transfer function in this specific case is

$$g(s) = \frac{s^2 + 7s + 1}{(3s+1)(2s+1)(5s+1)}$$

whose 2 zeros all lie in the LHP.

7.5 (a) The consolidated transfer function is

$$g(s) = \frac{K_1}{s} - \frac{K_2}{\tau_2 s + 1}$$

$$= \frac{K_1 - (K_2 - K_1\tau_2)s}{s(\tau_2 s + 1)}$$

with a single zero located at $s = \dfrac{K_1}{K_2 - K_1\tau_2}$

And now, given $K_1 > 0$, this zero will be positive only if

$$K_2 > K_1\tau_2$$

or
$$K_1 < \frac{K_2}{\tau_2} \qquad (7.12)$$

as required.

(b) The condition for inverse response is

$$K_1 < \frac{K_2}{\tau_2}$$

For the identified model, $K_1 = 5.3$ and $K_2/\tau_2 = 4.6$. Observe therefore that in this case K_1 is in fact greater than K_2/τ_2. Since it is a fact that the process in question does exhibit inverse response, then clearly some parameters have not been identified accurately enough.

Now, with the possibility that τ_2 could have been overestimated by as much as 20%, then a possibly more realistic value for τ_2 would be 1.833 as opposed to the reported value of 2.2.

With this new estimate,

$$\frac{K_2}{\tau_2} = \frac{10.1}{1.833} = 5.51$$

and now K_1 will now be less than K_2/τ_2, satisfying (7.12). It is therefore reasonable to conclude that τ_2 could have been overestimated by as much as 20%.

7.6
$$g(s) = \frac{K_1}{\tau_1 s + 1} - \frac{K_2}{(\tau_2 s + 1)^2}$$

or
$$g(s) = \frac{K_1\tau_2 s^2 + (2\tau_2 K_1 - K_2\tau_1)s + (K_1 - K_2)}{(\tau_1 s + 1)(\tau_2 s + 1)^2}$$

The condition for $as^2 + bs + c = 0$ to have <u>no</u> roots in the RHP is that $a, b, c > 0$. In this case, this requires

(i) $(K_1 - K_2) > 0 \Rightarrow K_1 > K_2$, which is given.

(ii) $K_1 \tau_2 > 0$ which is always true for positive K_1 (τ_2 is always > 0)

(iii) $2\tau_2 K_1 - K_2 \tau_1 > 0$

or $$2\left(\frac{K_1}{\tau_1}\right) > \frac{K_2}{\tau_2} \qquad (7.13)$$

Since $K_1 > K_2$, if $K_1 > 0$ then the only way the system in question will show inverse response is if the third condition in Eq (7.13) is violated, i.e. if

$$\frac{K_2}{\tau_2} > 2\left(\frac{K_1}{\tau_1}\right) \qquad (7.14)$$

as required. Note that Eq (7.7) in the text is

$$\frac{K_2}{\tau_2} > \frac{K_1}{\tau_1}.$$

which was the condition for two first order subsystems in opposition to show inverse response. The condition in Eq (P7.10), Eq (7.14) above, is for a <u>second</u> order system with 2 coincident poles in opposition to a first order system. It is interesting to note that in this case K_2/τ_2 is now required to be greater than twice K_1/τ_1 in order that the overall system exhibit inverse response.

7.7 (a) The transfer function relating y to u is obtained by consolidating the following expression:

$$y(s) = \left[\frac{-K_1}{|K_2-K_1|}\frac{1}{s(\tau_1 s+1)} + \frac{K_2}{|K_2-K_1|}\frac{1}{s(\tau_2 s+1)(\tau_3 s+1)}\right]u(s)$$

Since $K_2 > K_1$, $|K_2-K_1| = K_2-K_1$, and the implied transfer function is given by

$$g(s) = \frac{1}{K_2-K_1}\left\{\frac{-K_1\tau_2\tau_3 s^2 + [K_2\tau_1 - K_1(\tau_2+\tau_3)]s + (K_2-K_1)}{s(\tau_1 s+1)(\tau_2 s+1)(\tau_3 s+1)}\right\}$$

The numerator may be written as

$$as^2 + bs + 1 \qquad (7.15)$$

where

$$a = \frac{-K_1\tau_2\tau_3}{K_2-K_1} \qquad (7.16a)$$

$$b = \frac{K_2\tau_1 - K_1(\tau_2+\tau_3)}{K_2-K_1} \qquad (7.16b)$$

The conditions given in (P7.16) $\Rightarrow (K_2-K_1)$ is positive, and (P7.17) $\Rightarrow K_2\tau_1 - K_1(\tau_2+\tau_3)$ is negative; thus $b<0$.
Now, factoring (7.15) as

$$(\xi_1 s+1)(\xi_2 s+1) = as^2+bs+1 \qquad (7.17)$$

we now have that for (7.15) to have two roots in the RHP, $\xi_1\xi_2$ must be positive (i.e a in Eq (7.16a) above must be positive); in addition, $\xi_1+\xi_2$ (same as b in (7.15)) must be negative.

But from (7.16a) above, we see that for $K_1>0$, a will always be negative. Thus we conclude that $g(s)$ does not have two RHP zeros. Alternatively if either ξ_1 or ξ_2 is negative, so that $g(s)$ will have a single RHP zero (not two), (7.17) implies that a must be negative, which is in fact the case. Observe that for $g(s)$ to have no RHP zero, both a and b will have to be positive. But both a and b are negative. Thus the only conclusion is that under the given conditions, $g(s)$, the transfer function relating y to u, has a SINGLE RHP zero.

(b) $K_1 = 0.5 K_2$ so that clearly $K_2 > K_1$ and the individual in question has a healthy immune system. Furthermore,

$$\frac{\tau_1}{\tau_2 + \tau_3} = \frac{2}{8} = 0.25$$

and

$$\frac{K_1}{K_2} = 0.5$$

so that

$$\frac{\tau_1}{\tau_2 + \tau_3} < \frac{K_1}{K_3}$$

and the vaccine is effective.

Under these conditions the response to $u(t) = 5\delta(t)$ is obtained as (upon doing the required algebra)

$$y(t) = 5\left[1 + e^{-t/2} + 3e^{-t/3} - 5e^{-t/5}\right] \qquad (7.15)$$

a plot of which is shown in FIG S7.4

(c) y is a measure of "health" with respect to the amount of pathogen compared to the amount of corresponding antibodies:

$y = 0$: base case (no antibodies, no pathogens) or equal "number" of both.

$y < 0$ sicker system (fewer antibodies than pathogens)

$y > 0$ healthier system (more antibodies than pathogens)

when y is normalized by the amount of vaccine employed,

$y = 1$ implies complete immunization in which all the pathogens are eliminated, and only antibodies remain.

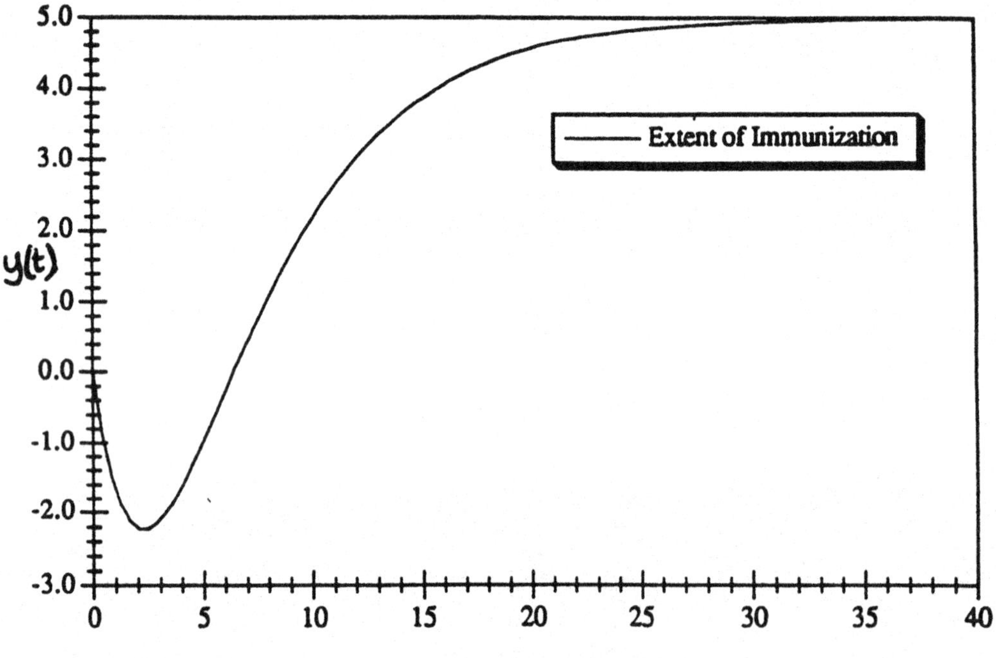

FIG S7.4

7.8 In general, the expression for the unit impulse response is

$$y(s) = \frac{1}{|K_2 - K_1|} \frac{\left\{-K_1\tau_2\tau_3 s^2 + \left[K_2\tau_1 - K_1(\tau_1+\tau_3)\right]s + (K_2-K_1)\right\}}{s(\tau_1 s+1)(\tau_2 s+1)(\tau_3 s+1)}$$

which rearranges to give

$$y(s) = \frac{\frac{(K_2 - K_1)}{|K_2 - K_1|}\left\{\left(\frac{-K_1}{K_2-K_1}\right)\tau_2\tau_3 s^2 + \frac{K_2\tau_1 - K_1(\tau_2-\tau_3)}{K_2 - K_1} s + 1\right\}}{s(\tau_1 s+1)(\tau_2 s+1)(\tau_3 s+1)}$$

This has the general form

$$y(s) = \frac{\mathcal{K}(as^2 + bs + 1)}{s(\tau_1 s+1)(\tau_2 s+1)(\tau_3 s+1)} \tag{7.16}$$

with

$$\mathcal{K} = \frac{(K_2 - K_1)}{|K_2 - K_1|} \tag{7.17}$$

$$a = \left(\frac{-K_1}{K_2 - K_1}\right)\tau_2\tau_3 \tag{7.18}$$

$$b = \frac{K_2\tau_1 - K_1(\tau_2 - \tau_3)}{K_2 - K_1} \tag{7.19}$$

The corresponding time response, $y(t)$, now has the following <u>general</u> characteristics:

1. <u>Final Value</u>

$$\lim_{t\to\infty} y(t) = \lim_{s\to 0} s\,y(s) = \lim_{s\to 0} \frac{\mathcal{K}(as^2+bs+1)}{(\tau_1 s+1)(\tau_2 s+1)(\tau_3 s+1)}$$

$$\boxed{\lim_{t\to\infty} y(t) = \mathcal{K} = \frac{K_2 - K_1}{|K_2 - K_1|}} \tag{7.20}$$

2. Nature of response

Primary determining factors: the nature of the numerator polynomial in (7.16); in turn determined by the signs associated with a and b.

From factoring $as^2 + bs + 1$ as $(\xi_1 s + 1)(\xi_2 s + 1)$ we have the following:

(i) $a < 0$; $b < 0$ $\Rightarrow$ one RHP zero $\Rightarrow$ inverse response
(ii) $a < 0$; $b > 0$ $\Rightarrow$ one RHP zero $\Rightarrow$ inverse response
(iii) $a > 0$, $b > 0$ $\Rightarrow$ NO RHP zero $\Rightarrow$ no inverse response
(iv) $a > 0$, $b < 0$ $\Rightarrow$ two RHP zeros $\Rightarrow$ two initial inversions

We may now use these facts to establish the required results

(a) For the unhealthy immune system, $K_1 > K_2$ implies

$$\mathcal{K} = \frac{K_2 - K_1}{|K_2 - K_1|} = -1$$

and from Eq (7.20) above, we observe that y will settle to a value of -1.

For an effective vaccine, the condition in (P7.17) implies that the numerator in Eq (7.19) above for b is negative. The unhealthy immune system implies that the denominator is also negative. Thus $b > 0$ in this case.

Also, a as given in Eq (7.18) above will be positive. Thus, with $a > 0$, $b > 0$, this corresponds to (iii) above with no RHP zero and therefore the response will show no inverse response. And the result is established.

(b) The healthy immune system means $\mathcal{K} = 1$ since $K_2 > K_1$. Thus y will settle to a value of $+1$.

The less-than-effective vaccine condition in (P7.19) means that the numerator of the expression for b is positive; the healthy immune system also means that the denominator is positive; so that in this case, $b > 0$.

Also, a as in (7.18) above will be <u>negative</u>. Thus with $a < 0$ and $b > 0$, this corresponds to (ii) above, with a single RHP zero so that the response shows inverse response as it approaches its final value of $+1$. And the result is established.

(c) The unhealthy immune system implies $\mathcal{K} = -1$ and hence that y will settle to an ultimate value of -1. The less-than-effective vaccine condition in (P7.19) means that the numerator of b is positive, but the unhealthy immune system means that the denominator is negative, the net effect: $b < 0$.

The unhealthy immune system also means that $a > 0$. Thus, with $a > 0$ and $b < 0$, this corresponds to (iv) above, with <u>two</u> RHP zeros. The response thus shows <u>two</u> inversions initially as it heads to its final value of -1.

(d) <u>CASE 1</u>: From the given parameters, we have:

$$K_1 > K_2 \quad \Rightarrow \quad \underline{\text{unhealthy}} \text{ immune system.}$$

and,

$$\frac{\tau_1}{\tau_2 + \tau_3} = \frac{2}{8} = 0.25 \; ; \quad \frac{K_1}{K_2} = 2, \quad \text{so that}$$

$$\frac{\tau_1}{\tau_2 + \tau_3} < \frac{K_1}{K_2} \quad \Rightarrow \quad \underline{\text{effective}} \text{ vaccine}$$

These are the same conditions as those investigated in part (a) above. The response is shown in the solid line in FIG S7.5, confirming the earlier result.

CASE 2 Here,

$$K_2 > K_1 \quad \Rightarrow \quad \text{healthy immune system}$$

and

$$\frac{K_1}{K_2}(\tau_2+\tau_3) = 0.5(3+2) = 0.25 \;; \text{ and } \tau_1 = 4$$

so that

$$\frac{K_1}{K_2}(\tau_2+\tau_3) < \tau_1 \quad \Rightarrow \quad \text{less effective vaccine}$$

These are the same conditions as those investigated in part (b) above. The specific response is shown in the dashed line in FIG S7.5, again confirming the earlier result: inverse response, final value of +1.

CASE 3 Here,

$$K_1 > K_2 \quad \Rightarrow \quad \text{unhealthy immune system};$$

and

$$\frac{K_1}{K_2}(\tau_2+\tau_3) = 3(1+1/3) = 4 \;; \text{ but } \tau_1 = 7$$

so that

$$\frac{K_1}{K_2}(\tau_2+\tau_3) < \tau_1 \quad \Rightarrow \quad \text{less effective vaccine.}$$

These are the same conditions as those investigated in part (c) above. The specific response in this case is shown in the dashed-dotted line in FIG S7.5: a double inversion followed by an approach to the final value of −1, confirming the earlier result.

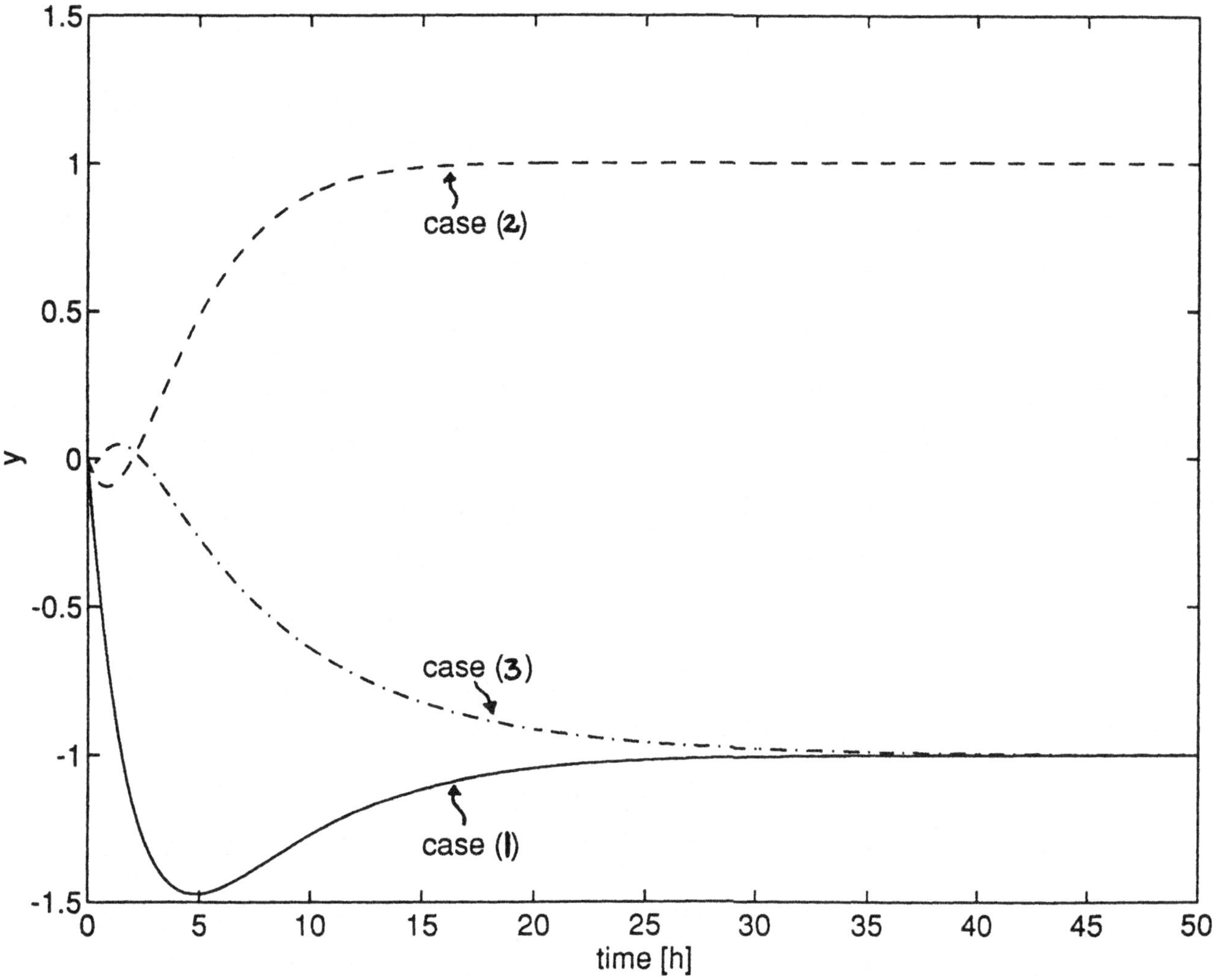
1.5
1
0.5
0
-0.5
-1
-1.5
y
0
5
10
15
20
25
30
35
40
45
50
time [h]
case (2)
case (3)
case (1)

FIG S 7.5

CHAPTER 8

8.1 The equation in question is of the form

$$\frac{dy}{dz} = -ay + bu \tag{8.1}$$

where
$$a = \frac{s+\beta}{v} \tag{8.2a}$$

$$b = \frac{\beta}{v} \tag{8.2b}$$

And by using the method of integrating factors, or from Eq (B.42) in Appendix B, the solution to (8.1) above is:

$$y(z,s) = e^{-az}y(0,s) + \int_0^z e^{-a(z-\sigma)}bu(s)\,d\sigma$$
$$= e^{-az}y(0,s) + bu(s)\,e^{-az}\int_0^z e^{a\sigma}d\sigma \tag{8.3}$$

since $u(s)$ is not a function of z and can therefore be taken out from under the integral sign. Carrying out the indicated integration in (8.3) above now leads to

$$y(z,s) = e^{-az}y(0,s) + \frac{b}{a}\left(1-e^{-az}\right)u(s)$$

If we now introduce (8.2a) and (8.2b) we obtain

$$y(z,s) = e^{-\left(\frac{s+\beta}{v}\right)z}y(0,s) + \left(\frac{\beta}{s+\beta}\right)\left[1-e^{-\left(\frac{s+\beta}{v}\right)z}\right]u(s)$$

precisely as obtained in Eq (8.37) in the text.

8.2 (a) The desired response is shown below in FIG S8.1

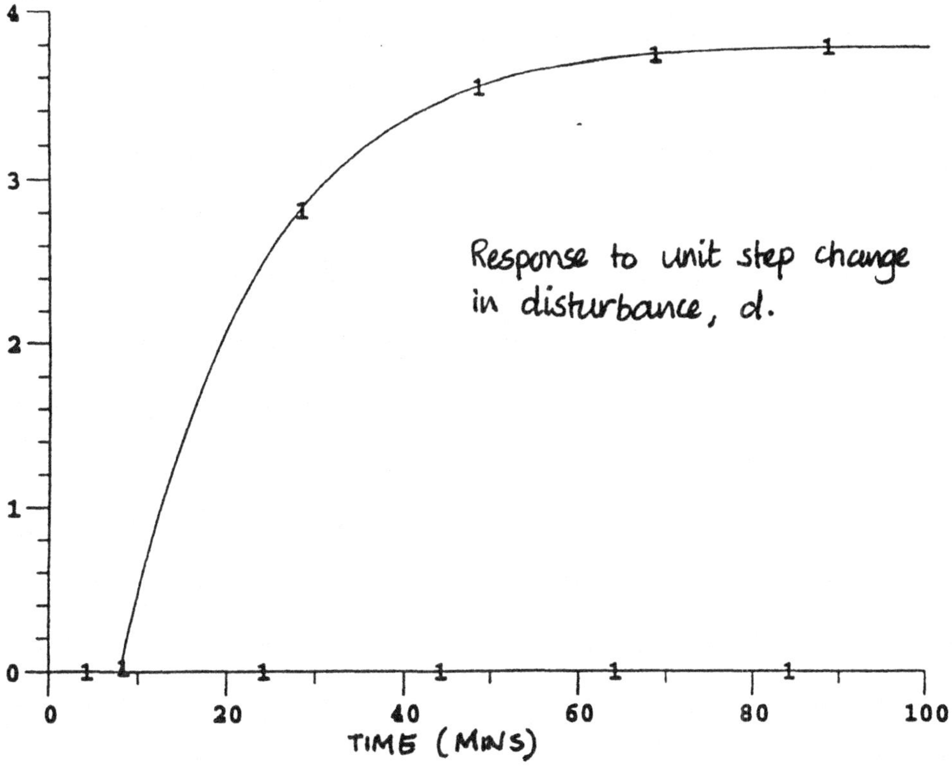

FIG S8.1

(b) Given

$$u(s) = g_{FF}(s)\, d(s)$$

with

$$g_{FF}(s) = -\frac{g_d}{g}$$

For this specific distillation column, we have

$$u(s) = \frac{-3.8}{12.8}\frac{(16.7s+1)}{(14.9s+1)} e^{-7s} d(s)$$

and the indicated transfer function between d and u is that of a lead/lag system with gain $K=-0.2969$, lead time constant $\xi = 16.7$, lag time constant $\tau = 14.9$, a

time delay of 7, and a lead-to-lag ratio $\rho = 1.121$
Thus, the response observed in u as a result of a step change in d will be given by

$$u(t) = \begin{cases} 0 & t \leq 7 \\ -0.2969\left(1 + 0.121\, e^{-(t-7)/14.9}\right); & t > 7 \end{cases}$$

(see Eq (5.66) in the main text).

A sketch of this response is shown in FIG S8.2 below.

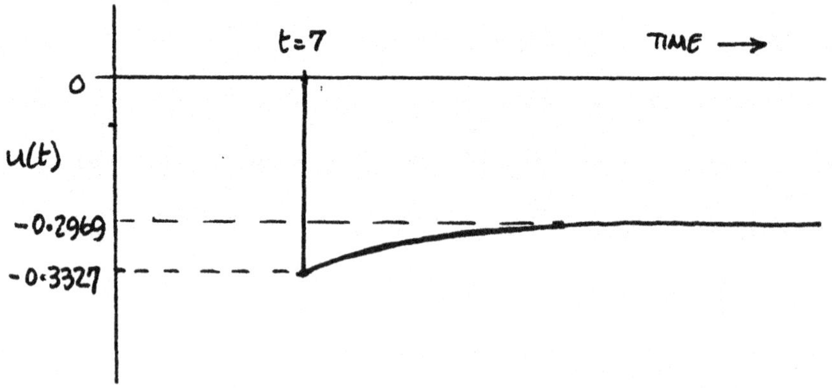

FIG S8.2

8.3 The "staircase" function is defined mathematically as:

$$u(t) = \begin{cases} 0 & t \leq 0 \\ 1 & 0 < t \leq 2 \\ 2 & 2 < t \leq 4 \\ 3 & 4 < t \leq 6 \\ 4 & t > 6 \end{cases}$$

By introducing the Heaviside function, $H(t)$ defined

as follows: (see Eq (3.37) in the text)

$$H(t) = \begin{cases} 0 & t \leq 0 \\ 1 & t > 0 \end{cases}$$

and also the delayed Heaviside function

$$H(t-\alpha) = \begin{cases} 0 & t \leq \alpha \\ 1 & t > \alpha \end{cases}$$

Observe that this "staircase" function may be compactly represented as:

$$u(t) = H(t) + H(t-2) + H(t-4) + H(t-6) \qquad (8.4)$$

Since now $\mathcal{L}\{H(t)\} = 1/s$, we observe that taking Laplace transforms of Eq (8.4) above yields

$$u(s) = \frac{1}{s} + \frac{e^{-2s}}{s} + \frac{e^{-4s}}{s} + \frac{e^{-6s}}{s}$$

or

$$u(s) = \frac{1}{s}\left[1 + e^{-2s} + e^{-4s} + e^{-6s}\right] \qquad (8.5)$$

And now, for the process represented by the given transfer function model, the required reactor temperature response is given by

$$y(s) = \left(\frac{0.32}{15.5s+1}\right)\frac{1}{s}\left[1 + e^{-2s} + e^{-4s} + e^{-6s}\right]$$

or

$$y(s) = \underbrace{\frac{0.32}{s(15.5s+1)}}_{y_0(s)} + \underbrace{\frac{0.32e^{-2s}}{s(15.5s+1)}}_{y_I(s)} + \underbrace{\frac{0.32e^{-4s}}{s(15.5s+1)}}_{y_{II}(s)} + \underbrace{\frac{0.32e^{-6s}}{s(15.5s+1)}}_{y_{III}(s)} \qquad (8.6)$$

Keeping in mind that the effect of the delays is merely to shift the undelayed responses, the Laplace inversion of each component of (8.6) now yields

$$y_0(t) = 0.32\left[1 - e^{-t/15.5}\right]$$

$$y_I(t) = \begin{cases} 0 & t \le 2 \\ 0.32\left[1 - e^{-(t-2)/15.5}\right]; & t > 2 \end{cases}$$

$$y_{II}(t) = \begin{cases} 0 & ; t \le 4 \\ 0.32\left[1 - e^{-(t-4)/15.5}\right]; & t > 4 \end{cases}$$

$$y_{III}(t) = \begin{cases} 0 & t \le 6 \\ 0.32\left[1 - e^{-(t-6)/15.5}\right]; & t > 6 \end{cases}$$

And now, the required response is given by

$$y(t) = y_0(t) + y_I(t) + y_{II}(t) + y_{III}(t)$$

and we obtain, after carrying out the simple algebra implied:

$$y(t) = \begin{cases} 0.32\left[1 - e^{-t/15.5}\right] & ; 0 < t \le 2 \\ 0.64\left[1 - \tfrac{1}{2}\left(e^{-t/15.5} + e^{-(t-2)/15.5}\right)\right. & ; 2 < t \le 4 \\ 0.96\left[1 - \tfrac{1}{3}\left(e^{-t/15.5} + e^{-(t-2)/15.5} + e^{-(t-4)/15.5}\right)\right. ; & 4 < t \le 6 \\ 1.28\left[1 - \tfrac{1}{4}\left(e^{-t/15.5} + e^{-(t-2)/15.5} + e^{-(t-4)/15.5} + e^{-(t-6)/15.5}\right)\right. ; & t > 6 \end{cases}$$

Which is precisely what was obtained earlier (in Prob 5.6).

8.4 The responses obtained from all 3 models are shown in FIG S8.3 where it is very difficult to distinguish one from the other. FIG S8.4 shows the initial portion of each of the 3 responses, and now we may be able to distinguish them.

MODEL 1 (The original)

$$g(s) = \frac{40e^{-20s}}{(900s+1)(25s+1)}$$

MODEL 2 (Neglecting the smaller time constant)

$$g(s) = \frac{40e^{-20s}}{900s+1}$$

MODEL 3 (Higher order model; no time delay)

$$g(s) = \frac{40}{(900s+1)(13s+1)^3}$$

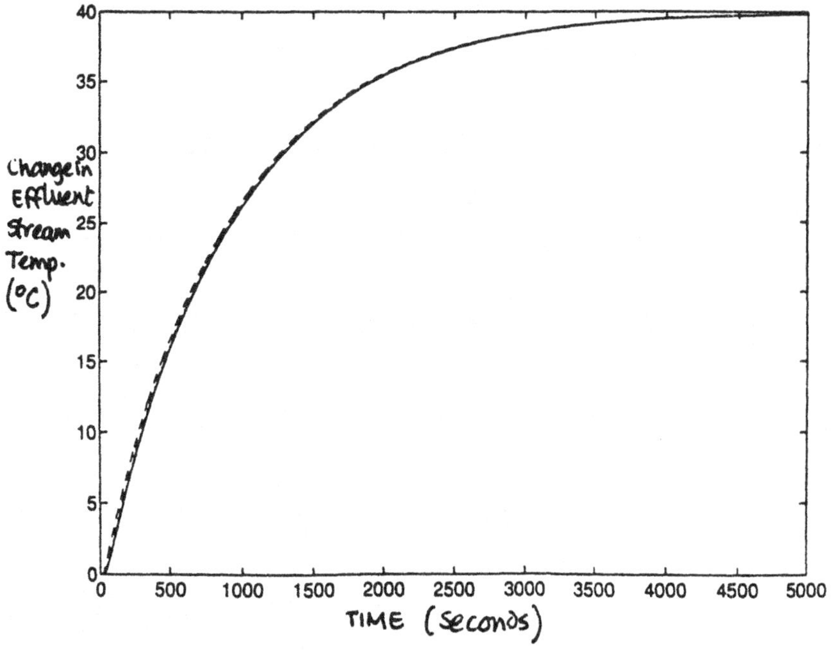

FIG S8.3

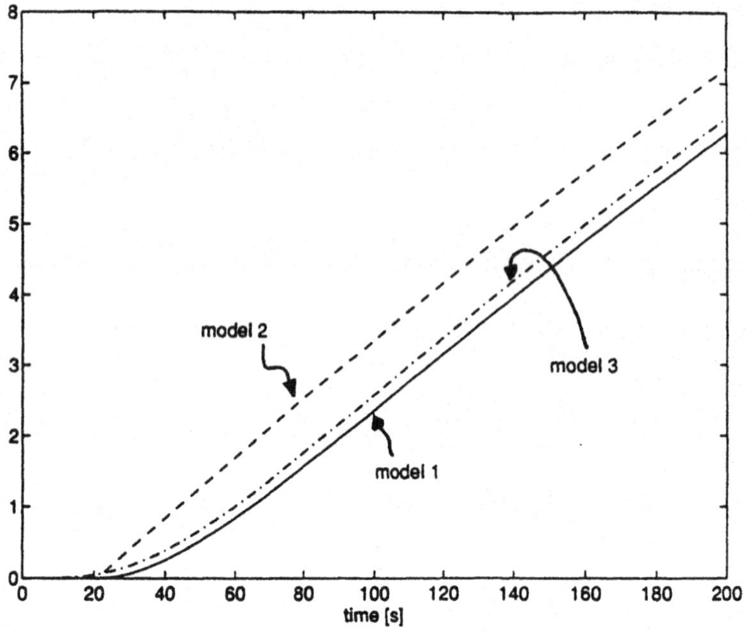

FIG S8.4

One of the main points of this exercise is that, since these different models have almost indistinguishable step responses (see FIG S8.3) whichever is more convenient for any specific application may therefore be employed.

For example, for the design of simple PID controllers, model 2 will be more convenient (See Chapter 15, sec. 15.4) On the other hand, for closed loop stability analysis (see Chapter 14), the presence of the time delay element is often problematic; Model 3 will be more convenient.

8.5 (a) The implications of the given models are:

$$v(s) = \frac{5s+1}{10s+1} u(s) \qquad \text{PREMIXER} \qquad (8.7)$$

$$w(s) = e^{-6s} v(s) \qquad \text{PIPE} \qquad (8.8)$$

and

$$y(s) = \frac{25}{12s+1} w(s) \qquad \text{REACTOR} \qquad (8.9)$$

From here, we immediately the following relationship between $y(s)$ and $u(s)$:

$$y(s) = \left(\frac{25}{12s+1}\right)\left(e^{-6s}\right)\left(\frac{5s+1}{10s+1}\right) u(s)$$

and the required transfer function is:

$$\boxed{g(s) = \frac{25(5s+1)e^{-6s}}{(12s+1)(10s+1)}} \qquad (8.10)$$

(b) The response of $v(t)$ to $u(t)$ is obtained from

$$v(s) = \frac{5s+1}{10s+1} u(s)$$

and with $u(s) = \frac{0.02}{s}$, we see that the required response is that of a lead-lag system with $\rho = 5/10 = 0.5$; $\tau = 10$, and a step size of 0.02, i.e.

$$\boxed{v(t) = 0.02\left[1 - 0.5e^{-0.1t}\right]} \qquad (8.11)$$

The response observed in $w(t)$ is as a result of changes in $v(t)$, and $w(s)$ is related to $v(s)$ according to Eq (8.8) above: thus, we easily obtain from (8.11) above

$$w(t) = \begin{cases} 0 & t \leq 6 \\ 0.02\left[1 - 0.5e^{-0.1(t-6)}\right]; & t > 6 \end{cases} \qquad (8.12)$$

Alternatively, obtain $w(t)$ from the fact that $w(s)$ is related to $u(s)$ by combining Eqs (8.7) and (8.8) above to give

$$w(s) = \frac{(5s+1)e^{-6s}}{(10s+1)} u(s)$$

The same result as in (8.12) will be obtained.

Finally, the response $y(t)$ is obtained from the overall transfer function deduced in part (a); i.e.

$$y(s) = \frac{25(5s+1)e^{-6s}}{(12s+1)(10s+1)} \cdot \frac{0.02}{s}$$

Using the results in Section 6.6.1 of the text (particularly Eq (6.87) in the text, with $\xi = 5$, $\tau_1 = 10$, $\tau_2 = 12$) and recalling the effect of the delay term, the desired result is:

$$y(t) = \begin{cases} 0 ; & t \leq 6 \\ 0.5\left[1 + \frac{5}{2}e^{-(t-6)/10} - \frac{7}{2}e^{-(t-6)/10}\right]; & t > 6 \end{cases}$$

Each response is sketched below in FIG S8.5(a) (b) and (c)

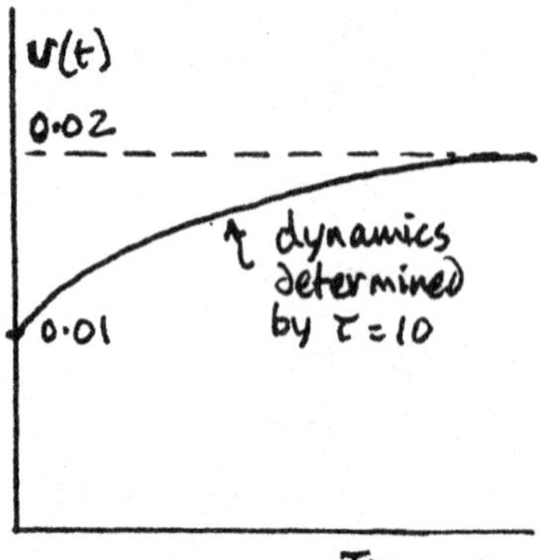

FIG S8.5(a)

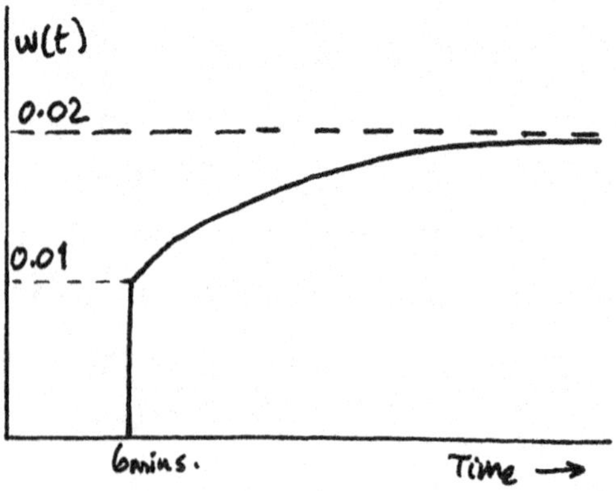

FIG S8.5(b).

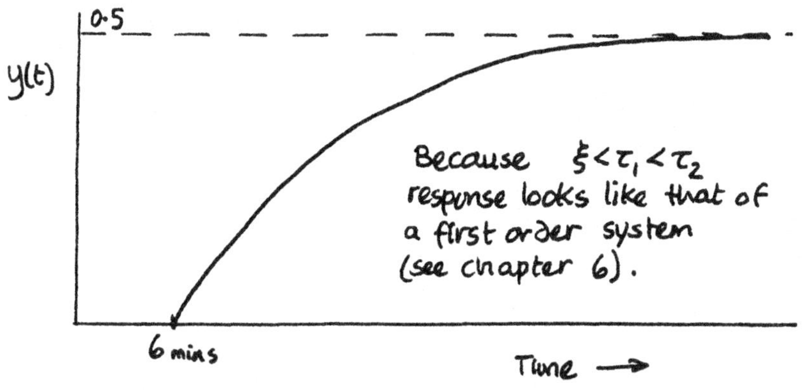

FIG S8.5 (c)

(c) The implication of the alteration is that the transfer function $g_1(s)$ is now given by

$$g_1(s) = \frac{20s+1}{10s+1}$$

As a result, we now have

$$v(s) = \frac{20s+1}{10s+1}\, u(s) \tag{8.13}$$

$$w(s) = \frac{(20s+1)\,e^{-6s}}{(10s+1)}\, u(s) \tag{8.14}$$

and

$$y(s) = \frac{25(20s+1)\,e^{-6s}}{(10s+1)}\, u(s) \tag{8.15}$$

And now, for $v(t)$, $p=2$, everything else remains the same; for $w(t)$, it is merely $v(t)$ shifted by 6 mins; for $y(t)$, $\xi = 20 > \tau_2 = 12 > \tau_1 = 10$ — and such a strong zero now implies there will be an overshoot in the step response.

The new responses are sketched in FIG S8.6 (a), (b) and (c).

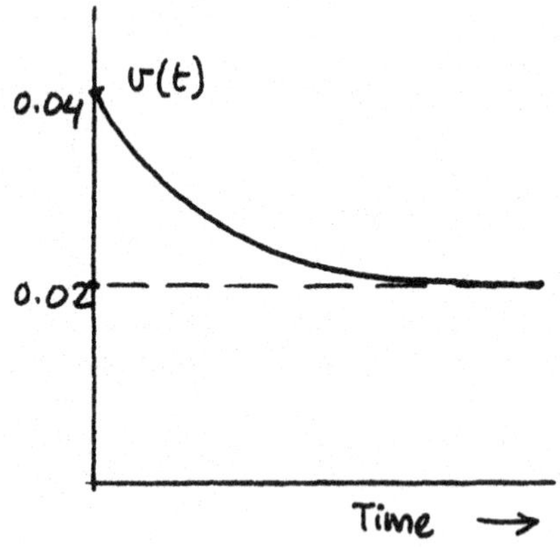

FIG S8.6 (a)

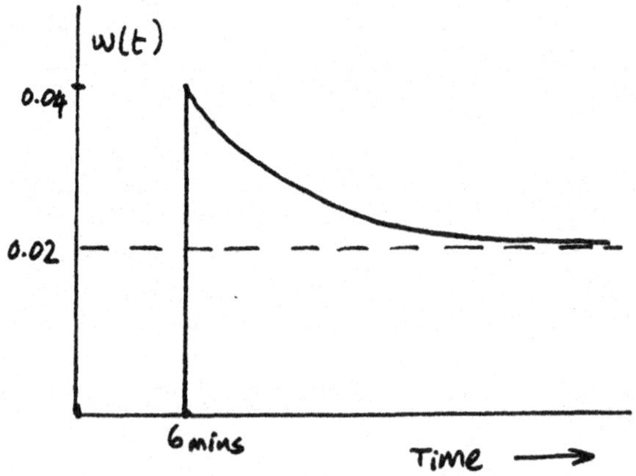

FIG S 8.6(b).

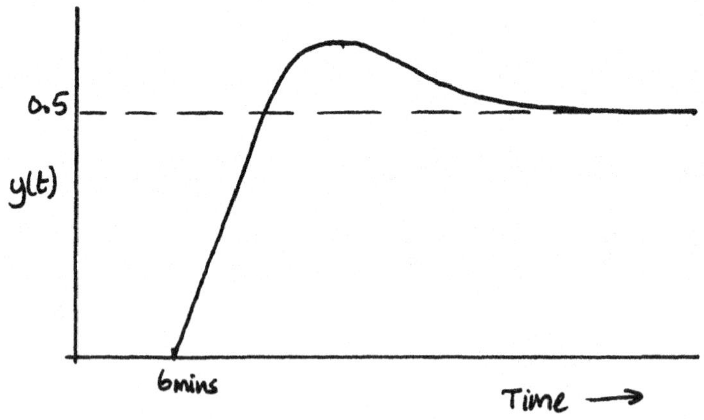

FIG S 8.6(c).

8.6 The unit step response obtained from the 6th order model is shown in FIG S8.7 in the solid line. The corresponding response of the first-order-plus-time delay system is shown in the dashed-dotted line.

A comparison of these two responses indicates that, but for a short initial period in which the simpler model shows a delay, the two responses are virtually identical.

The simpler model employs the delay term $e^{-7.5s}$ to replace the 5th order term $(1.5s+1)^5$. Note that $5 \times 1.5 = 7.5$.

The response indicates that, as far as step responses are concerned, the essence captured by the "more complicated" model, which requires 7 parameters, is also more or less captured just as effectively by the simpler model which employs only 3 parameters. In essence, Eq (P8.5) indicates that a single delay term, $e^{-7.5s}$ might suffice to capture the part of the process represented by $(1.5s+1)^5$.

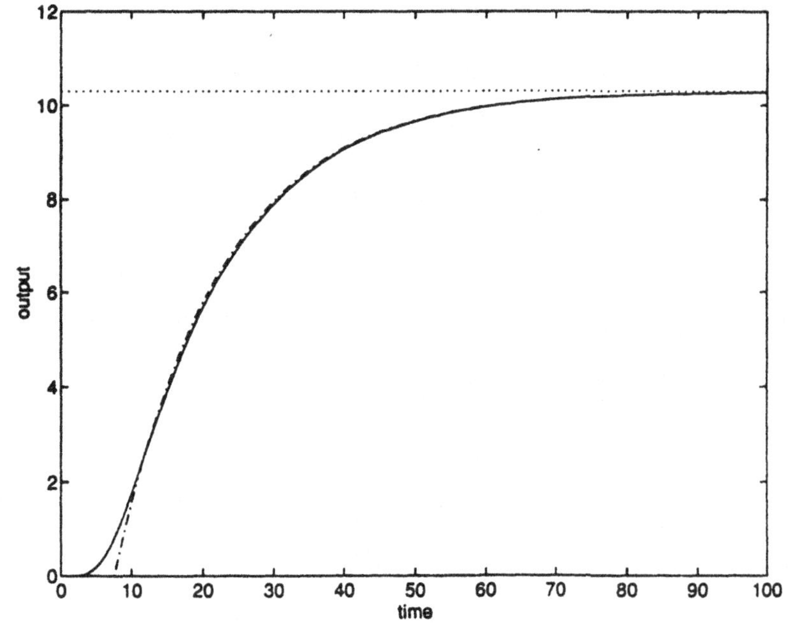

FIG S8.7

8.7 (a) The difference between Tank A and Tank B appears to be that of proximity to the reactor; u appears to be the reactant composition in the feed stream; y appears to be the composition of the reactant in the reactor. The reasoning goes as follows:

A change in reactant composition in the feed stream will show a delay in affecting reactant composition in the reactor <u>if</u> feed tank B is farther away from the reactor than tank A is. Furthermore, because of transportation lag through the recycle stream, what will be returning to the reactor in this recycle stream will be a delayed version of the reactant composition in the reactor itself: this will then be the $y(t-\alpha_0)$ term.

(b) Taking Laplace transforms of Eq (P8.6), we obtain

$$\left(s - a - a_1 e^{-\alpha_0 s}\right) y(s) = b u(s) + \gamma d(s)$$

so that

$$y(s) = \underbrace{\left(\frac{b}{s - a - a_1 e^{-\alpha_0 s}}\right)}_{g(s)} u(s) + \underbrace{\left(\frac{\gamma}{s - a - a_1 e^{-\alpha_0 s}}\right)}_{g_d(s)} d(s)$$

with the indicated transfer functions.

For the model in Eq (P8.7), we obtain

$$g(s) = \frac{b e^{-\alpha_1 s}}{s - a - a_1 e^{-\alpha_0 s}}$$

with the same $g_d(s)$ as above; i.e.

$$g_d(s) = \frac{\gamma}{s - a - a_1 e^{-\alpha_0 s}}.$$

The time delay systems in Section 8.3 had the $e^{-\alpha s}$ terms only in the transfer function numerator; the transfer functions obtained here have the additional $e^{-\alpha_0 s}$ term in the denominator. (See Eq (8.58) in the text.)

(c) Under the specified circumstances, the transfer functions g_A and g_B, (respectively process transfer functions when feed comes from Tank A, and Tank B) are given by:

$$g_A(s) = \frac{3}{s+2-0.5e^{-2s}} \quad ; \quad g_B(s) = \frac{3e^{-5s}}{s+2-0.5e^{-2s}}$$

The step responses are shown in FIG S8.8 below. As expected one is merely a shifted version of the other.

Tank A is to be preferred: for then the process responds instantaneously to changes in feed composition. The significant delay introduced by using Tank B (see Figure) can be problematic for control. (See Chapters 15 and 17)

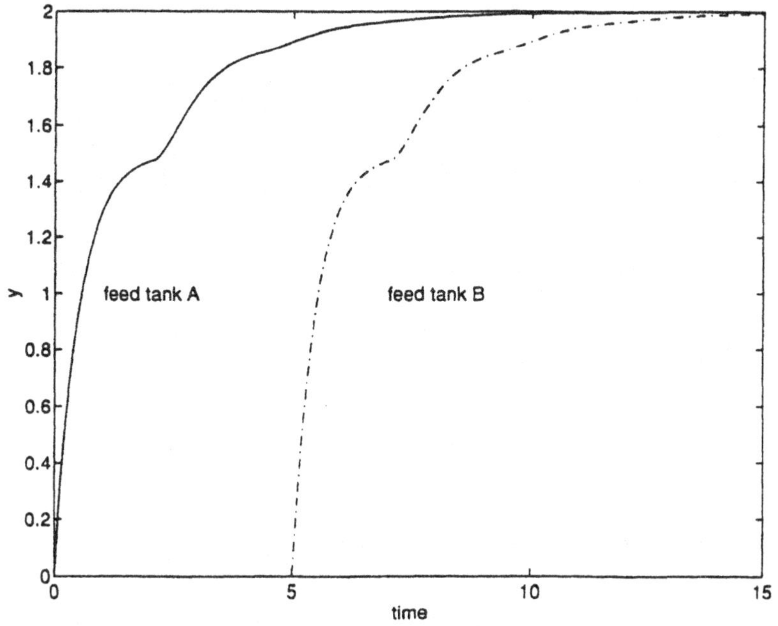

FIG S8.8

8.8 By taking Laplace transforms of (P8.9) along with the boundary conditions in (P8.10) and (P8.11) we obtain

$$sy(z,s) = \frac{d^2y(z,s)}{dz^2} \tag{8.16}$$

along with:

$$\frac{dy}{dz} = \beta(y-u) \quad ; \quad z=0 \tag{8.17}$$

$$\frac{dy}{dz} = 0 \quad ; \quad z=1 \tag{8.18}$$

The second order ordinary differential equation in (8.16)

$$\frac{d^2y}{dz^2} - sy = 0$$

has the characteristic equation

$$r^2 - s = 0$$

whose roots are

$$r = \pm\sqrt{s} \tag{8.19}$$

The solution to (8.16) is therefore given by

$$y(z,s) = A_1 e^{+\sqrt{s}\,z} + A_2 e^{-\sqrt{s}\,z} \tag{8.20}$$

and the constants A_1 and A_2 are to be determined using the boundary conditions given in Eqs (8.17) and (8.18) above. From (8.20) we obtain by differentiation, that

$$\frac{dy}{dz} = A_1\sqrt{s}\,e^{z\sqrt{s}} - A_2\sqrt{s}\,e^{-z\sqrt{s}} \tag{8.21}$$

so that for $z=0$, with the aid of (8.17) above we obtain

$$\beta(y-u) = A_1\sqrt{s} - A_2\sqrt{s} \tag{8.22}$$

Similarly, with $z=1$, and from (8.18), we obtain

$$0 = A_1\sqrt{s}\,e^{\sqrt{s}} - A_2\sqrt{s}\,e^{-\sqrt{s}} \tag{8.23}$$

From (8.23) we have that

$$A_1 = A_2 e^{-2\sqrt{s}} \quad (8.24)$$

which upon introducing into (8.22) and rearranging yields

$$A_2 = \frac{\beta}{\sqrt{s}} \frac{(y-u)}{(e^{-2\sqrt{s}}-1)} \quad (8.25)$$

and therefore

$$A_1 = \frac{\beta (y-u)}{\sqrt{s}\,(e^{-2\sqrt{s}}-1)} e^{-2\sqrt{s}} \quad (8.26)$$

Returning to Eq. (8.20) we now have

$$y(z,s) = \frac{\beta (y-u)}{\sqrt{s}\,(e^{-2\sqrt{s}}-1)} \left[e^{-2\sqrt{s}} e^{z\sqrt{s}} + e^{-z\sqrt{s}} \right]$$

or

$$y(z,s) = \xi (y-u) \quad (8.27)$$

where

$$\xi = \frac{\beta}{\sqrt{s}} \frac{\left[e^{-2\sqrt{s}} e^{z\sqrt{s}} + e^{-z\sqrt{s}} \right]}{(e^{-2\sqrt{s}} - 1)} \quad (8.28)$$

Rearranging (8.27) gives

$$y = \left(\frac{\xi}{\xi - 1} \right) u \quad (8.29)$$

so that

$$g(z,s) = \frac{\xi}{\xi - 1} \quad (8.30)$$

From (8.28) we have

$$\xi - 1 = \frac{\beta \left[e^{-2\sqrt{s}} e^{z\sqrt{s}} + e^{-z\sqrt{s}} \right] - \left(\sqrt{s}\, e^{-2\sqrt{s}} - \sqrt{s} \right)}{\sqrt{s}\,(e^{-2\sqrt{s}} - 1)} \quad (8.31)$$

Thus, from (8.28) and (8.31), (8.30) now rearranges to give

$$g(z,s) = \frac{e^{-2\sqrt{s}} e^{z\sqrt{s}} + e^{-z\sqrt{s}}}{\left(e^{-2\sqrt{s}} e^{z\sqrt{s}} + e^{-z\sqrt{s}}\right) + \frac{\sqrt{s}}{\beta}\left(1 - e^{-2\sqrt{s}}\right)} \qquad (8.32)$$

so that with

$$A = e^{-2\sqrt{s}} e^{z\sqrt{s}} + e^{-z\sqrt{s}}$$

$$B_1 = e^{-2\sqrt{s}} e^{z\sqrt{s}} + e^{-z\sqrt{s}} = A$$

$$B_2 = \frac{\sqrt{s}}{\beta}\left(1 - e^{-2\sqrt{s}}\right)$$

$$g(s) = \frac{A(z,s)}{B_1 + B_2}$$

as required.

In particular, setting $z=0$ in (8.32), yields

$$g(0,s) = \frac{e^{-2\sqrt{s}} + 1}{e^{-2\sqrt{s}} + 1 + \frac{\sqrt{s}}{\beta}\left(1 - e^{-2\sqrt{s}}\right)}$$

$$g(0,s) = \frac{1}{1 + \frac{\sqrt{s}}{\beta} \tanh\sqrt{s}}$$

where

$$\tanh\sqrt{s} = \frac{1 - e^{-2\sqrt{s}}}{1 + e^{-2\sqrt{s}}}$$

as required.

8.9 The reverse Padé approximation for a RHP zero is:

$$1 - \frac{\alpha}{2}s \approx \left(1 + \frac{\alpha}{2}s\right)e^{-\alpha s}$$

so that for the transfer function given in Eq (P8.14) we have the following approximation:

$$g(s) \approx \frac{(3s+1)e^{-6s}}{(2s+1)(5s+1)} \qquad (8.33)$$

A plot of the original system's response to a unit step input is shown in FIG S8.9, in the solid line. The approximating time delay system response is shown in the dashed-dotted line. The approximation appears reasonable.

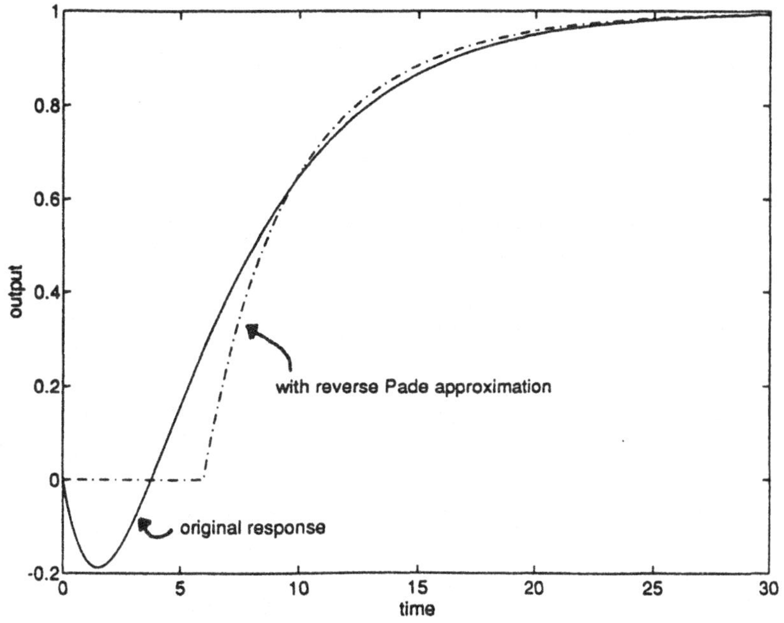

FIG S 8.9

8.10 By definition of e,

$$e = \lim_{m \to \infty}\left(1 + \frac{1}{m}\right)^m \qquad (8.34)$$

so that

$$e^{\alpha} = \lim_{m \to \infty} \left(1 + \frac{1}{m}\right)^{\alpha m}$$

Now let $N = \alpha m$; then $m = N/\alpha$; so that

$$e^{\alpha} = \lim_{N \to \infty} \left(1 + \frac{\alpha}{N}\right)^{N} \qquad (8.35)$$

Now for $\alpha = -\beta s$, we obtain

$$e^{-\beta s} = \lim_{N \to \infty} \left(1 + \frac{-\beta s}{N}\right)^{N}$$ as required.

Thus the time delay may be represented as an infinite set of RHP zeros.

From the identity

$$e^{-\alpha s} = \frac{e^{-\frac{\alpha s}{2}}}{e^{\frac{\alpha s}{2}}} \qquad (8.36)$$

now use Eq (P8.15) for the numerator in Eq (8.36) above and use Eq (8.21) in the text (p251) for the denominator,

we have first, from (P8.15) that

$$e^{-\frac{\alpha s}{2}} = \lim_{N \to \infty} \left(1 + \frac{-\alpha s}{2N}\right)^{N}$$

and from Eq 8.21, p251 of the text, we have

$$e^{+\frac{\alpha s}{2}} = 1 \Big/ \left\{ \lim_{N \to \infty} \left[\frac{1}{\left(1 + \frac{\alpha s}{2N}\right)^{N}} \right] \right\}$$

so that

$$e^{-\alpha s} = \frac{e^{-\frac{\alpha s}{2}}}{e^{\frac{\alpha s}{2}}} = \frac{\lim_{N\to\infty}\left(1+\frac{-\alpha s}{2N}\right)^N}{1\Big/\left\{\lim_{N\to\infty}\left[\frac{1}{\left(1+\frac{\alpha s}{2N}\right)^N}\right]\right\}}$$

$$e^{-\alpha s} = \lim_{N\to\infty}\left(1+\frac{-\alpha s}{2N}\right)^N \cdot \lim_{N\to\infty}\left(\frac{1}{1+\frac{\alpha s}{2N}}\right)^N \qquad (8.37)$$

And now, since each limit converges, it can be shown that

$$\left(\lim_{N\to\infty} A_N\right)\left(\lim_{N\to\infty} B_N\right) = \lim_{N\to\infty}\left(A_N \cdot B_N\right)$$

so that Eq. (8.37) now becomes

$$e^{-\alpha s} = \lim_{N\to\infty}\left[\frac{\left(1+\frac{-\alpha s}{2N}\right)}{\left(1+\frac{\alpha s}{2N}\right)}\right]^N \qquad (8.38)$$

as required.

8.11 From the given transfer function model,

$$y(s) = \frac{\left(1-0.5e^{-10s}\right)}{(40s+1)(15s+1)}\, u(s)$$

we obtain,

$$\left(600s^2+55s+1\right)y(s) = u(s) - 0.5e^{-10s}u(s) \qquad (8.39)$$

Now, the most obvious differential equation from which Eq. (8.39) above could have been obtained by Laplace transform is

$$600\frac{d^2y}{dt^2} + 55\frac{dy}{dt} + y = u(t) - 0.5u(t-10) \qquad (8.40)$$

To obtain the impulse response model, we need to obtain the impulse response function $g(t)$ from

$$g(s) = \frac{1 - 0.5e^{-10s}}{(40s+1)(15s+1)}$$

By partial fraction expansion, we obtain

$$g(s) = \left(1 - 0.5e^{-10s}\right)\left[\frac{8/5}{40s+1} - \frac{3/5}{15s+1}\right]$$

and from here, we obtain

$$g(t) = \begin{cases} \frac{1}{25}\left(e^{-t/40} - e^{-t/15}\right); & t \leq 10 \\ \frac{1}{25}\left(e^{-t/40} - 0.5e^{-(t-10)/40} - e^{-t/15} + 0.5e^{-(t-10)/15}\right); & t > 10 \end{cases} \qquad (8.41)$$

The impulse response model is now

$$y(t) = \int_0^t g(t-\sigma)u(\sigma)d\sigma = \int_0^t g(t)u(t-\sigma)d\sigma$$

where $g(t)$ is given in Eq (8.41) above.

8.12 The consolidated transfer function is

$$g(s) = \frac{K_1 e^{-\alpha s} - K_2}{(\tau s + 1)} \qquad (8.42)$$

By definition, the transfer function zero is the value of s for which the numerator vanishes; in this case, therefore we find the zero by solving the following equation for s:

$$K_1 e^{-\alpha s} - K_2 = 0$$

or

$$e^{-\alpha s} = \frac{K_2}{K_1}$$

$$-\alpha s = \ln \frac{K_2}{K_1}$$

yielding

$$s = \frac{-\left(\ln \frac{K_2}{K_1}\right)}{\alpha} \quad \text{as required.}$$

For this value of s to be > 0 requires

$$\ln\left(\frac{K_2}{K_1}\right) < 0$$

and since $e^0 = 1$, that

$$\frac{|K_2|}{|K_1|} < 1$$

or

$$|K_2| < |K_1| \quad \text{as required.}$$

8.13 (a) The simple reason is the delays: observe that if $K_1 > K_2$ (each assumed positive), the ultimate value of the response, $(K_1 - K_2)$, will be positive. However, because the delay α_1 associated with g_1 is longer, the first response to be registered will be that due to g_2, which is in the opposite direction.

For the specific situation given, the step response is shown in FIG S8.10 where the inverse response is clear.

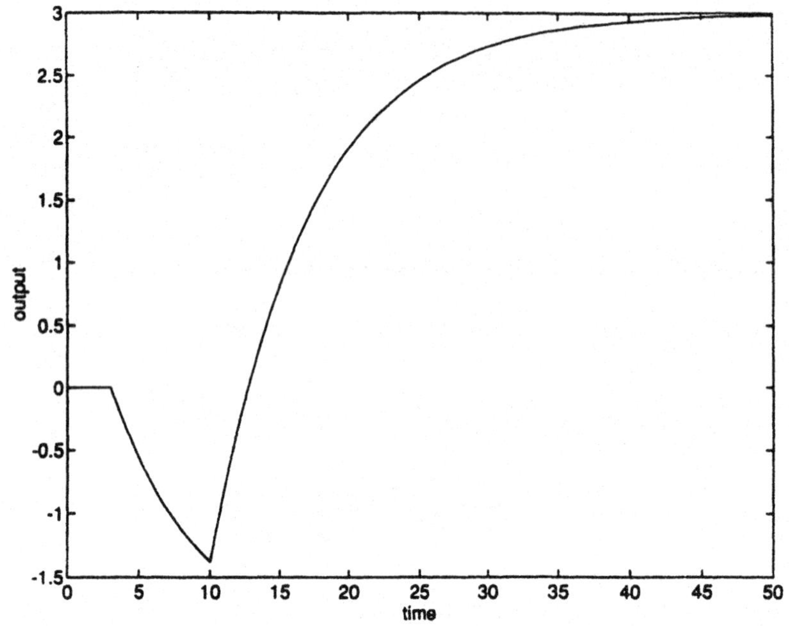

FIG S8.10

(b) when $\tau_2 = \tau_1 = \tau$, the composite transfer function is given by

$$g(s) = \frac{K_1 e^{-\alpha_1 s} - K_2 e^{-\alpha_2 s}}{(\tau s + 1)}$$

and its zero is obtained by solving

$$K_1 e^{-\alpha_1 s} - K_2 e^{-\alpha_2 s} = 0$$

for s; i.e.

$$K_1 e^{-\alpha_1 s} = K_2 e^{-\alpha_2 s}$$

$$\frac{K_1}{K_2} = e^{(\alpha_1 - \alpha_2) s}$$

from where we obtain

$$(\alpha_1 - \alpha_2) s = \ln \left(K_1 / K_2 \right)$$

which now yields

$$s = \frac{\ln K_1 - \ln K_2}{\alpha_1 - \alpha_2}$$

as required.

For the composite system to exhibit inverse response requires:

Condition 1: $\ln K_1 > \ln K_2$ (i.e. $K_1 > K_2$)

<u>and</u> $\alpha_1 > \alpha_2$

or

Condition 2: $\ln K_1 < \ln K_2$ (i.e. $K_1 < K_2$)

<u>and</u> $\alpha_1 < \alpha_2$

CHAPTER 9

9.1 Representing the complex number

$$z = a + bj \tag{9.1}$$

in an Argand diagram as shown below

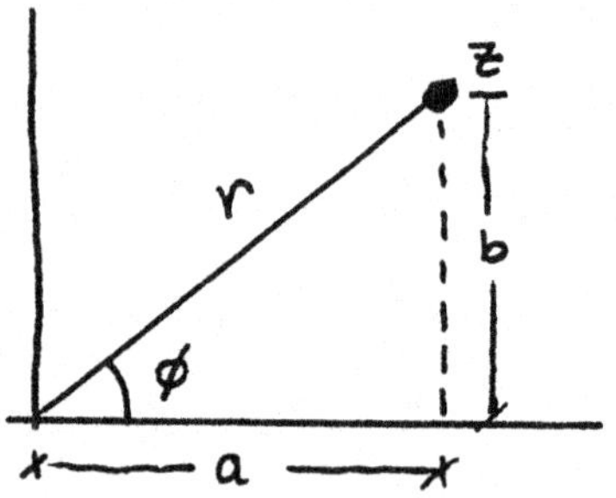

indicates that

$$r = \sqrt{(a^2+b^2)} = |z| \quad \text{as defined in Eq (P9.2a)}$$

and

$$a = r\cos\phi = |z|\cos\phi \tag{9.2}$$
$$b = r\sin\phi = |z|\sin\phi \tag{9.3}$$

so that $\tan\phi = b/a$

or $\phi = \tan^{-1}(b/a)$ as in Eq (P9.2b).

Introducing (9.2) and (9.3) into (9.1) we now have

$$z = |z|(\cos\phi + j\sin\phi) \tag{9.4}$$

and recalling Euler's identity (Eq (9.8a), p 278 in text) Eq (9.4) above becomes

$$z = |z|e^{j\phi} \text{ , as required}$$

9.2 For this current situation in which

$$g(s) = g^*(s)e^{-\alpha s}$$

recalling the effect of the delay on a system's response, we have from Eq. (9.7) on p 278 in the text that $y(t)$ in this case will be

$$y(t) = \begin{cases} 0 & t \le \alpha \\ \sum A_i e^{r_i(t-\alpha)} + B_1 e^{j\omega(t-\alpha)} + B_2 e^{-j\omega(t-\alpha)} ; & t > \alpha \end{cases}$$

Thus,

$$y(t)\Big|_{t\to\infty} = B_1 e^{j\omega(t-\alpha)} + B_2 e^{-j\omega(t-\alpha)}$$

$$= B_1 e^{-\alpha j\omega} \cdot e^{j\omega t} + B_2 e^{\alpha j\omega} e^{-j\omega t}$$

or

$$y(t)\Big|_{t\to\infty} = \tilde{B}_1 e^{j\omega t} + \tilde{B}_2 e^{-j\omega t} \quad (9.5a)$$

with

$$\tilde{B}_1 = B_1 e^{-\alpha j\omega} \quad (9.5b)$$

$$\tilde{B}_2 = B_2 e^{\alpha j\omega} \quad (9.5c)$$

As in the text, the constants B_1 and B_2 are determined from the delay-free transfer function. Since in this case this refers to $g^*(s)$, we now have that

$$B_1 = \frac{A g^*(j\omega)}{2j} \quad (9.6a)$$

$$B_2 = \frac{A g^*(-j\omega)}{2j} \quad (9.6b)$$

Thus, from Eqs (9.5b) and (9.5c) above, Eqs (9.6) now imply

$$\tilde{B}_1 = \frac{A g(j\omega) e^{-\alpha j\omega}}{2j} \tag{9.7a}$$

$$\tilde{B}_2 = \frac{A g(-j\omega) e^{\alpha j\omega}}{2j} \tag{9.7b}$$

Now, by definition, $g(s) = g^*(s) e^{-\alpha s}$ for all s; in particular, for $s = j\omega$, we have

$$g(j\omega) = g^*(j\omega) e^{-\alpha j\omega} \tag{9.8a}$$

similarly

$$g(-j\omega) = g^*(-j\omega) e^{\alpha j\omega} \tag{9.8b}$$

and by substituting (9.8) into (9.7) above, we obtain

$$\tilde{B}_1 = \frac{A g(j\omega)}{2j}$$

$$\tilde{B}_2 = \frac{A g(-j\omega)}{2j}$$

so that (9.5a) becomes

$$y(t)\Big|_{t\to\infty} = \left[\frac{A\, g(j\omega)}{2j}\right] e^{j\omega t} + \left[\frac{A g(-j\omega)}{2j}\right] e^{-j\omega t}$$

From where we now observe that Eq. (9.13) on p. 279 in the text holds for all $g(s)$ whether it contains a delay or not. All the results in the text from this point on therefore hold regardless of the presence or absence of a time delay.

9.3 (a) We begin with the expression

$$\frac{AR}{K} = \frac{1}{\sqrt{(1-\omega^2\tau^2)^2 + (2\zeta\omega\tau)^2}}$$

To simplify the algebra, let $z = AR/K$, and $\theta = \omega\tau$: thus yielding

$$z = \left[(1-\theta^2)^2 + (2\zeta\theta)^2\right]^{-1/2}$$

From here,

$$\frac{\partial z}{\partial \theta} = 0 \quad \text{implies that}$$

$$-\tfrac{1}{2}\left[(1-\theta^2)^2 + (2\zeta\theta)^2\right]^{-3/2}\left\{2(1-\theta^2)(-2\theta) + 2\theta(2\zeta)^2\right\} = 0$$

or,

$$\cancel{2\theta} \times 2(1-\theta^2) = \cancel{2\theta}(2\zeta)^2$$

$$\theta^2 = 1 - 2\zeta^2$$

$$\theta = \sqrt{1-2\zeta^2} \qquad (9.9)$$

And for $\theta = \omega\tau$ to be <u>real</u> we now require

$$1 - 2\zeta^2 > 0$$

or

$$\zeta^2 < \tfrac{1}{2}$$

$$\zeta < \frac{1}{\sqrt{2}}$$

Establishing Eq (9.56) in the text.

From Eq (9.9) above, when resonance occurs,

$$\theta = \omega_r\tau = \sqrt{1-2\zeta^2}$$

From where we obtain immediately that the resonance frequency is given by

$$\omega_r = \frac{1}{\tau}\sqrt{1 - 2\zeta^2} \quad \text{establishing Eq (9.57).}$$

(b) The Nyquist plots are shown in Fig S9.1 below, using the same scale to emphasize the differences. Nevertheless the effect of resonance exhibited by g_2 is not so obvious.

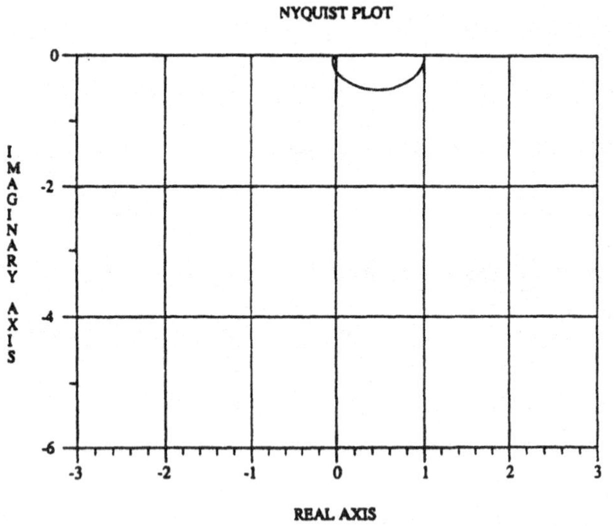

$$g_1(s) = \frac{1}{s^2 + 4s + 1}$$

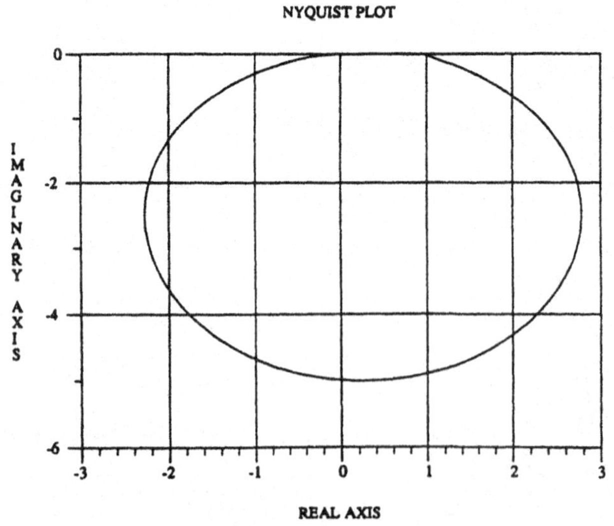

$$g_2(s) = \frac{1}{s^2 + 0.2s + 1}$$

FIG S9.1

9.4 For the given sinusoidal input,

$$u(s) = \frac{A\omega}{s^2+\omega^2}$$

and therefore the response of the given pure capacity system to this input will be obtained from

$$y_c(s) = \frac{A\omega K}{s(s^2+\omega^2)} \qquad (9.10)$$

The general transfer function for the second order system is:

$$g_2(s) = \frac{K_2}{\tau^2 s^2 + 2\zeta\tau s + 1}$$

For the undamped system, $\zeta = 0$, and the unit step response in this case will be obtained from

$$y_2(s) = \frac{K_2}{(\tau^2 s^2 + 1)} \cdot \frac{1}{s}$$

or

$$y_2(s) = \frac{K_2/\tau^2}{s(s^2 + 1/\tau^2)} \qquad (9.11)$$

Observe now that (9.10) and (9.11) are exactly equivalent with $\tau = 1/\omega$, and

$$\boxed{K_2 = \frac{AK}{\omega}} \qquad (9.12)$$

[9.5] The Bode plot for $g(s) = 1$ is as given in Fig 9.6 on p. 290 of the text (with $K=1$). Both AR, and ϕ are identically 1 for all frequencies.

The Bode plot for

$$g(s) = \frac{1}{bs}\left(1 - e^{-bs}\right) \quad \text{for } b = 1 \text{ and } b = 0.1$$

are shown below in Figs S9.2(a) and 9.2 (b) respectively. The plot in FIG S9.2 (b) (for $b = 0.1$) is a better approximation to the plot in FIG 9.6 of the text. Implication: A narrower rectangular pulse (Laplace transform as given in Eq (P9.3)) is a better approximation to an ideal impulse ($u(s) = 1$) than a wider rectangular pulse.

However, for $b = 0.01$, a narrower pulse still, the Bode plot shown in Fig S9.2(c) is not necessarily better than the one in Fig S9.2(b). The pulse height (= 1 in all cases) needs to increase

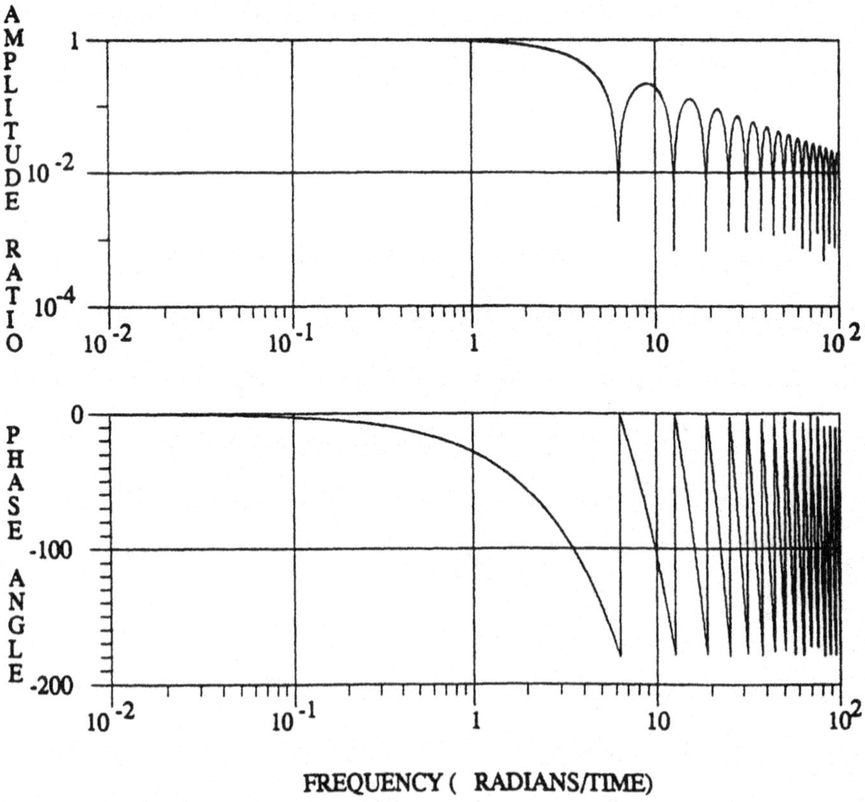

FIG S 9.2(a) $b=1$

as the pulse width is reduced in order to obtain better approximations.

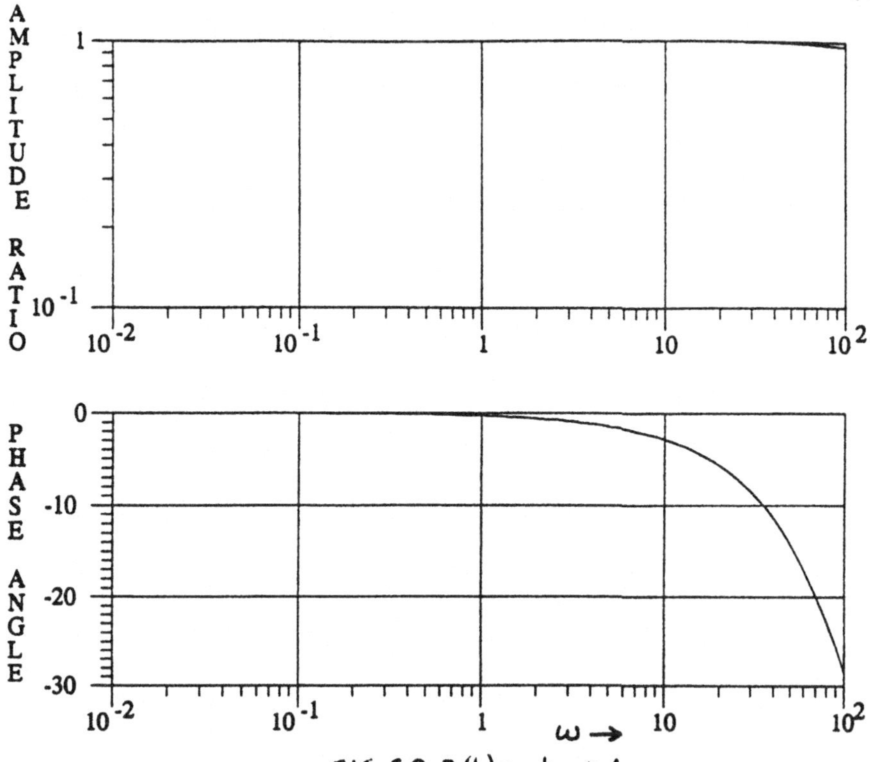

FIG S9.2(b): b = 0.1

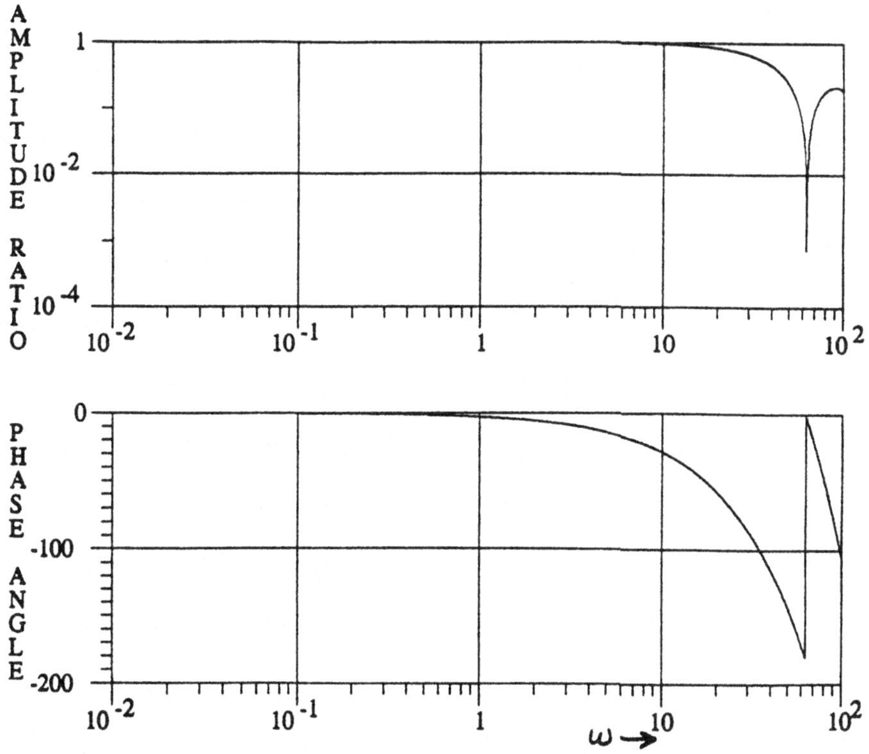

FIG S9.2(c): b = 0.01

9.6 (a)

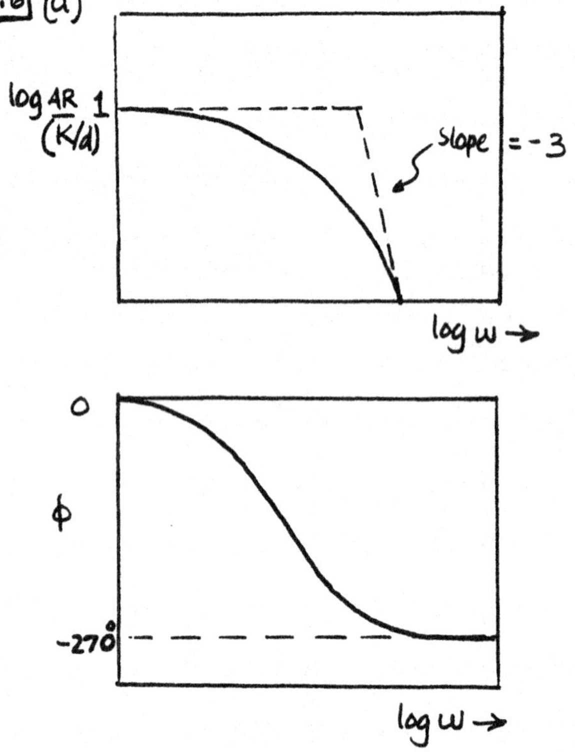

$$g_1(s) = \frac{K}{as^3 + bs^2 + cs + d}$$

FIG S9.3

For the given $g_2(s)$, system is equivalent to a 2nd order system with $\tau = 1$, $\zeta = 0.5$ ($\therefore$ underdamped) and since $\zeta < 0.707$, the AR will show resonance. The sketch is now shown below.

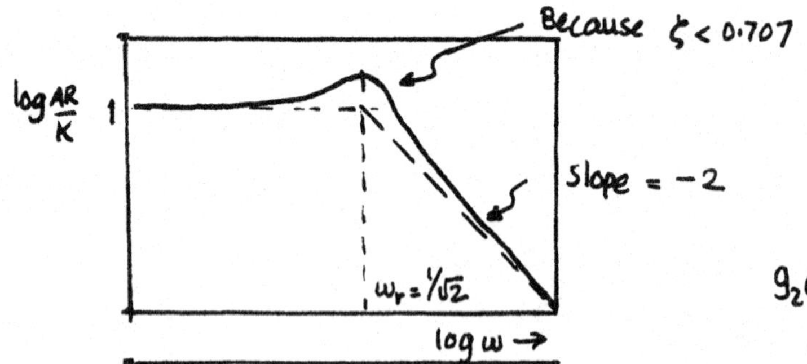

$$g_2(s) = \frac{Ke^{-\alpha s}}{s^2 + s + 1}$$

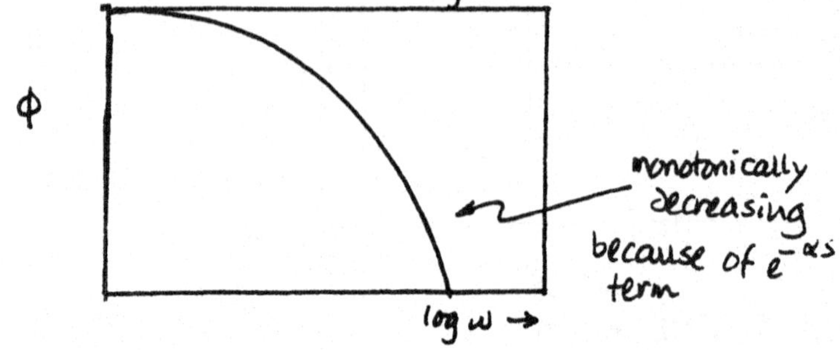

FIG S9.4

(b) The Bode diagram has the following peculiarities which provide us with the required clues:

1. Phase characteristics show no finite limit as $\omega \to \infty$
   Implying: Presence of a time delay;
2. High Frequency Asymptote for the MR plot has slope = −1
   Implying: A pole-zero excess of 1.
3. "Overshoot" in both MR and ϕ plots
   Implying: At least one "strong" zero. (see Fig 9.15, p.301)

The simplest transfer function form which would give rise to such a Bode diagram is

$$g(s) = \frac{K(\xi s+1)e^{-\alpha s}}{(\tau_1 s+1)(\tau_2 s+1)}$$

with the added requirement that

$$\xi > \tau_1 \text{ and } \tau_2$$

because only then can we see the indicate "overshoot".

9.7 The required transfer functions are

$$g_1(s) = \frac{3}{s+2-0.5e^{-2s}} \qquad (9.13)$$

for case 1 (Eq (P9.6)), and

$$g_2(s) = \frac{3e^{-5s}}{s+2-0.5e^{-2s}} \qquad (9.14)$$

for case 2 (Eq (P9.6)). The Bode diagrams are shown in

FIG S9.5 below: solid lines for case 1, dashed-dotted lines for case 2.

Note that the AR plots are identical. The plot in the solid line differs from that of a standard first order system in the intermediate frequency range: the LF and HF behavior are similar. The additional e^{-2s} term in the denominator of $g_1(s)$ in (9.13) above is responsible for the "ripples" and accounts for the single most important difference between FIG S9.5 and the standard 1st order system Bode plot.

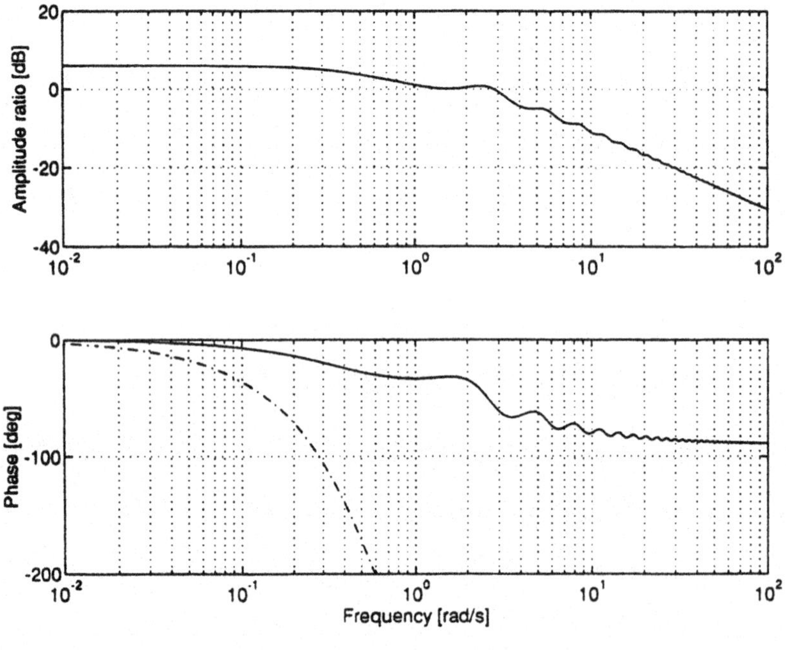

FIG S9.5

9.8 To understand properly the Bode diagram for the system

$$g(s) = \frac{1 - 0.5e^{-10s}}{(40s+1)(15s+1)} \qquad (9.15)$$

let

$$g_1(s) = 1 - 0.5e^{-10s} \qquad (9.16a)$$

$$g_2(s) = \frac{1}{40s+1} \qquad (9.16b)$$

$$g_3(s) = \frac{1}{15s+1} \qquad (9.16c)$$

Then, in particular,

$$g_1(j\omega) = 1 - 0.5e^{-10j\omega}$$

$$g_1(j\omega) = 1 - 0.5\cos 10\omega + 0.5j\sin 10\omega$$

From here we obtain

$$AR_1 = |g_1(j\omega)|$$

$$AR_1 = \sqrt{(1.25 - \cos 10\omega)} \qquad (9.17)$$

and

$$\phi_1 = \angle g_1(j\omega)$$

$$\phi_1 = \tan^{-1}\left(\frac{0.5\sin 10\omega}{1 - 0.5\cos 10\omega}\right) \qquad (9.18)$$

From $g_2(j\omega)$ and $g_3(j\omega)$ we obtain (see p288, Eq(9.34))

$$AR_2 = \left[1 + (40\omega)^2\right]^{-1/2}; \quad \phi_2 = -\tan^{-1}(40\omega)$$

$$AR_3 = \left[1 + (15\omega)^2\right]^{-1/2}; \quad \phi_3 = -\tan^{-1}(15\omega)$$

So that for the overall system, we have

$$\log AR = \tfrac{1}{2}\log(1.25 - \cos 10\omega) - \tfrac{1}{2}\log\left[1 + (40\omega)^2\right] - \tfrac{1}{2}\log\left[1 + (15\omega^2)\right] \qquad (9.19)$$

and

$$\phi = \tan^{-1}\left(\frac{0.5\sin 10\omega}{1 - 0.5\cos 10\omega}\right) - \tan^{-1}(40\omega) - \tan^{-1}(15\omega) \qquad (9.20)$$

Observe now that the AR_1 and ϕ_1 contributions introduce periodicity into the overall AR and ϕ behavior.

The Bode plots are shown in FIG S9.6. The key difference between this Bode diagram and that of a typical time-delay system (apart from the periodic components) is that in this case ϕ does not go to ∞ as $\omega \to \infty$, as would be the case with the time delay system.

The difference arises because $1 - 0.5e^{-10s}$ has a fundamentally different dynamic characteristic. Note that the unit step response of g_1 in (9.16a) shows no initial delay; the unit step response of a time-delay system, on the other hand shows a delay. For such time delay systems $\phi \to \infty$ as $\omega \to \infty$; for this system with $1 - 0.5e^{-10s}$, $0 > \phi > -200°$.

9.9 (a) PI controller

$$g(j\omega) = K_c\left(1 + \frac{1}{\tau_I j\omega}\right) = K_c\left(\frac{\tau_I j\omega + 1}{\tau_I j\omega}\right)$$

from which we obtain

$$AR = |g(j\omega)| = K_c\left[\frac{(\tau_I\omega)^2 + 1}{(\tau_I\omega)^2}\right]^{1/2} \qquad (9.20)$$

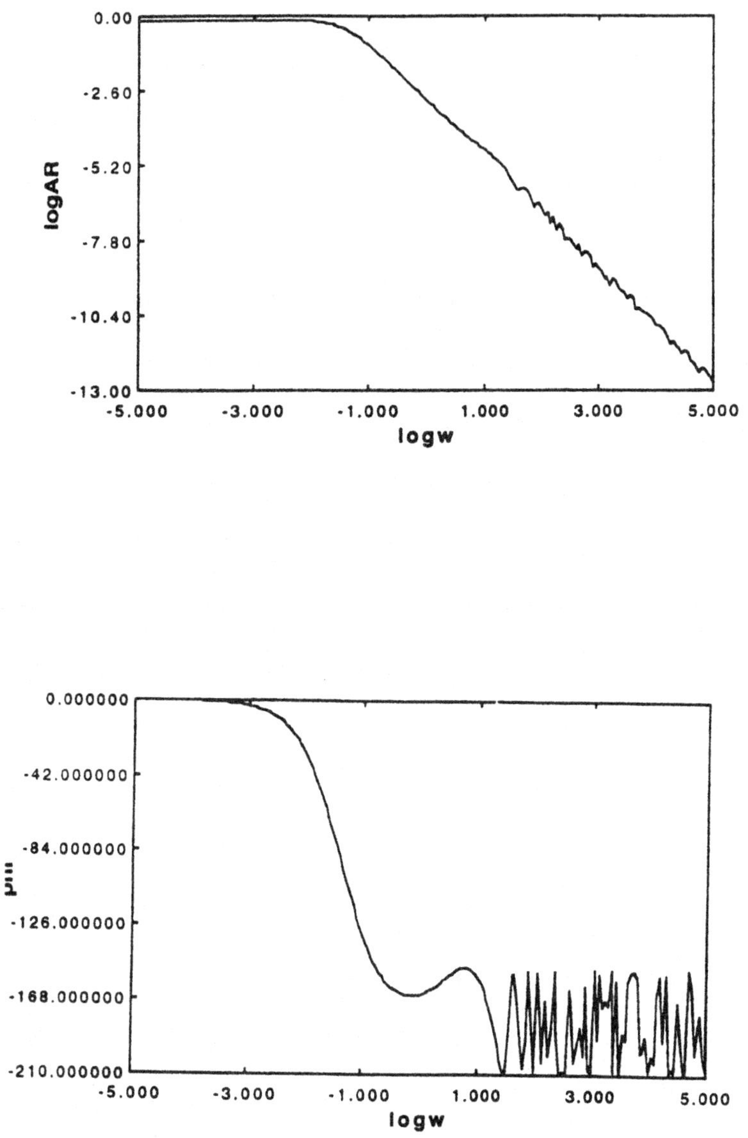
0.00
-2.60
-5.20
-7.80
-10.40
-13.00
logAR
-5.000
-3.000
-1.000
1.000
3.000
5.000
logw
0.000000
-42.000000
-84.000000
-126.000000
-168.000000
-210.000000
-5.000
-3.000
-1.000
1.000
3.000
5.000
logw

FIG S 9.6

and

$$\phi = \tan^{-1}\left(\frac{-1}{\tau_I \omega}\right) \qquad (9.21)$$

PID Controller

$$g(j\omega) = K_c\left(1 + \frac{1}{\tau_I j\omega} + \tau_D j\omega\right).$$

$$AR = |g(j\omega)| = K_c\sqrt{1 + \frac{(1-\tau_I\tau_D\omega^2)^2}{(\tau_I\omega)^2}} \qquad (9.22)$$

$$\phi = \angle g(j\omega) = \tan^{-1}\left(\tau_D\omega - \frac{1}{\tau_I\omega}\right)$$

$$\phi = \tan^{-1}\left(\frac{\tau_I\tau_D\omega^2 - 1}{\tau_I\omega}\right). \qquad (9.23)$$

(b) Asymptotic Behavior

PI Controller

From Eq(9.20) above, observe that

$$\log\left(\frac{AR}{K_c}\right) = \frac{1}{2}\log\left[1 + \frac{1}{(\tau_I\omega)^2}\right] \qquad (9.24)$$

Thus as $\omega \to 0$,

$$\log\left(\frac{AR}{K_c}\right) \to \frac{1}{2}\log(\tau_I\omega)^{-2} = -\log\tau_I\omega$$

A LFA of slope −1.

From (9.21), as $\omega \to 0$

$$\phi \to \tan^{-1}(-\infty) = -90^\circ$$

As $\omega \to \infty$, $\log(AR/K_c) \to \frac{1}{2}\log(1) = 0$

and $\phi \to \tan^{-1}(0) = 0$

These characteristics are as shown in Fig 9.19a, p-305.

PID controller

$$\log \frac{AR}{K_c} = \frac{1}{2}\log\left[1 + \left(\tau_D\omega - \frac{1}{\tau_I\omega}\right)^2\right]$$

$$\phi = \tan^{-1}\left(\tau_D\omega - \frac{1}{\tau_I\omega}\right).$$

As $\omega \to 0$ $\quad \log \frac{AR}{K_c} \to \frac{1}{2}\log(\tau_I\omega)^{-2} = -\log(\tau_I\omega) \quad (9.25a)$

$\phi \to \tan^{-1}\left(-\frac{1}{\tau_I\omega}\right) = -90^\circ \quad (9.25b)$

As $\omega \to \infty$ $\quad \log \frac{AR}{K_c} \to \frac{1}{2}\log(\tau_D\omega)^2 = \log(\tau_D\omega) \quad (9.26a)$

$\phi \to \tan^{-1}(\tau_D\omega) = \tan^{-1}(\infty) = +90^\circ \quad (9.26b)$

Eq (9.25a) above indicates a LFA with a negative slope while Eq (9.26a) indicates a HFA with a positive slope. Furthermore, the LFA meets the line $\frac{AR}{K_c} = 1$ at the "corner frequency" $\omega = 1/\tau_I$, while the HFA meets this line at $\omega = 1/\tau_D$.

Also, Eq (9.25b) and Eq (9.26b) indicate phase behavior commencing at -90° and tending asymptotically to $+90^\circ$ at high frequencies.

These characteristics are consistent with what is shown in Fig 9.19b, p.305.

9.10 Given the anaerobic digester's transfer function as

$$g(s) = \frac{0.52}{(45.5s+1)(12.2s+1)}$$

we observe that this implies a system with the characteristics of an overdamped 2nd order system.

For such a system, from Eq (9.50a) on p295 of the text we have that the AR is given by

$$AR = \frac{0.52}{\left(\sqrt{(45.5\omega)^2+1}\right)\left(\sqrt{(12.2\omega)^2+1}\right)} \tag{9.27}$$

Thus if the input sine wave amplitude is 5 units, then the output sine wave amplitude will be 5 × AR. And for the specific frequency $\omega = 0.2$ radians/min, we obtain from Eq (9.27) above that the digester output amplitude, A_o, is given by

$$A_o = 0.108$$

At $\omega = 2$ radians/min, we obtain

$$A_o(\omega=2) = 0.00117$$

And we find that the output amplitude is reduced by a factor of about 10^{-2} as a result of a one order of magnitude increase in input frequency.

9.11 The Bode diagrams are shown in FIGS 9.7a, b, c. respectively. In each case g_1 always has the minimum phase angle behavior. For (a), g_2, because of the time delay shows a phase behavior which increase with time. (The observed high frequency behavior is a computational artifact).

For (b) g_1 has an ultimate ϕ of $-90°$; g_2 has an ultimate ϕ of $-270°$. For (c), g_2 starts at $-180°$; g_1 ends at $-90°$.

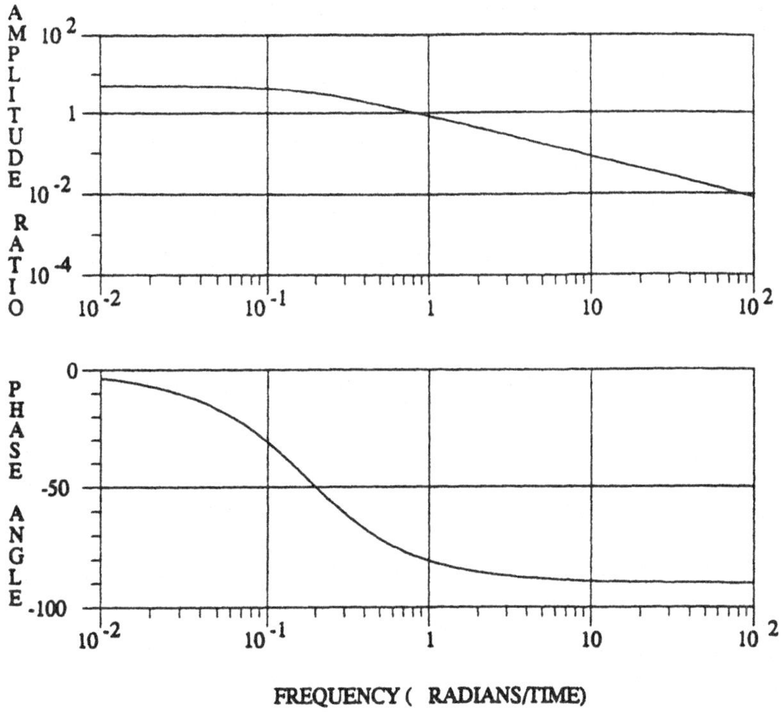

$$g_1(s) = \frac{5}{6s+1}$$

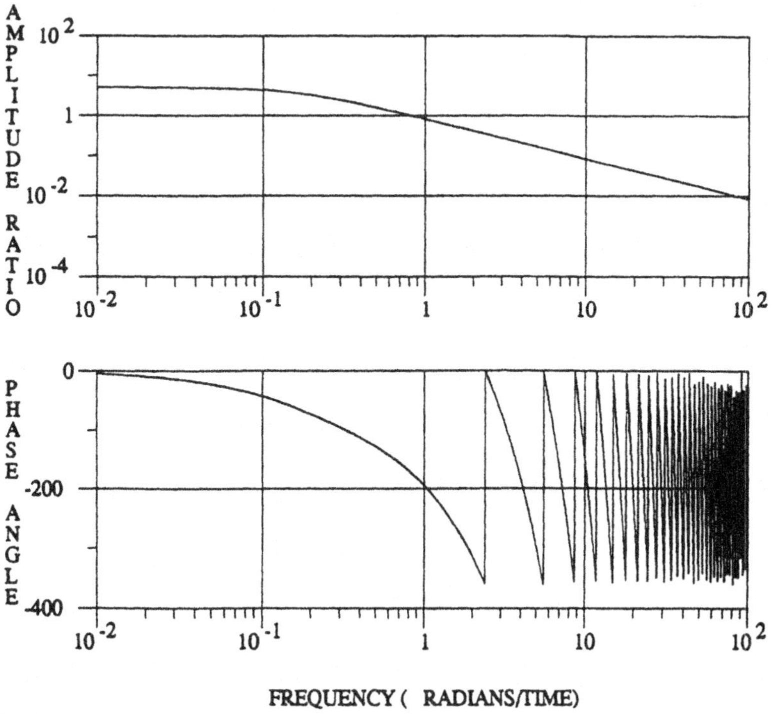

$$g_2(s) = \frac{5e^{-2s}}{6s+1}$$

FIG S 9.7(a)

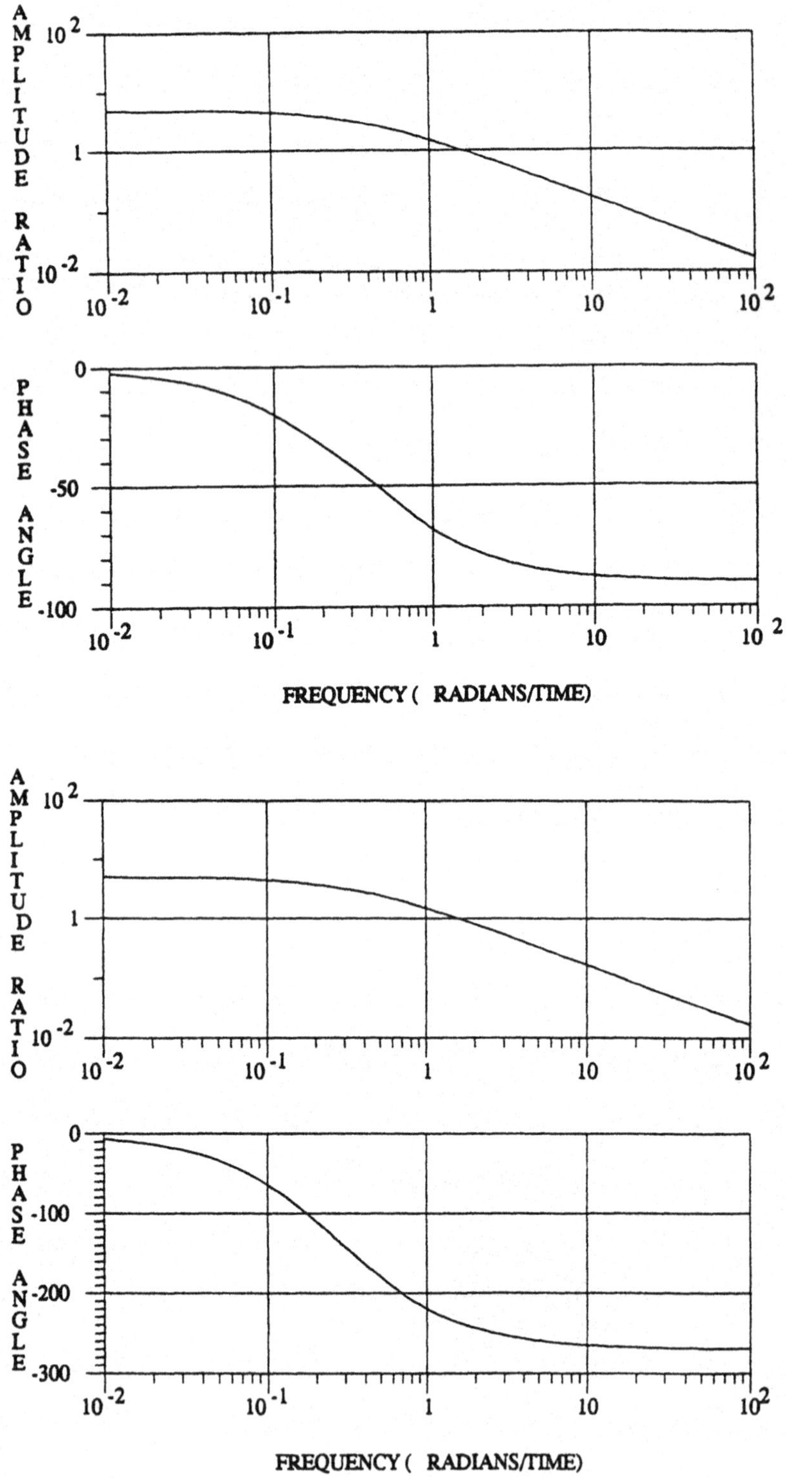

$$g_1(s) = \frac{5(4s+1)}{(2s+1)(6s+1)}$$

$$g_2(s) = \frac{5(-4s+1)}{(2s+1)(6s+1)}$$

FIG S 9-7-(b)

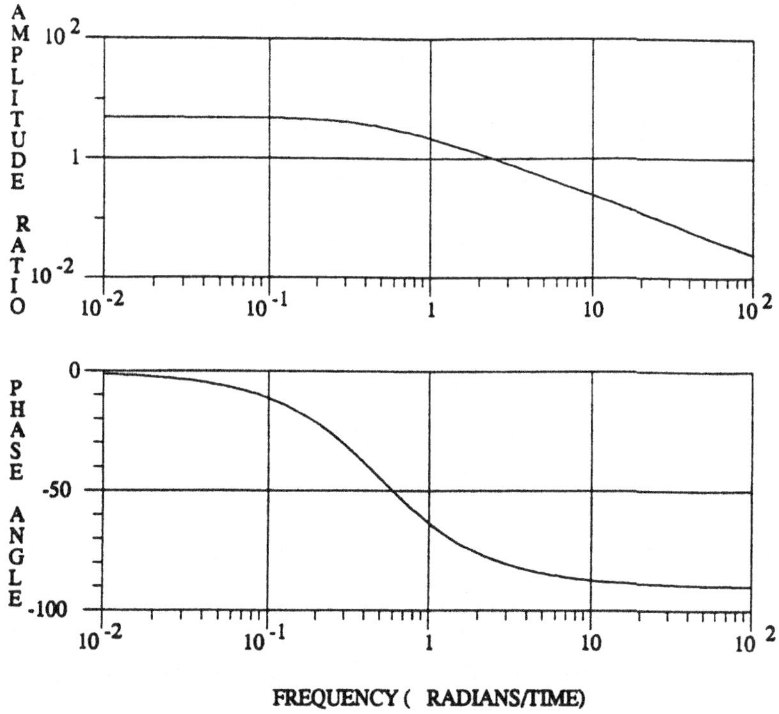

$$g_1(s) = \frac{5}{2s+1}$$

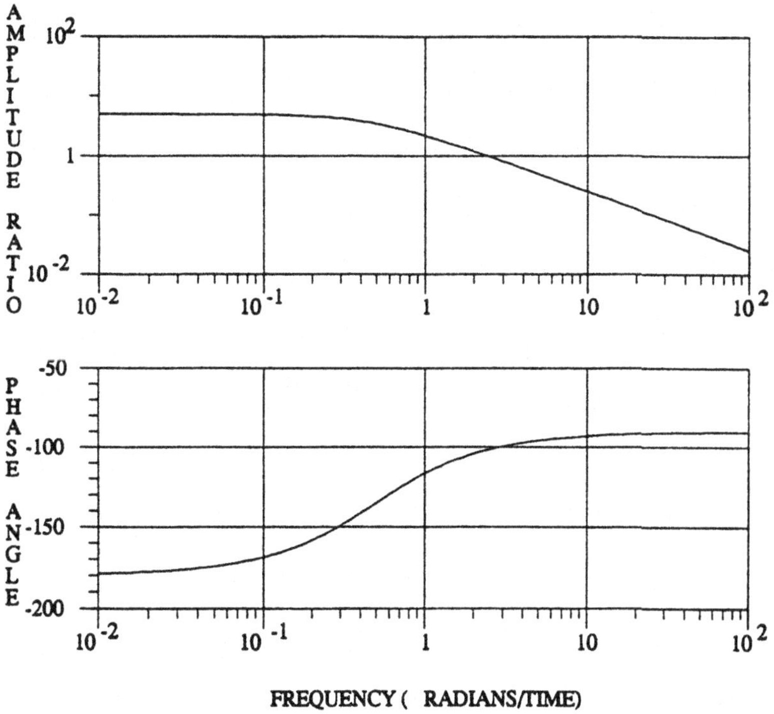

$$g_2(s) = \frac{5}{2s-1}$$

FIG S9.7 (c)

9.12 The Nyquist plots are shown in FIGS S9.8(a),(b) and (c).

NYQUIST PLOT

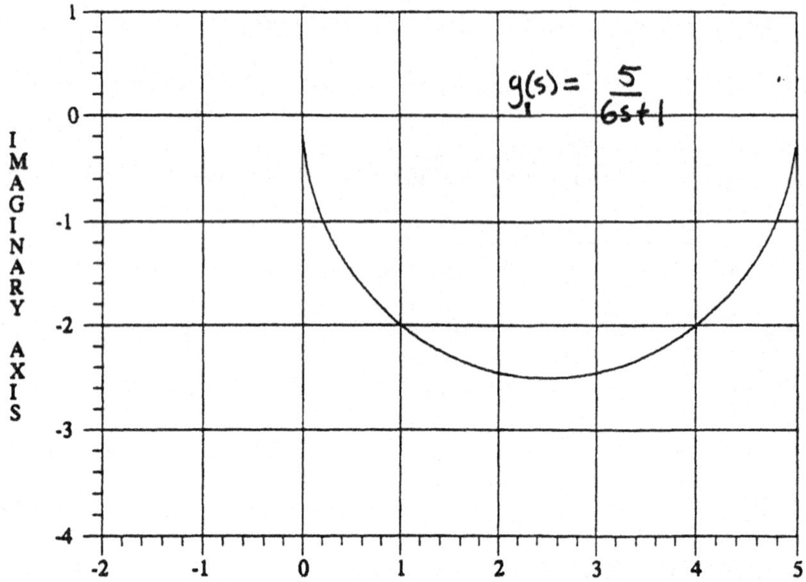

REAL AXIS

NYQUIST PLOT

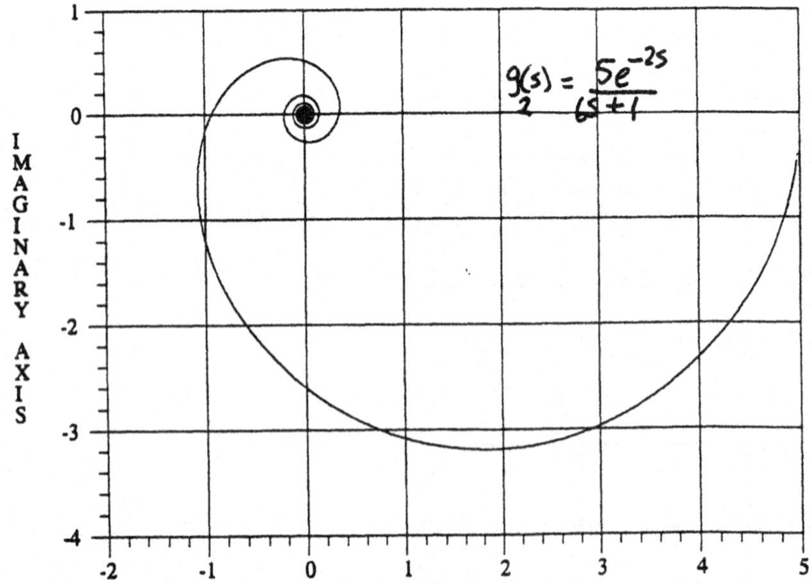

REAL AXIS

FIG S9.8(a)

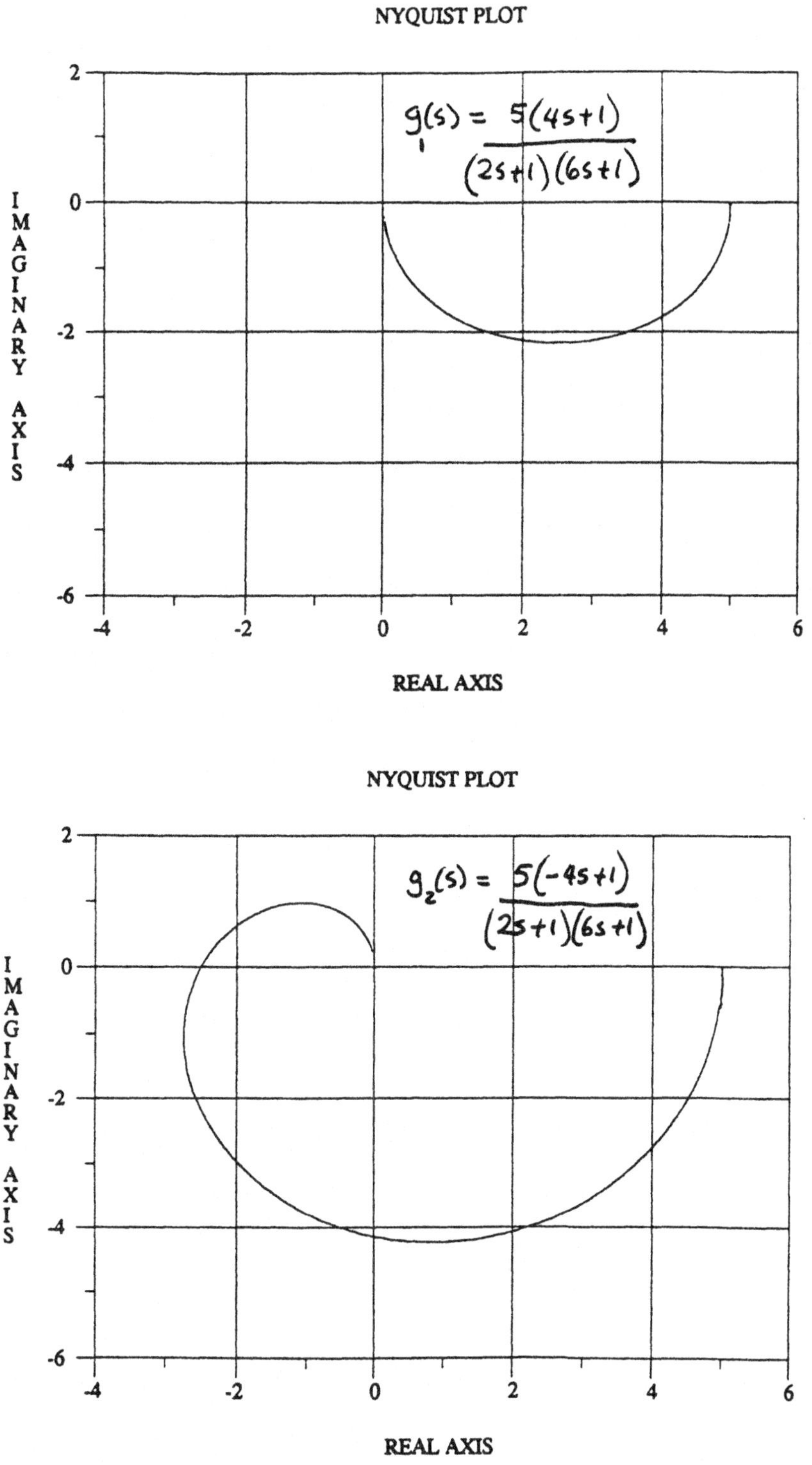
NYQUIST PLOT
$g_1(s) = \frac{5(4s+1)}{(2s+1)(6s+1)}$
IMAGINARY AXIS
REAL AXIS
2
0
-2
-4
-6
-4
-2
0
2
4
6
NYQUIST PLOT
$g_2(s) = \frac{5(-4s+1)}{(2s+1)(6s+1)}$
IMAGINARY AXIS
REAL AXIS
2
0
-2
-4
-6
-4
-2
0
2
4
6

FIG S 9.8 (b)

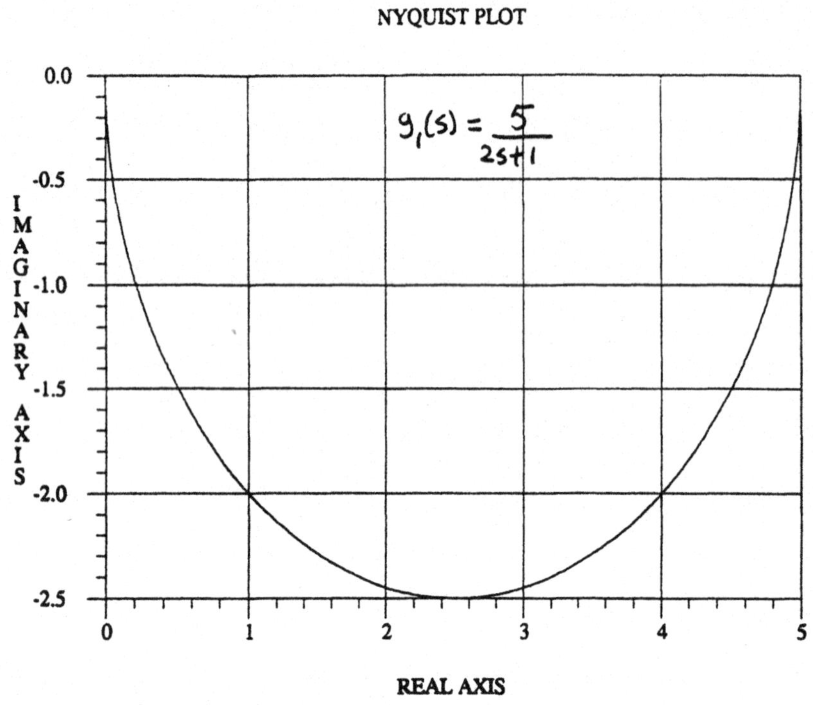
NYQUIST PLOT
$g_1(s) = \frac{5}{2s+1}$
IMAGINARY AXIS
REAL AXIS
0.0
-0.5
-1.0
-1.5
-2.0
-2.5
0
1
2
3
4
5

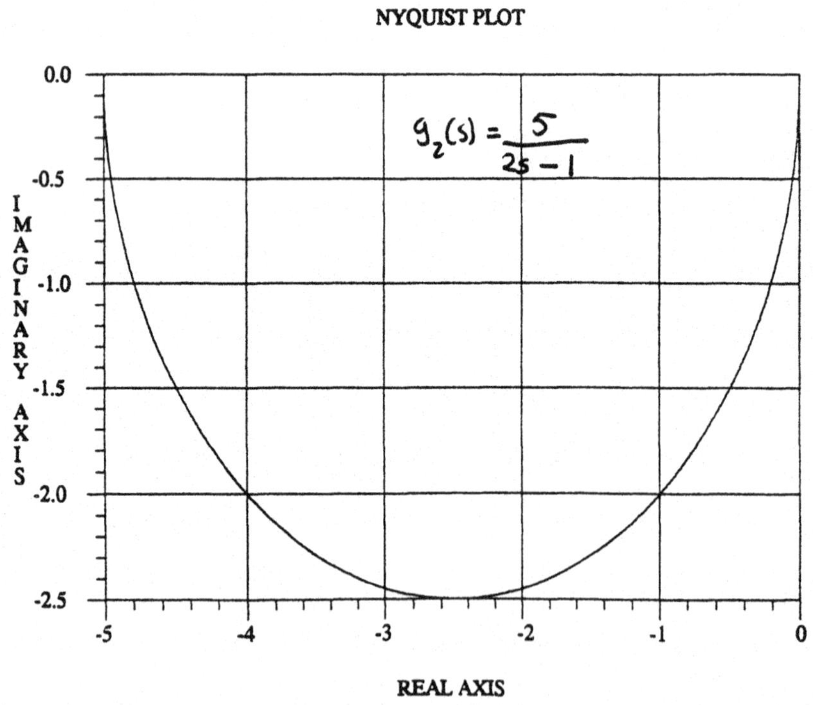
NYQUIST PLOT
$g_2(s) = \frac{5}{2s-1}$
IMAGINARY AXIS
REAL AXIS
0.0
-0.5
-1.0
-1.5
-2.0
-2.5
-5
-4
-3
-2
-1
0

FIG S 9.8(c)

Chapter 10

10.1 When the function $f(\xi)$ is linearized around the point $\xi = \xi_s$ using a first order Taylor approximation, the pertinent expression is:

$$f(\xi) \approx f(\xi_s) + f'(\xi)\Big|_{\xi_s}(\xi - \xi_s) \qquad (10.1)$$

where $f'(\xi) = \dfrac{\partial f}{\partial \xi}$

and if we now define the following deviation variables:

$$\left.\begin{aligned} y &= f(\xi) - f(\xi_s) \\ x &= \xi - \xi_s \end{aligned}\right\} \qquad (10.2)$$

Eq. (10.1) becomes

$$y = \frac{\partial f}{\partial \xi}\Big|_{\xi_s} x \qquad (10.3)$$

or

$$y = J(\xi_s)\,x \qquad (10.4)$$

where $J(\xi_s) = \dfrac{\partial f}{\partial \xi}\Big|_{\xi_s}$

These results may now be applied to the expressions to be linearized.

(a) $$k(T) = k_0 e^{-E/RT}$$

In which case $J(T_s) = \dfrac{\partial k}{\partial T}\Big|_{T_s} = \dfrac{k_0 E}{R T_s^2} e^{-E/RT_s}$

so that

$$y = k(T) - k(T_s) = \left(\frac{k_0 E}{R T_s^2} e^{-E/RT_s}\right)(T - T_s)$$

or, with $x = T - T_s$, we have

$$y = k(T) - k(T_s) = \left(\frac{k_0 E}{R T_s^2} e^{-E/RT_s}\right) x \qquad (10.5)$$

(b) Let $x = T - T_s$

and $y = H(T) - H(T_s)$

Then since $H(T) = b_0 + b_1 T + b_2 T^2 + b_3 T^3 + b_4 T^4$

$$\left.\frac{\partial H}{\partial T}\right|_{T=T_s} = b_1 + 2b_2 T_s + 3b_3 T_s^2 + 4b_4 T_s^3$$

and the required linearized expression is:

$$y = \left(b_1 + 2b_2 T_s + 3b_3 T_s^2 + 4b_4 T_s^3\right) x \qquad (10.6)$$

(c) $$p^o(T) = e^{[A - B/(T+C)]}$$

From here, we obtain

$$\left.\frac{\partial p^o(T)}{\partial T}\right|_{T=T_s} = \frac{B}{(T_s+C)^2} e^{[A - B/(T_s+C)]}$$

Thus, the required linearized equation is:

$$y = p^o(T) - p^o(T_s) = \left\{\frac{B}{(T_s+C)^2} e^{[A - B/(T_s+C)]}\right\} \underbrace{(T - T_s)}_{x} \qquad (10.7)$$

(d) $$y(x) = \frac{\alpha x}{1 + (\alpha - 1)x}$$

From here,

$$\left.\frac{\partial y}{\partial x}\right|_{x_s} = \frac{\alpha}{[1 + (\alpha-1)x_s]^2}$$

So that the required linearized expression is:

$$\bar{y} = y(x) - y(x_s) = \left\{\frac{\alpha}{[1 + (\alpha-1)x_s]^2}\right\}(x - x_s) \qquad (10.8)$$

where $(x - x_s) \to \bar{x}$

(e) $$q(T) = \epsilon\sigma A T^4$$

$$\left.\frac{\partial q}{\partial T}\right|_{T_s} = 4\epsilon\sigma A T_s^3$$

and therefore the linearized expression is:

$$y = q(T) - q(T_s) = \left(4\epsilon\sigma A T_s^3\right) x \qquad (10.9)$$

where $x = (T - T_s)$

10.2 (a) The original bilinear equation

$$\frac{dY(t)}{dt} = aY(t) + nY(t)U(t) + bU(t) \qquad (P10.6)$$

contains $Y(t)U(t)$ as the only nonlinear term: this may be linearized as follows.

$$Y(t)U(t) \approx Y_sU_s + Y_s(U - U_s) + U_s(Y - Y_s) \qquad (10.10)$$

At steady state (P10.6) becomes

$$0 = aY_s + nY_sU_s + bU_s \qquad (10.11)$$

Subtracting Eq (10.11) from (P10.6), introducing the approximation in (10.10) and using the given deviation variables, we obtain immediately

$$\frac{dy}{dt} = (a + U_s)y + (b + Y_s)u \qquad (10.12)$$

as the corresponding approximate linear model; by taking Laplace transforms now and rearranging, we obtain the following transfer function model:

$$y(s) = \left\{ \frac{(b + Y_s)}{[s - (a + U_s)]} \right\} u(s) \qquad (10.13)$$

with the required transfer function indicated in winged brackets.

(b) In terms of the given deviation variables, Eq (P10.7) is

$$\frac{dy}{dt} = ay + bu$$

and upon Laplace transform the transfer function model is

$$y(s) = \frac{b}{(s - a)} u(s) \qquad (10.14)$$

indicating a first order transfer function with gain K and time constant τ given by

$$K = -b/a \; ; \; \tau = -1/a$$

By comparison, the linear approximation of the bilinear model has the following parameters:

$$K_b = -(b + Y_s)/(a + U_s) \qquad (10.15a)$$

$$\tau_b = \frac{-1}{(a + U_s)} \qquad (10.15b)$$

Observe therefore that the effect of the bilinear term is to make the "apparent system time constant" dependent on U_s, while the "apparent system gain" is dependent on both Y_s and U_s as shown in Eq (10.15).

10.3 Given a general nonlinear model

$$\frac{d\eta}{dt} = f(\eta, \mu)$$

a first order Taylor series approximation is given by:

$$\frac{dy}{dt} = \left.\frac{\partial f}{\partial \eta}\right|_{(\eta^*,\mu^*)} y + \left.\frac{\partial f}{\partial \mu}\right|_{(\eta^*,\mu^*)} u$$

where $y = \eta - \eta^*$; $u = \mu - \mu^*$

Applying these results to the model:

$$A\frac{dh}{dt} = F_i - ch^{2/3} \qquad (P10.8)$$

we obtain

$$A\frac{dy}{dt} = \left(-\frac{2}{3}ch_s^{-1/3}\right)y + u \qquad (10.16)$$

where $y = h - h_s$; $u = F_i - F_{i_s}$

Upon taking Laplace transforms in (10.16) and rearranging we obtain

$$y(s) = \frac{\left(\frac{3}{2C} h_s^{1/3}\right)}{\left[\left(\frac{3A}{2C} h_s^{1/3}\right)s + 1\right]} u(s); \quad (10.17)$$

the first order transfer function indicated has a gain K, and time constant τ, explicitly given by

$$K = \frac{3}{2C} h_s^{1/3} \quad (10.18a)$$

$$\tau = \frac{3A}{2C} h_s^{1/3} \quad (10.18b)$$

(b) With the given process parameters and operating conditions, the linearized model's gain and time constant are given by

$$K = 2.4$$

$$\tau = 0.6$$

For a change in flowrate of 0.2 m^3/min (from 0.32 to 0.52) by definition of the gain of a linear system, the linearized model will predict a <u>change</u> in the liquid level of

$$\Delta h = (2.4) \times 0.2 = 0.48.$$

Since the initial liquid level was 0.512 m, the linearized model predicts a final steady state liquid level of

$$h_{fs} = 0.512 + 0.48$$

or

$$h_{fs} = 0.992 \text{ m}.$$

And for a tank with total height of 1 m, this model therefore indicates <u>no overflow</u>.

Using the nonlinear model directly, however, we observe that at ss,

$$A\frac{dh}{dt} = 0 = F_{i_s} - ch_s^{2/3}$$

$$0 = 0.52 - (0.5)(h_s)^{2/3}$$

or $h_s = 1.06\,m$

indicating that the tank will in fact overflow.

10.4 From the given transformation

$$z = \exp\left[h^{3/2}\left(\frac{4R}{3} - \frac{2h}{5}\right)\right] \qquad (P10.10)$$

we obtain that since

$$\frac{dz}{dt} = \frac{dz}{dh}\cdot\frac{dh}{dt}$$

then

$$\frac{dz}{dt} = \left\{\exp\left[h^{3/2}\left(\frac{4R}{3} - \frac{2h}{5}\right)\right]\right\}\left\{2Rh^{1/2} - h^{3/2}\right\}\frac{dh}{dt}$$

$$\frac{dz}{dt} = z\left[h^{-1/2}\left(2Rh - h^2\right)\right]\cdot\frac{dh}{dt} \qquad (10.19)$$

If we now introduce the RHS of the original model in Eq (P10.9) into (10.19) above for dh/dt, we obtain, upon some simple rearrangement that.

$$\frac{dz}{dt} = -\frac{c}{\pi}z + \frac{z}{\pi}F_i h^{-1/2}$$

and finally we introduce Eq (P10.11) to obtain

$$\frac{dz}{dt} = -\frac{c}{\pi} z + \frac{1}{\pi} v$$

which is of the required form with

$$a = -c/\pi \quad ; \quad b = 1/\pi .$$

10.5 (a) The original process model

$$\frac{dh}{dt} = \frac{1}{\pi} \frac{(F_i - ch^{1/2})}{(2Rh - h^2)}$$

when linearized around (F_{i_s}, h_s) becomes

$$\frac{dy}{dt} = \left.\frac{\partial f}{\partial F_i}\right|_{ss} u + \left.\frac{\partial f}{\partial h}\right|_{ss} y \tag{10.20}$$

where $y = h - h_s$; $u = F_i - F_{i_s}$, and

$$f(F_i, h) = \frac{1}{\pi} \frac{(F_i - ch^{1/2})}{(2Rh - h^2)} \tag{10.21}$$

From Eq (10.21) we now obtain

$$\left.\frac{\partial f}{\partial F_i}\right|_{ss} = \frac{1}{\pi} \frac{1}{(2Rh_s - h_s^2)} = 0.1232 \tag{10.22}$$

when the appropriate numerical values are introduced.
Similarly, we obtain

$$\left.\frac{\partial f}{\partial h}\right|_{ss} = \frac{1}{\pi} \left[\frac{(2Rh_s - h_s^2)(-\tfrac{1}{2} c h_s^{-1/2}) - \overbrace{(F_{i_s} - ch_s^{1/2})}^{0}(2R - 2h_s)}{(2Rh_s - h_s^2)^2} \right]$$

simplifying to give

$$\left.\frac{\partial f}{\partial h}\right|_{ss} = -\frac{c}{2\pi}\left[\frac{h_s^{-1/2}}{2Rh_s - h_s^2}\right] = -0.1027$$

Thus Eq.(10·20) now becomes

$$\frac{dy}{dt} = 0.1232u - 0.1027y$$

The required response obtained using this approximate linear model is shown in the dashed line in FIG S10.1.

(b) The more accurate response is shown in the solid line in FIG S10.1. The linearized model underestimates the process gain, clearly: however, observe how closely it approximates the true response in the initial portion (up till about 25 minutes and $h \approx 1.3$m)

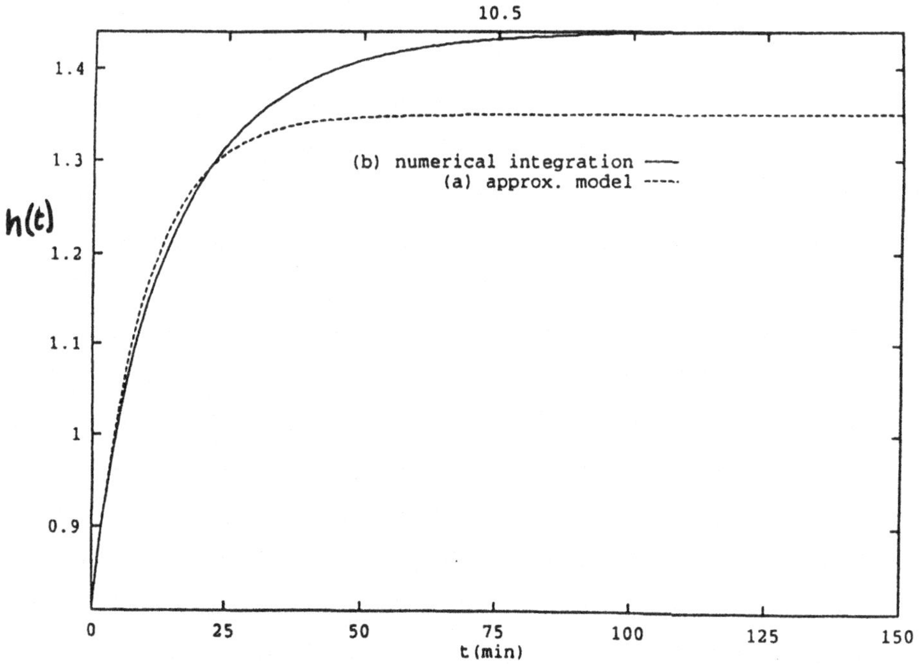

FIG S10.1

10.6 Starting from

$$\frac{dV}{dt} = F_i - ch^{1/2} \qquad (P10.13)$$

if $\mathcal{V} = V/V_s$; $\mathcal{H} = h/h_s$; $\mathcal{F}_i = \dfrac{F_i}{F_{i_s}}$

then from (P10.13), we have

$$V_s \frac{d}{dt}\left(V/V_s\right) = F_i - ch^{1/2}$$

and

$$\frac{V_s}{F_{i_s}} \frac{d}{dt}\left(\frac{V}{V_s}\right) = \frac{F_i}{F_{i_s}} - \frac{ch^{1/2}}{F_{i_s}} \qquad (10.23)$$

But since $F_{i_s} = ch_s^{1/2}$ (from (P10.13) at SS conditions)
Eq (10.23) above becomes

$$\frac{V_s}{F_{i_s}} \frac{d}{dt}\left(\frac{V}{V_s}\right) = \frac{F_i}{F_{i_s}} - \frac{h^{1/2}}{h_s^{1/2}}$$

from where we immediately obtain

$$\tau_s \frac{d\mathcal{V}}{dt} = \mathcal{F}_i - \mathcal{H}^{1/2} \qquad (10.24)$$

as required, with $\tau_s = V_s/F_{i_s}$.

Now let
$$\tilde{\mathcal{V}} = \mathcal{V} - \mathcal{V}_s = \mathcal{V} - 1$$
$$\tilde{\mathcal{H}} = \mathcal{H} - \mathcal{H}_s = \mathcal{H} - 1$$
$$\tilde{\mathcal{F}}_i = \tilde{\mathcal{F}} - \mathcal{F}_s = \mathcal{F} - 1$$

we may now linearize (10.24) above in terms of these deviation variables to obtain

$$\tau_s \frac{d\tilde{\mathcal{V}}}{dt} = \tilde{\mathcal{F}}_i - \tfrac{1}{2}\tilde{\mathcal{H}}\left(\mathcal{H}_s^{-1/2}\right)$$

but since $\mathcal{H}_s = 1$, we now have

$$\tau_s \frac{d\tilde{\mathcal{V}}}{dt} = \tilde{\mathcal{F}}_i - \frac{1}{2}\tilde{\mathcal{H}} \qquad (10.25)$$

It now remains to eliminate $\tilde{\mathcal{V}}$ and write it in terms of $\tilde{\mathcal{H}}$.

By definition of the volume in the conical tank,

$$V = Ah$$

$$= \pi\left(\frac{Rh}{H}\right)^2 h$$

And now

$$\mathcal{V} = \frac{V}{V_s} = \frac{\pi\left(\frac{Rh}{H}\right)^2 h}{\pi\left(\frac{Rh_s}{H}\right)^2 h_s}$$

i.e

$$\mathcal{V} = \left(\frac{h}{h_s}\right)^3 = \mathcal{H}^3 \qquad (10.26)$$

thus,

$$\frac{d}{dt}(\tilde{\mathcal{V}}) = \frac{d}{dt}(\mathcal{H}^3 - 1)$$

and since

$$\mathcal{H}^3 \approx \mathcal{H}_s^3 + 3\mathcal{H}_s^2(\tilde{\mathcal{H}})$$

($\mathcal{H}_s^3 = 1$)

we obtain

$$\frac{d}{dt}(\tilde{\mathcal{V}}) \approx 3\mathcal{H}_s^2 \frac{d}{dt}(\tilde{\mathcal{H}})$$

and finally Eq. (10.25) becomes

$$3\mathcal{H}_s^2 \tau_s \frac{d}{dt}(\tilde{\mathcal{H}}) = \tilde{\mathcal{F}}_i - \frac{1}{2}\tilde{\mathcal{H}}$$

or, since $\mathcal{H}_s = 1$

$$6\tau_s \frac{d\tilde{\mathcal{H}}}{dt} = 2\tilde{\mathcal{F}}_i - \tilde{\mathcal{H}}$$ as required.

10.7 When the original nonlinear model is linearized using a first order Taylor series approximation, the result is:

$$\frac{dx_1}{dt} = \left.\frac{\partial f_1}{\partial \xi_1}\right|_{ss} x_1 + \left.\frac{\partial f_1}{\partial \xi_2}\right|_{ss} x_2 + \left.\frac{\partial f_1}{\partial \mu}\right|_{ss} u$$

$$\frac{dx_2}{dt} = \left.\frac{\partial f_2}{\partial \xi_1}\right|_{ss} x_1 + \left.\frac{\partial f_2}{\partial \xi_2}\right|_{ss} x_2$$

with $f_1 = a_1\xi_1 - a_{12}\xi_1\xi_2 + b\mu$

and $f_2 = -a_2\xi_2 + a_{21}\xi_1\xi_2$

we have

$$\frac{dx_1}{dt} = \left(a_1 - a_{12}\xi_2^*\right)x_1 - \left(a_{12}\xi_1^*\right)x_2 + bu \qquad (10.27a)$$

$$\frac{dx_2}{dt} = +a_{21}\xi_2^* + \left(a_{21}\xi_1^* - a_2\right)x_2 \qquad (10.27b)$$

which is of the given form with

$$\alpha_{11} = (a_1 - a_{12}\xi_2^*) \; ; \; \alpha_{12} = a_{12}\xi_1^*$$

$$\alpha_{21} = a_{21}\xi_2^* \; ; \; \alpha_{22} = \left(a_2 - a_{21}\xi_1^*\right)$$

10.8 The required linearization may be achieved either
(i) by simply linearizing the nonlinear terms in isolation, and reintroducing into the original equation or,
(ii) by obtaining a complete linear approximation in terms

deviation variables.
We will do the latter.

By defining

$$x_1 = C_A - C_{As}; \quad x_2 = T - T_s; \quad d_1 = C_{Af} - C_{Af_s}$$

$$d_2 = T_f - T_{f_s} \qquad u = \bar{T}_c - \bar{T}_{c_s}$$

then the complete approximate linear model takes the form

$$\frac{dx_1}{dt} = \left.\frac{\partial f_1}{\partial C_A}\right|_s x_1 + \left.\frac{\partial f_1}{\partial T}\right|_s x_2 + \left.\frac{\partial f_1}{\partial C_{Af}}\right|_s d_1$$

$$\frac{dx_2}{dt} = \left.\frac{\partial f_2}{\partial C_A}\right|_s x_1 + \left.\frac{\partial f_2}{\partial T}\right|_s x_2 + \left.\frac{\partial f_2}{\partial T_f}\right|_s d_2 + \left.\frac{\partial f_2}{\partial \bar{T}_c}\right|_s u$$

with f_1 as the RHS of the given Eq. (4.26) and f_2 as the RHS of the accompanying equation, we obtain

$$\frac{dx_1}{dt} = -\left[\frac{1}{\theta} + k_0 e^{-(E/RT_s)}\right]x_1 + \left[\frac{E}{RT_s^2} C_{As} k_0 e^{-(E/RT_s)}\right]x_2 + \frac{1}{\theta} d_1 \tag{10.28a}$$

$$\frac{dx_2}{dt} = \left[\beta k_0 e^{-(E/RT_s)}\right]x_1 - \left[\frac{1}{\theta} - \frac{\beta E}{RT_s^2} C_{As} k_0 e^{-(E/RT_s)} + \frac{hA}{\rho C_p V}\right]x_2$$

$$+ \frac{1}{\theta} d_2 + \frac{hA}{\rho C_p V} u. \tag{10.28b}$$

Alternatively, the Taylor series approximation:

$$k_0 e^{(-E/RT)} C_A \approx k_0 e^{-(E/RT_s)} C_{As} + k_0 e^{-(E/RT_s)} (C_A - C_{As})$$

$$+ k_0 e^{-(E/RT_s)} \frac{E}{RT_s^2} C_{As} (T - T_s) \tag{10.29}$$

could simply be introduced for the nonlinear $k_0 e^{-(E/RT)}$ term in the original equation. This will yield the same set of equations as those boxed above, upon simplification.

10.9 (a) The linearized model takes the form

$$\frac{dx_1}{dt} = \left[-3 - \left(1.36594\sqrt{\xi_2^*}\right)\right]x_1 - \left[\frac{1}{2}(1.36594)\frac{\xi_1^*}{\sqrt{\xi_2^*}}\right]x_2$$

$$\frac{dx_2}{dt} = 60u - 3.03161\,x_2$$

$$\frac{dx_3}{dt} = \left(0.00134106\sqrt{\xi_2^*}\right)x_1 + \left[\frac{1}{2}\left(0.00134106\frac{\xi_1^*}{\sqrt{\xi_2^*}}\right) + 0.0034677\right]x_2 - 10x_3$$

$$\frac{dx_4}{dt} = -\left(\frac{\xi_4^*}{\xi_3^{*2}}\right)x_3 + \frac{1}{\xi_3^*}x_4$$

where we have made use of the fact that if

$$\frac{d\xi}{dt} = f(\xi, \mu) \quad \text{then linearization yields}$$

$$\frac{d\underset{\sim}{x}}{dt} = \left.\frac{\partial f}{\partial \underset{\sim}{\xi}}\right|_* \underset{\sim}{x} + \left.\frac{\partial f}{\partial \underset{\sim}{\mu}}\right|_* \underset{\sim}{u}.$$

(b) The linearized equations may be written, for simplicity in the following form:

$$\frac{dx_1}{dt} = a_{11}x_1 + a_{12}x_2$$

$$\frac{dx_2}{dt} = a_{22}x_2 + bu$$

$$\frac{dx_3}{dt} = a_{31}x_1 + a_{32}x_2 + a_{33}x_3$$

$$\frac{dx_4}{dt} = a_{41}x_1 + a_{42}x_2 + a_{44}x_4$$

$$y = c_3x_3 + c_4x_4 \qquad (10.30)$$

Laplace transforms and subsequent rearrangement yields

$$(s - a_{11})\,x_1 = a_{12}x_2 \qquad (10.31)$$

$$(s - a_{22})\,x_2 = bu \qquad (10.32)$$

$$(s - a_{33})\,x_3 = a_{31}x_1 + a_{32}x_2 \qquad (10.33)$$

$$(s - a_{44})\,x_4 = a_{41}x_1 + a_{42}x_2 \qquad (10.34)$$

From (10.32) we obtain

$$x_2 = \left(\frac{b}{s - a_{22}}\right)u \qquad (10.35)$$

and from here, and (10.31), we obtain

$$x_1 = \left[\frac{a_{12}b}{(s-a_{11})(s-a_{22})}\right]u \qquad (10.36)$$

Substituting (10.35) and (10.36) into (10.33) and (10.34) and simplifying leads to

$$x_3 = \left[\frac{a_{31}a_{12}b + (s-a_{11})a_{32}b}{(s-a_{11})(s-a_{22})(s-a_{33})}\right]u \qquad (10.37)$$

and

$$x_4 = \left[\frac{a_{41}a_{12}b + (s-a_{11})a_{42}b}{(s-a_{11})(s-a_{22})(s-a_{44})}\right]u \qquad (10.38)$$

From the given steady state values, we obtain

$$a_{11} = -3.498; \quad a_{12} = -10.3164; \quad a_{22} = -3.03161$$

$$a_{31} = 0.000489; \quad a_{32} = 0.044805; \quad a_{33} = -10$$

$$a_{41} = 49.8569; \quad a_{42} = 1032.8741; \quad a_{44} = -10$$

$$b = 60; \quad c_3 = -1.266 \times 10^7; \quad c_4 = 506.3291$$

Introducing these values into (10.37) and (10.38), and observing that $(s-a_{33}) = (s-a_{44}) = (s+10)$, the result may be introduced into (10.30) and rearranged to give the required transfer function model:

$$y(s) = b\left\{\frac{c_3\left[a_{31}a_{12} + (s-a_{11})a_{32}\right] + c_4\left[a_{41}a_{12} + (s-a_{11})a_{42}\right]}{(s-a_{11})(s-a_{22})(s-a_{44})}\right\}u(s)$$

and with the numerical values, we obtain, finally that the required transfer function is given by

$$g(s) = \frac{-1.98803 \times 10^5 (0.126s+1)}{(0.286s+1)(0.330s+1)(0.1s+1)} \qquad (10.39)$$

CHAPTER 11

11.1 (a) Transfer function poles are located at

$$s = -1/\tau \; ; \quad s = \pm \omega j$$

The purely imaginary poles indicate system is BIBO stable but <u>not</u> asymptotically stable. (Unit step response will oscillate continuously.)

(b) Transfer function poles are located at

$$s = -3 \; ; \quad s = -1$$

implying <u>both</u> BIBO stability and asymptotic stability.

(c) Transfer function poles: $s = -3$; $s = +1$

The positive pole implies <u>BIBO instability</u> as well as asymptotic instability.

(d) Transfer function poles: $s = 0$; $s = -3$; (the additional pole at $s = -1$ is at the same location at the lone zero. But since this is a "stable pole", the pole-zero cancellation does not "obscure" a problem).

The pole at the origin indicates only <u>conditional</u> BIBO stability, since the system response to such bounded inputs as steps is <u>unbounded</u> while the response to such inputs as rectangular pulses is bounded. The pole $s = 0$ implies asymptotic instability.

11.2 The model in question is:

$$\frac{d^2\theta}{dt^2} + \frac{k}{m}\frac{d\theta}{dt} + \frac{g}{R}\sin\theta = 0$$

the nonlinear term, $\sin\theta$ has the following 1st order Taylor series approximation:

$$\sin\theta \approx \sin\theta_s + (\cos\theta_s)(\theta - \theta_s) \qquad (11.1)$$

(i) For $\theta_s = 0$,

$\sin\theta_s = 0$ and $\cos\theta_s = 1$ so that

$$\sin\theta \approx (\theta - \theta_s)$$

So that at this steady state, the linearized model is

$$\frac{d^2\theta}{dt^2} + \frac{k}{m}\frac{d\theta}{dt} + \frac{g}{R}(\theta - \theta_s) = 0$$

or

$$\frac{d^2x}{dt^2} + \frac{k}{m}\frac{dx}{dt} + \frac{g}{R}x = 0 \qquad (11.2)$$

where $x = \theta - \theta_s$.

Eq (11.2) is a linear, second order ODE with characteristic equation

$$r^2 + \frac{k}{m}r + \frac{g}{R} = 0 \qquad (11.3)$$

And now because k/m is positive as is g/R, observe that Eq (11.3) has both roots in the LHP. Thus the steady state $\theta_s = 0$ is STABLE.

(ii) For $\theta_s = +\pi$,

$\sin\theta_s = 0$, and $\cos\theta_s = -1$

so that, around this steady state

$$\sin\theta \approx -(\theta - \theta_s)$$

The approximate linear model is now

$$\frac{d^2\theta}{dt^2} + \frac{k}{m}\frac{d\theta}{dt} - \frac{g}{R}(\theta - \theta_s) = 0$$

or

$$\frac{d^2x}{dt^2} + \frac{k}{m}\frac{dx}{dt} - \frac{g}{R}x = 0 \qquad (11\text{-}4)$$

The characteristic equation for this 2nd order ODE is:

$$r^2 + \frac{k}{m}r - \frac{g}{R} = 0 \qquad (11\text{-}5)$$

with a negative coefficient, $\left(-g/R\right)$, indicating that there is precisely one root in the RHP. The steady state $\theta = \pi$ is therefore UNSTABLE.

11.3 (a) From the original nonlinear process model,

$$\frac{dY}{dt} = -2.5Y + 0.5YU + 1.5U$$

we obtain that at steady state,

$$0 = -2.5Y^* + 0.5Y^*U^* + 1.5U^* \qquad (11\text{-}6)$$

which is easily rearranged to give

$$Y^* = \frac{3U^*}{5 - U^*} \qquad (11\text{-}7)$$

as the required steady state relationship.

A sketch of this functional relation is shown below in FIG S 11.1. Note the asymptote at $U^* = 5$ (vertical) and the horizontal asymptote at $Y^* = -3$

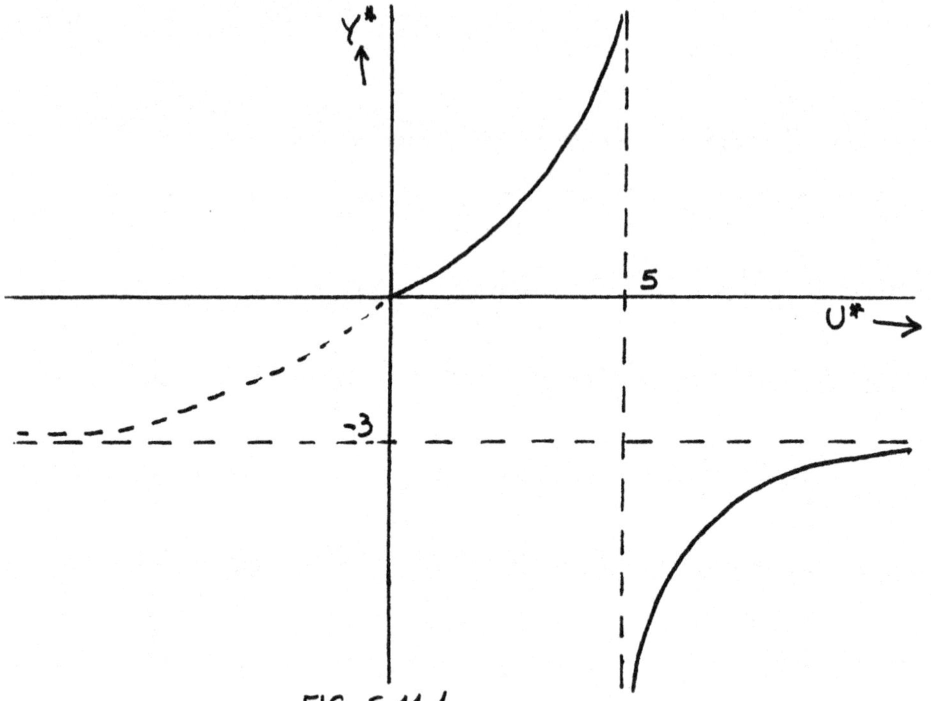

FIG S 11.1

(b) Defining $y = Y - Y^*$, $u = U - U^*$, the original model is linearized to obtain

$$\frac{dy}{dt} = (-2.5 + 0.5U^*)\,y + (1.5 + 0.5Y^*)\,u \qquad (11.8)$$

and the stability of the steady state is determined by the sign associated with the coefficient of y. For the specific case of $U^* = 3.0$, this coefficent is -1.0 and therefore we conclude that this steady state is STABLE. The bioreactor will thus operate stably in the vicinity of this operating condition.

(c) From Eq (11.7) above, we obtain that $U^* = 6$ corresponds to $Y^* = -18$. A negative value is physically impossible for Y, hence the contemplated conditions correspond to an infeasible steady state. Furthermore, from Eq (11.8) we observe that with $U^* = 6$, the coefficient of y becomes $+0.5$, indicating an <u>unstable</u> steady state.

11.4 As was obtained in Eq (10.39) of the solution to Problem 10.9, the approximate linear transfer function for the polymerization reactor is:

$$g(s) = \frac{-1.98803 \times 10^5 (0.126s+1)}{(0.286s+1)(0.330s+1)(0.1s+1)}$$

with 3 poles located at:

$$s = -3.498 ; \quad s = -3.03161 ; \quad s = -10$$

(Recall that a 4th pole at $s = -10$ was "cancelled" by a corresponding zero at the same location; since this is a <u>stable</u> pole, the implied cancellation is not problematic) Because all poles of $g(s)$ lie in the LHP, we conclude that the indicated steady state is STABLE.

11.5 From Problem 10.8, the linearized equation is as given in (10.28b). For this specific problem, with the additional assumption that C_A is held constant, Eq (10.28b) becomes (with $x_1 = 0$ as a consequence of this assumption of constant C_A)

$$\frac{dx_2}{dt} = -\left[\frac{1}{\theta} - \frac{\beta E}{RT_s^2} C_{A_s} k_0 e^{-(E/RT_s)} + \frac{hA}{\rho C_p V}\right] x_2 + \frac{1}{\theta} d_2 + \frac{hA}{\rho C_p V} u \tag{11.9}$$

For (11.9) to be stable now requires that

$$\frac{1}{\theta} - \frac{\beta E}{RT_s^2} C_{As} k_0 e^{-(E/RT_s)} + \frac{hA}{\rho C_p V} > 0$$

or :

$$\boxed{\frac{hA}{\rho C_p V} > \frac{\beta E}{RT_s^2} C_{As} k_0 e^{-(E/RT_s)} - \frac{1}{\theta}.} \qquad (11.10)$$

11.6 When the original process model is linearized around the steady state (F_{i_s}, h_s), we obtain

$$\frac{dy}{dt} = \left.\frac{\partial f}{\partial h}\right|_{ss} y + \left.\frac{\partial f}{\partial F_i}\right|_{ss} u \qquad (11.11)$$

with $y = h - h_s$; $u = F_i - F_{i_s}$, and

$$f(F_i, h) = \frac{1}{\pi}\frac{(F_i - ch^{1/2})}{(2Rh - h^2)} \qquad (11.12)$$

The stability or otherwise of (11.11) is determined solely by the sign associated with $\left.\frac{\partial f}{\partial h}\right|_{ss}$.

From Eq (11.12) we obtain

$$\left.\frac{\partial f}{\partial h}\right|_{ss} = \frac{1}{\pi}\left[\frac{(2Rh_s - h_s^2)(-\frac{1}{2} c h_s^{-1/2}) - (F_{i_s} - ch_s^{1/2})(2R - 2h_s)}{(2Rh_s - h_s^2)^2}\right]$$

(with $F_{i_s} - ch_s^{1/2} \to 0$)

$$= -\frac{c}{\pi}\left[\frac{h_s^{-1/2}}{2Rh_s - h_s^2}\right]$$

$$\boxed{\left.\frac{\partial f}{\partial h}\right|_{ss} = -\frac{c}{\pi}\left[\frac{1}{h_s^{3/2}(2R - h_s)}\right]} \qquad (11.13)$$

We now observe that $c > 0$, $\pi > 0$, $h_s > 0$ always, and, for a hemispherical tank $2R > h_s$ always.

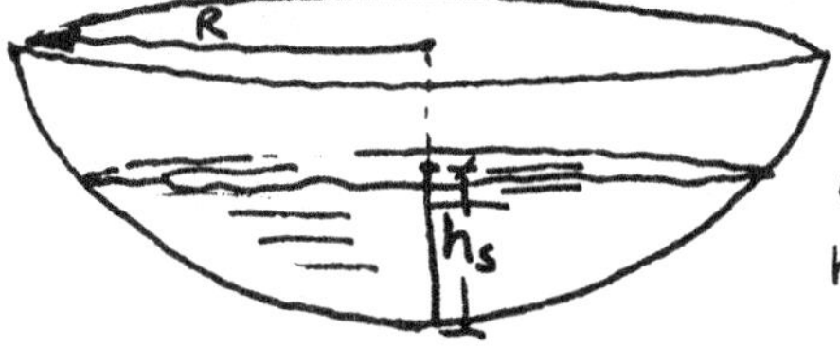

observe: $h_s \leq R$
hence $h_s < 2R$ always.

The implication is that the RHS of (11.13) will <u>always</u> be negative for <u>all</u> values of h_s. Thus the system is stable at all steady states.

Physical Justification:

The Modeling equation is based on flow through outlet valve at a rate proportional to the square root of liquid level in the tank. For any inlet flow rate F_{i_s}, there will always be a value h_s for which

$$F_{i_s} = c h_s^{1/2}$$

for a fixed valve resistance c. Any disturbance away from this equilibrium results in process adjustments that will ultimately restore the process.

For example, an increase in F_i will cause an increase in liquid level until such a value of h such that $ch^{1/2}$ now equals the new, increased, F_i, at which point the process equilibrium is restored. (See p. 348 in the text) This process is an example of a self-regulating process.

11.7 (a) The linearized process model is

$$\frac{dy}{dt} = \left.\frac{\partial f}{\partial Y}\right|_{ss} y + \left.\frac{\partial f}{\partial U}\right|_{ss} u \qquad (11.14)$$

where $y = Y - Y^*$, $u = U - U^*$ and

$$f(Y,U) = -Y(t) + 2.6U(t) - 2.0U(t)^2 + F(t) \qquad (11.15)$$

(since $F(t)$ is constant, (11.14) does not include $\frac{\partial f}{\partial F}$ which is zero).

From (11.15), we obtain

$$\left.\frac{\partial f}{\partial Y}\right|_{ss} = -1 \; , \quad \left.\frac{\partial f}{\partial U}\right|_{ss} = (2.6 - 4U^*)$$

so that the linearized model is

$$\frac{dy}{dt} = -y + (2.6 - 4U^*)u \qquad (11.16)$$

Upon taking Laplace transforms and rearranging, we obtain the following transfer function model

$$y(s) = \frac{(2.6 - 4U^*)}{s+1}\, u$$

indicating a first order transfer function with

$$K = 2.6 - 4U^* \; ; \quad \tau = 1.$$

At $U^* = 0.65$, $K = 0$. i.e. the process gain vanishes at this operating steady state.

(b) At ss, $\frac{dy}{dt} = 0$, and from the original process model we obtain the steady state relationship

$$2.0U^{*2} + 2.6U^* + Y^* = 0 \qquad (11.17)$$

A plot of Y^* versus U^* is shown in FIG S11.2.

Since the operational definition of the steady-state gain is the ratio of the change in Y, (at steady-state) to the change in U responsible for the response, in this case then, the steady state "gain" at any point is defined as $\partial Y^*/\partial U^*$ — the slope of the tangent to the curve in FIG S11.2 at the point in question.

Observe that for $U^* < 0.65$, $\partial Y^*/\partial U^*$ is positive; while for $U^* > 0.65$, $\partial Y^*/\partial U^*$ is negative. Thus for operating conditions corresponding to $U^* < 0.65$, the system has a positive gain; the process gain switches sign when $U^* > 0.65$. The gain therefore becomes identically zero when $U^* = 0.65$.

Note from (11.17) that

$$\frac{dY^*}{dU^*} = 4U^* - 2.6$$

which we had earlier identified as the gain of the linearized system model.

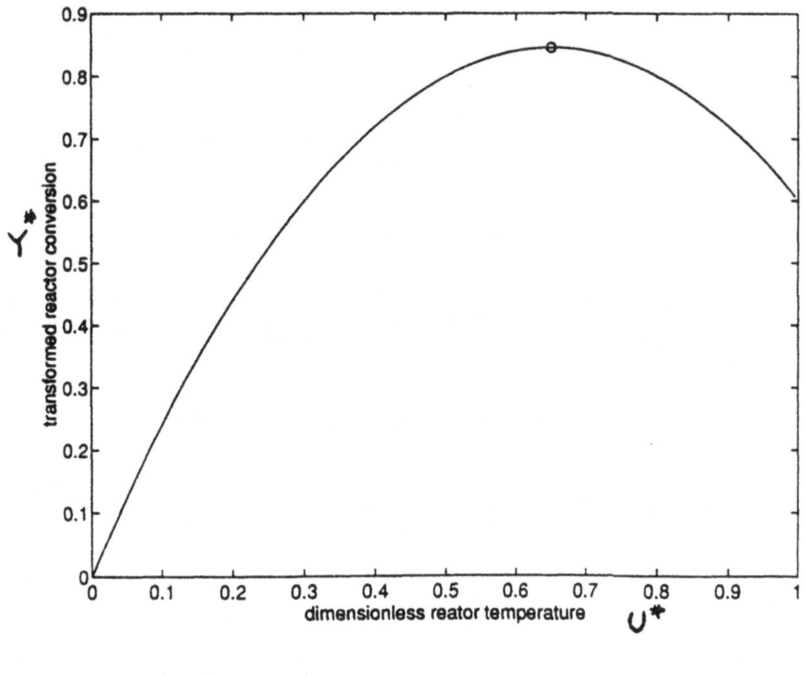

FIG S 11.2

(c) At the point $U^* = 0.65$ a "singularity" occurs. If we exclude this point, for now, we observe, from Eq. (11.16) that the system is always stable, since the coefficient of y is -1 regardless of the steady state under consideration.

11.8 (a) With $g(s) = \frac{6}{2s+1}$, $g^+(s)$ is now given by

$$g^+(s) = \frac{\left(\frac{6}{2s+1}\right)K}{1 + \left(\frac{6}{2s+1}\right)K}$$

or
$$g^+(s) = \frac{6K}{2s + (1+6K)}$$

which is a transfer function with a single pole located at

$$s = \frac{-(1+6K)}{2}$$

Observe now that for all $K > 0$, this pole will always lie in the LHP: the system represented by $g^+(s)$ is therefore <u>always</u> stable for $K > 0$. Thus <u>no</u> value of $K > 0$ will make $g^+(s)$ unstable.

(b) With $g(s) = \frac{6}{(2s+1)(4s+1)}$, we now have

$$g^+(s) = \frac{6K}{(2s+1)(4s+1) + 6K} = \frac{6K}{8s^2 + 6s + (1+6K)}$$

A transfer function whose poles are located at

$s = r_1$, $s = r_2$ where r_1 and r_2 are the roots of the indicated denominator quadratic.

There are several ways of establishing that r_1 and r_2 will always lie in the LHP for $K > 0$. The most straightforward is to note that, starting from the factorization

$$(s+p_1)(s+p_2) \qquad p_1 > 0,\ p_2 > 0$$

and expanding, we will obtain the quadratic

$$s^2 + (p_1+p_2)s + p_1p_2$$

a quadratic with all coefficients positive, and whose roots are $-p_1, -p_2$, both negative. Thus, so long as all the coefficients of a quadratic are positive (as is the case with $8s^2 + 6s + (1+6K)$ for $K>0$) its two roots will always lie in the LHP.

Conversely we may proceed to evaluate these roots and obtain

$$r_1, r_2 = \frac{-6 \pm 2\sqrt{1-48K}}{16}$$

and hence observe that there is no $K>0$ for which any of these roots will be in the RHP.

(c) Under the given circumstances, we have

$$g^+(s) = \frac{6K}{(2s+1)(4s+1)(6s+1)+6K}$$

$$= \frac{6K}{48s^3 + 44s^2 + 12s + (1+6K)}$$

and the roots of the denominator cubic polynomial will

determine the system stability.
we will now use the given results with

$$a_0 = 48\ ;\quad a_1 = 44\ ;\quad a_2 = 12\ ;\quad a_3 = (1 + 6K)$$

(i) For $K = 1$, $\quad a_3 = 7$

and all coefficients a_0, a_1, a_2, a_3 are > 0

Next, $\quad a_1 a_2 = 44 \times 12$

$\qquad\quad a_0 a_3 = 7 \times 48$

and $a_1 a_2 > a_0 a_3$, hence the system is stable.

(ii) For $K = 2$, this time $a_3 = 13$, and now

$$a_0 a_3 = 48 \times 13$$

so that $\quad a_1 a_2 \not> a_0 a_3$
and hence the system is not stable for $K = 2$.

11.9 (a) For $g(s)$ as given in (P11.11), g_{CL} is given by

$$g_{CL} = \frac{4.5 K_c}{3s + (4.5K_c - 1)}$$

a transfer function with a single pole located at

$$s = -\frac{(4.5K_c - 1)}{3}$$

For this pole to lie in the LHP (i.e. for the system to be stable) $\quad 4.5K_c - 1 > 0$

i.e.

$$K_c > 1/4.5 \qquad (11.18)$$

If the open loop unstable system had been specified with an <u>arbitrary</u> RHP pole at $s = 1/\tau$, i.e.

$$g(s) = \frac{4.5}{\tau s - 1}$$

observe that

$$g_{CL}(s) = \frac{4.5K_c}{\tau s + (4.5K_c - 1)}$$

and the closed loop pole, located at

$$s = -\frac{(4.5K_c - 1)}{\tau}$$

will lie strictly in the LHP if $K_c > 1/4.5$, which is the same condition given in Eq (11.18), entirely independent of the location of the RHP open loop pole.

(b) With the open loop unstable system given by

$$g(s) = \frac{4.5}{(3s-1)(\tau_2 s + 1)}$$

$$g_{CL}(s) = \frac{4.5K_c}{3\tau_2 s^2 + (3 - \tau_2)s + (4.5K_c - 1)}$$

and stability now requires that the two roots of

$$3\tau_2 s^2 + (3 - \tau_2)s + (4.5K_c - 1) = 0 \qquad (11.19)$$

all lie strictly in the LHP.

Noting that both roots of a quadratic will lie in the LHP if <u>all</u> the coefficients are positive (or all the coefficients are negative), we now observe that stability in this case requires

$$3 - \tau_2 > 0 \tag{11.20}$$

$$4.5K_c - 1 > 0 \tag{11.21}$$

Eq (11.21) is recognizable as giving rise to the same condition obtained earlier in part a; Eq (11.20) on the other hand requires that $\tau_2 < 3$ if the system is to be stable. It is this requirement which makes the closed loop system stability dependent on the location of τ_2 relative to 3.

Thus, in addition to requiring $K_c > 1/4.5$, stability in this case also requires that τ_2 be less than 3.

CHAPTER 12

12.1 (a) A sectional view of the thermometer's bulb is shown below

From the given assumptions, an appropriate mathematical model may be obtained from the following energy balance:

Rate of Accumulation = Rate of Input − Rate of output.

$$\frac{d}{dt}\left[mC_p(T-T_0)\right] = hA(T_s - T) - 0$$

where T_0 = Reference (Datum) Temperature; h = H.T. coeff.; A = Effective area of H.T.; T_s = Surrounding Temp.; 0 → No heat losses.

Since m, C_p are constant, the energy balance equation simplifies to:

$$\tau \frac{dT}{dt} = T_s - T \qquad (12.1)$$

where

$$\tau = \frac{mC_p}{hA} \qquad (12.2)$$

(b) Observe from the supplied data that $T(0) = 78°F$; Defining now the deviation variable

$$y = T - T(0) \qquad (12.3)$$

the model in Eq (12.1) becomes

$$\tau \frac{dy}{dt} = y_s - y \qquad (12.4)$$

with

$$y_s = T_s - T(0)$$

Eq (12.4) has the following closed form solution, given a step change A in y_s:

$$y = A(1 - e^{-t/\tau}) \qquad (12.5)$$

From the supplied data $A = 139.8 - 78 = 61.8$; and the parameter τ must be estimated using Eq (12.5), i.e.

$$(T - 78) = 61.8(1 - e^{-t/\tau}) \qquad (12.6)$$

Either by employing the logarithmic transformation of (12.6), i.e.

$$\ln\left[1 - \left(\frac{T-78}{61.8}\right)\right] = -t/\tau \qquad (12.7)$$

and linear regression techniques, or, by employing nonlinear regression directly on Eq (12.6), one easily obtains

$$\hat{\beta} = 1/\hat{\tau} = 0.033 \Rightarrow \hat{\tau} = 30.$$

A plot of the model equation:

$$(T - 78) = 61.8(1 - e^{-0.033t})$$

along with the data is shown in FIG S12.1. The indication is that the model is adequate, and the parameter estimate is reasonable.

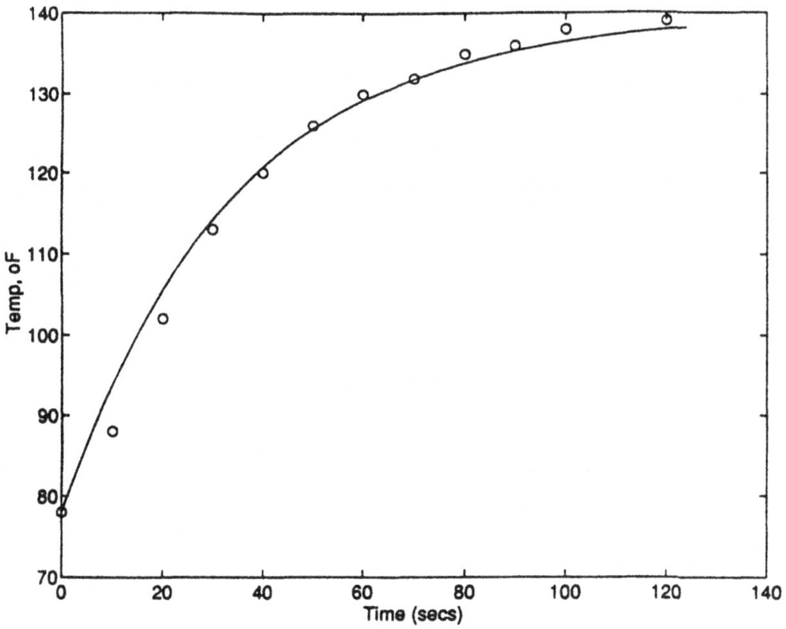

FIG S 12·1

12.2 A reasonable theoretical model for the water heater based on the given assumptions is obtained from energy balances for the tank, and for the coil:

Overall energy balance for tank:

$$\rho_\ell V_\ell C_{p_\ell} \frac{dT}{dt} = UA(T_c - T) + wC_{p_\ell}(T_i - T^*) - wC_{p_\ell}(T - T^*)$$

Rate of Accumulation. (constant physical properties)	Rate of heat transfer from coil	Rate of heat input with inflowing liquid	Rate of heat output with outflowing liquid.

which simplifies to

$$\rho_\ell V_\ell C_{p_\ell} \frac{dT}{dt} = UA(T_c - T) + wC_{p_\ell}(T_i - T). \qquad (12.8)$$

Overall energy balance for coil.

$$m_c C_{p_c} \frac{dT_c}{dt} = K_c(T_d - T) - UA(T_c - T) \qquad (12.9)$$

Rate of accumulation — Rate of input through controller — Rate of output to tank liquid

Define the following deviation variables:

$$x_1 = T - T_s, \quad x_2 = T_c - T_{cs}; \quad d = T_i - T_{is}; \quad y_d = T_d - T_{d_s}$$

also define the following parameters:

$$\alpha_1 = \frac{UA}{\rho_l V_l C_{p_l}}; \quad \alpha_2 = \frac{w}{\rho_l V_l}$$

$$\beta_1 = \frac{UA}{m_c C_{p_c}}, \quad \beta_2 = \frac{K_c}{m_c C_{p_c}}$$

then Eqs (12.8) and (12.9) become

$$\frac{dx_1}{dt} = -(\alpha_1 + \alpha_2)x_1 + \alpha_1 x_2 + \alpha_2 d \qquad (12.10)$$

$$\frac{dx_2}{dt} = (\beta_2 - \beta_1)x_1 - \beta_2 x_2 + \beta_1 y_d \qquad (12.11)$$

And in conjunction with

$$y = x_1 \qquad (12.12)$$

(since the tank temperature is the sole measured variable), we have the complete water heater model in terms of deviation variables, in state-space form.

12.3 (a) An appropriate mathematical model may be obtained for this process by carrying out material balances on salt, and on water, as follows:

<u>Salt Material Balance</u>

$$\frac{d}{dt}(VC_B) = F_{B_f}C_{B_f} - FC_B$$

and since $V = Ah$, with A constant, and $F = k\sqrt{h}$, we have

$$A\frac{d}{dt}(hC_B) = F_{B_f}C_{B_f} - k\sqrt{h}\,C_B \qquad (12.13)$$

<u>Water Material Balance</u>

Assume that mixing ξ gms of salt with ϕ litres of water does not alter appreciably the volume of the resulting mixture: This implies that F_{B_f} litres of brine contains F_{B_f} liters of water.

$$\frac{d}{dt}(\rho_w V) = \rho_w(F_{B_f} + F_w) - \rho_w F$$

ρ_w = constant density of water. Thus, we have

$$A\frac{dh}{dt} = F_{B_f} + F_w - k\sqrt{h} \qquad (12.14)$$

Now, from (12.13) we have that

$$A\left\{C_B\frac{dh}{dt} + h\frac{dC_B}{dt}\right\} = F_{B_f}C_{B_f} - C_B k\sqrt{h}$$

Introducing (12.14) for $A\frac{dh}{dt}$ and rearranging gives the desired process model incorporating <u>both</u> material balances:

$$\frac{dC_B}{dt} = \frac{1}{Ah}\left\{-C_B\left(F_{B_f} + F_W\right) + F_{B_f}C_f\right\} \qquad (12.15)$$

along with

$$\frac{dh}{dt} = \frac{1}{A}\left\{F_{B_f} + F_W - k\sqrt{h}\right\} \qquad (12.16)$$

Observe the following about this model:

(i) It is a set of 2 coupled, nonlinear ODE's, even though the coupling is "one-way" – C_B is affected by changes in h, but h does not depend on C_B.

(ii) At steady state Eq (12.15) indicates that

$$C_B^* = \frac{F_{B_f}C_{Bf}}{F_{B_f} + F_W} \qquad (12.17)$$

which is consistent with well known mixing results.

(b) By redesigning the process to operate at constant volume, h becomes constant; and the process model simplifies to one equation:

$$\frac{dC_B}{dt} = \alpha\left[-C_B\left(F_{B_f} + F_W\right) + F_{B_f}C_{B_f}\right] \qquad (12.18)$$

with no need for (12.16), in principle. If the constant volume is achieved by fixing F_{B_f}, and F_W, and manipulating only C_{B_f}, then observe that Eq (12.18) will now be a linear ODE; if on the other hand constant volume is achieved by allowing both F_{B_f} and F_W to vary but in such a way that $F_{B_f} + F_W = F\,(k\sqrt{h})$ then Eq (12.18) will be nonlinear. Regardless, the constant volume model is still simpler.

12.4 (a) An appropriate model may be obtained from an organism material balance and a substrate balance.

Organism Balance.

$$\frac{d}{dt}(VC_x) = FC_{x_0} - FC_x + \mu C_x V \qquad (12.19)$$

Rate of Accumulation. — Rate In — Rate out — Growth rate (Rate of production)

Substrate Balance

The implication of Eq (P12.4) is that the rate of "production" of the substrate is

$$-\frac{\mu C_x V}{S_{x_s}}$$

since the rate of organism growth is $\mu C_x V$. Thus the substrate balance is:

$$\frac{d}{dt}(VC_s) = F(C_{s_0} - C_s) - \frac{\mu C_x V}{S_{x_s}} \qquad (12.20)$$

Since V is constant, the model simplifies to:

$$\frac{dC_x(t)}{dt} = \frac{F}{V}\left[C_{x_0} - C_x(t)\right] + \mu(t)C_x(t) \qquad (12.21a)$$

$$\frac{dC_s(t)}{dt} = \frac{F}{V}\left[C_{s_0} - C_s(t)\right] - \frac{\mu(t)C_x(t)}{S_{x_s}} \qquad (12.21b)$$

$$\mu(t) = \mu_0\left[\frac{C_s(t)}{K_s + C_s(t)}\right] \qquad (12.21c)$$

(b) For the specific digester, the model is:

$$\frac{dC_x}{dt} = 0.1(C_{x_0} - C_x) + \mu(t) C_x$$

$$\frac{dC_s}{dt} = 0.1(C_{s_0} - C_s) - 50\mu(t) C_x$$

$$\mu = 0.4\left(\frac{C_s}{3.33\times10^{-3} + C_s}\right)$$

With $C_{x_0} = 0.3$ mole/m^3, $C_{s_0} = 167$ mole/m^3

The resulting behavior of C_x, C_s and μ are shown below in FIG S12.2.

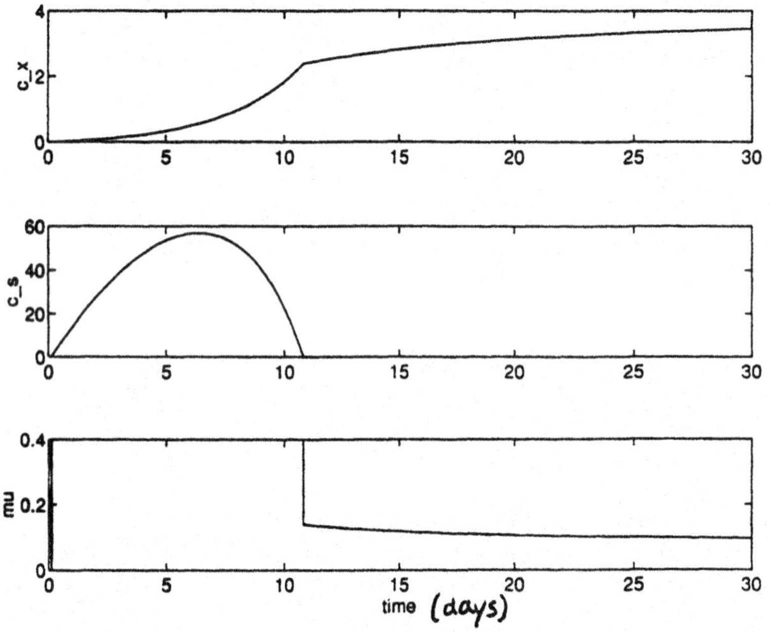

FIG S12.2

12.5 Assume that the given R, E are on the basis of a unit of time; i.e. Flows in and out of each stage are R moles/unit time, E moles/unit time.

Total Material balance (Raffinate) stage n:

$$\frac{d(h_n)}{dt} = R - R \quad \Rightarrow \quad h_n = \text{constant}$$

h_n = raffinate holdup in stage n is CONSTANT.

Similarly, for the Extract,

$$\frac{d}{dt}(H_n) = E - E \quad \Rightarrow \quad H_n = \text{constant}.$$

Material Balance on A:

$$\frac{d}{dt}\left(h_n y_n + H_n x_n\right) = R y_{n-1} + E x_{n+1} - R y_n - E x_n$$

- $\frac{d}{dt}(h_n y_n)$: Rate of Acc. in Raffinate
- $\frac{d}{dt}(H_n x_n)$: Rate of Acc. in Extract
- $R y_{n-1}$: In with raffinate from stage n-1
- $E x_{n+1}$: In with extract from stage n+1
- $R y_n$: Out with Raffinate
- $E x_n$: Out with Extract

Apply equilibrium $x_n = k y_n$, and simplify to obtain

$$\left(h_n + kH_n\right)\frac{dy_n}{dt} = R y_{n-1} - (R + kE) y_n + kE y_{n+1} \qquad (12.22)$$

$$n = 1, 2, \ldots, N.$$

12.6 Solution not provided.

12.7 From the given model, and the deviation variables, the model becomes

$$\frac{dx_1}{dt} = -\left(k_1 + \frac{1}{\theta}\right)x_1 + \frac{1}{\theta}u$$

$$\frac{dx_2}{dt} = -\left(k_2 + \frac{1}{\theta}\right)x_2 - \frac{\alpha}{\theta}u$$

$$y = -x_1 - x_2$$

since c_f is constant. This model is in the stipulated form with

$$a_1 = -\left(k_1 + \frac{1}{\theta}\right)$$

$$a_2 = -\left(k_2 + \frac{1}{\theta}\right)$$

$$b_1 = 1/\theta \; ; \quad b_2 = -\alpha/\theta$$

a_1 represents rate of A consumption – by reaction and by net flow out of the reactor; a_2 represents the same thing for B; b_1 is the reciprocal residence time for A; b_2 represents the reduction in c_B concentration as a result of any increase in the amount of A in the feed.

(a) In this case, it is possible to obtain a closed form solution to the process model given $u = -1$. We have

$$x_1 = -\frac{b_1}{a_1}\left(1 - e^{a_1 t}\right)$$

$$x_2 = -\frac{b_2}{a_2}\left(1 - e^{a_2 t}\right)$$

so that

$$y = \left[\left(\frac{b_1}{a_1} + \frac{b_2}{a_2}\right) - \frac{b_1}{a_1}e^{a_1 t} - \frac{b_2}{a_2}e^{a_2 t}\right] \qquad (12.23)$$

we may now estimate the four parameters a_1, a_2 ζ_1 and ζ_2 from

$$y(t) = \left[(\zeta_1 + \zeta_2) - \zeta_1 e^{a_1 t} - \zeta_2 e^{a_2 t} \right] \qquad (12.24)$$

from where we may also deduce b_1, b_2.

From data, obtain, via nonlinear estimation

$$a_1 = -0.5$$

$$a_2 = -0.2$$

$$\zeta_1 = -1.667$$

$$\zeta_2 = 2.667$$

From which we obtain

$$b_1 = 5/6 \; ; \qquad b_2 = -8/15.$$

(b) A plot of the model response and the data is shown below. The model fit appears adequate.

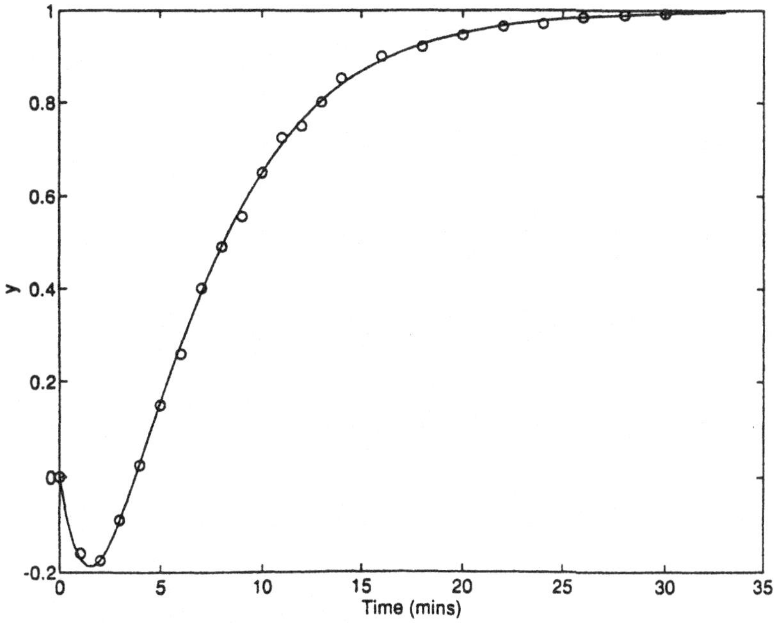

FIG S12.3

CHAPTER 13

13.1 (a) The first-order-plus-time-delay transfer function given in Eq (13.1) may be obtained from Eq (13.3) by setting $\xi = 0$, $\tau_1 = \tau$, and $\tau_2 = 0$. Thus, by introducing these values into (P13.2) and (P13.3) we obtain

$$\left.\begin{aligned} a_1 &= e^{-\Delta t/\tau} \\ a_2 &= 0 \\ b_1 &= K[1 - e^{-\Delta t/\tau}] \\ b_2 &= 0 \end{aligned}\right\} \quad (13.1)$$

so that the corresponding expression for this "special case" is:

$$y(k) = a_1 y(k-1) + b_1 u(k-m-1) \qquad (13.2)$$

with parameters as specified above.

The second "special case"

$$g(s) = \frac{Ke^{-\alpha s}}{(\tau_1 s+1)(\tau_2 s+1)}$$

is obtained by setting $\xi = 0$. The corresponding expression is therefore

$$y(k) = a_1 y(k-1) + a_2 y(k-2) + b_1 u(k-m-1) + b_2 u(k-m-2)$$

with a_1 and a_2 remaining as in Eq (P13.2), but with

$$b_1 = K\left[1 - \left(\frac{\tau_1}{\tau_1-\tau_2}\right)e^{-\Delta t/\tau_1} - \left(\frac{\tau_2}{\tau_1-\tau_2}\right)e^{-\Delta t/\tau_2}\right]$$

$$b_2 = K\left[e^{-\Delta t/\tau_1}\,e^{-\Delta t/\tau_2} - \left(\frac{\tau_1}{\tau_1-\tau_2}\right)e^{-\Delta t/\tau_2} - \left(\frac{\tau_2}{\tau_2-\tau_1}\right)e^{-\Delta t/\tau_1}\right]$$

(b) The equations (P13-2a,b) and (P13-3a,b) must be solved for K, τ_1, τ_2 and ξ.

Observe first that (P13-2a,b) involve only a_1 and a_2 in terms of τ_1 & τ_2; thus we first solve for τ_1 and τ_2 from here.

We may use the fact that

$$(s-\phi_1)(s-\phi_2) = s^2 - (\phi_1+\phi_2)s + \phi_1\phi_2$$

directly to solve for τ_1 & τ_2. Alternatively, let

$\phi_1 = e^{-\Delta t/\tau_1}$; $\phi_2 = e^{-\Delta t/\tau_2}$, so that $a_1 = \phi_1 + \phi_2$; and $a_2 = -(\phi_1\phi_2)$. Then

$$a_1^2 + 4a_2 = (\phi_1 - \phi_2)^2$$

or $\sqrt{a_1^2 + 4a_2} = \phi_1 - \phi_2$
in conjunction with
$a_1 = \phi_1 + \phi_2$

(using the positive square root. Result unchanged because of symmetry, if negative sq. root chosen.)

solve simultaneously to yield

$$\phi_1 = \frac{1}{2}\left(a_1 + \sqrt{a_1^2 + 4a_2}\right) = e^{-\Delta t/\tau_1}$$

$$\phi_2 = \tfrac{1}{2}\left(a_1 - \sqrt{a_1^2 + 4a_2}\right) = e^{-\Delta t/\tau_2}$$

The explicit expressions for τ_1 and τ_2 are now easily obtained as:

$$\tau_1 = \frac{-\Delta t}{\ln\left[\tfrac{1}{2}\left(a_1 + \sqrt{a_1^2 + 4a_2}\right)\right]} \qquad (13.3)$$

$$\tau_2 = \frac{-\Delta t}{\ln\left[\tfrac{1}{2}\left(a_1 - \sqrt{a_1^2 + 4a_2}\right)\right]} \qquad (13.4)$$

Thus, given a_1 and a_2 (and Δt, time between each sampled data) τ_1 and τ_2 are obtained from these equations.

Next, let

$$\left.\begin{aligned} C_{11} &= \left(\frac{\tau_1}{\tau_1 - \tau_2}\right)\phi_1 \; ; \quad C_{12} = \left(\frac{\tau_1}{\tau_1 - \tau_2}\right)\phi_2 \\ C_{21} &= \left(\frac{\tau_2}{\tau_2 - \tau_1}\right)\phi_1 \; ; \quad C_{22} = \left(\frac{\tau_2}{\tau_2 - \tau_1}\right)\phi_2 \end{aligned}\right\} \qquad (13.5)$$

all known once τ_1 and τ_2 are known. Then (P13.3a, b) become

$$b_1 = K\left[(1 - C_{11} - C_{22}) + (\phi_1 + \phi_2)\xi\right]$$

$$b_2 = K\left[(a_2 - C_{12} - C_{21}) + (\phi_1 + \phi_2)\xi\right]$$

These equations may be solved simultaneously for K and ξ to obtain:

$$K = \frac{b_1 - b_2}{(1 - C_{11} - C_{22}) + (a_2 + C_{12} + C_{21})} \qquad (13.6)$$

$$\xi = \frac{(a_2 + C_{12} + C_{21})(b_1 - b_2 C_{11} - b_2 C_{22})}{(b_1 - b_2)(\phi_1 + \phi_2)} \qquad (13.7)$$

which may be simplified further if desired or use as is. The required explicit expressions are therefore as given in Eqs (13.3), (13.4) and (13.6) and (13.7) along with Eq (13.5).

13.2 By definition of the process gain,

$$K = \frac{\Delta\ \text{output}}{\Delta\ \text{input}} = \frac{20\,°C}{0.2\ \text{gal/min}} \quad \text{(From the diag)}$$

$$K = 100\ °C/\text{gal/min}.$$

As shown in Fig S13.1 below, α is estimated as 1 min, and $\tau = 4.5 - 1 = 3.5$ mins.

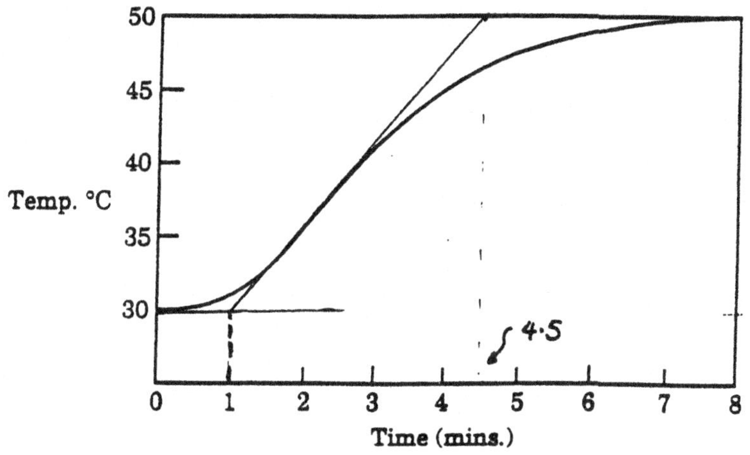

FIG S13.1

Thus approximate transfer function is:

$$g(s) = \frac{100\,°C/\text{gal/min}\ e^{-1s}}{(3.5s + 1)}$$

13.3 By reading off some data from FIG P13.1 and employing the method of section 3.3 in the text one possible set of results is illustrated in FIG S 13.2 below. (Process gain = 100 °C/gal/min was used)

$$\text{Slope} = -\frac{1}{\tau} = -0.81374 \Rightarrow \hat{\tau} = 1.2289 \text{ min.}$$

$$\text{Intercept} = \frac{\alpha}{\tau} = 1.6861 \Rightarrow \hat{\alpha} = 2.072 \text{ min}$$

observe:

(i) These values are very different from those obtained in Prob 13.2 above;

(ii) FIG S 13.2 shows significant departure from a "good" straight line fit to the data. This is clearly due to the fact that the actual process response in FIG P13.1 is obviously of higher order: the first-order-plus-time-delay approximation may thus be a poor one; the lack of fit may also be due to errors introduced into the data by "reading off" from paper.

(iii) Another set of "read off" data would probably produce a different set of results.

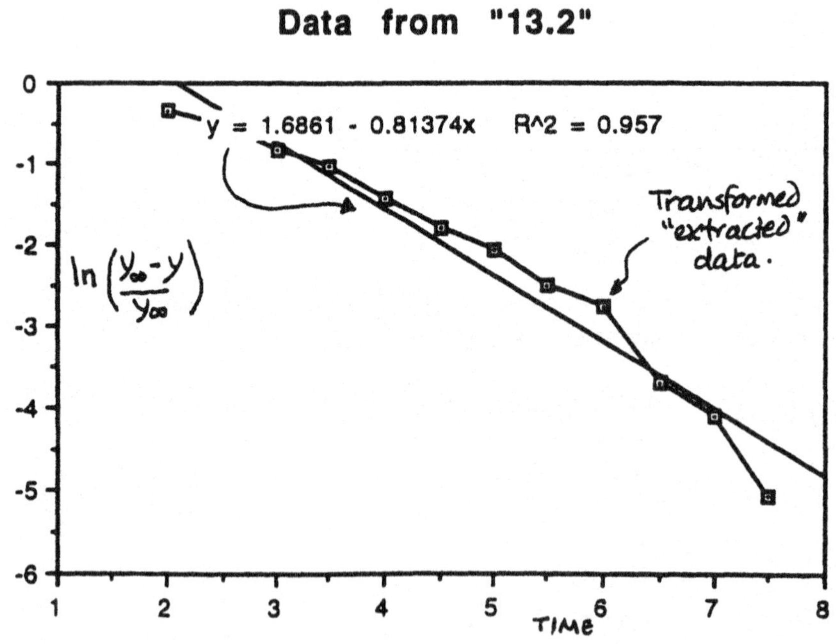

FIG S 13.2

13.4 (a) The data set strongly suggests a first-order-plus-time-delay process model will be adequate. Using any of the several methods presented in the text (or by employing a model identification software such as MATLAB or MIDENT in CONSYD) a reasonable set of model parameter estimates is obtained as:

$$K = -0.0094 \text{ ppm/gpm} \; ; \; \tau = 55.0s \; ; \; \alpha = 40s.$$

(b) The input is obviously a step of size −100 gpm; and the response of a system whose transfer function is

$$g(s) = \frac{-0.0094e^{-40s}}{55s+1}$$

when such an input is implemented, is shown in the solid line in FIG S 13.3; the supplied data is shown in circles. Observe that the model fit is adequate.

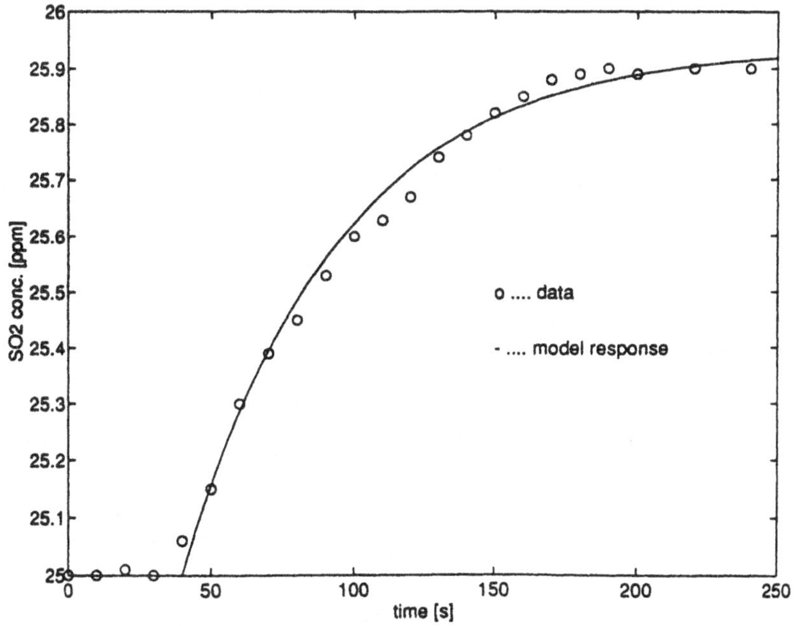

FIG S 13.3

13.5 The strategy is to solve (13.53b,c) simultaneously for a_1 and a_2, substitute the expression for a_1 into (13.53a) and obtain the required expression for ξ.

Eqns (13.53b,c) may be rewritten as:

$$a_1\mu_1 - a_2 = \frac{\mu_2}{2} \qquad (13.8)$$

$$a_1\frac{\mu_2}{2} - a_2\mu_1 = \frac{\mu_3}{6} \qquad (13.9)$$

[Eq (13.8) $\times\, \mu_1$] − [Eq (13.9)] gives

$$a_1\left(\mu_1^2 - \frac{\mu_2}{2}\right) = \frac{\mu_1\mu_2}{2} - \frac{\mu_3}{6}$$

which simplifies to give

$$a_1 = \frac{3\mu_1\mu_2 - \mu_3}{3(2\mu_1^2 - \mu_2)}. \qquad (13.10)$$

From (13.8) we obtain

$$a_2 = a_1\mu_1 - \frac{\mu_2}{2}$$

and from Eq (13.53a) in the text we have

$$\xi = a_1 - \mu_1$$

as required. If desired, Eq (13.10) above could be substituted for a_1 in each of the two expressions given above for a_2 and ξ.

13.6 Solution not provided.

13.7 For this problem, a plot of the data shown the time delay clearly : $\alpha = 2.00$ mins.

Thus, it is necessary to estimate the gain K, and the time constant τ, only. For this reason, it is advisable to "shift" the data by 2.00 mins, and estimate K and τ from the time shifted data, $\tilde{t}_k = t_k - 2$; $\tilde{g}(\tilde{t}_k)$.

With such a shift, $\tilde{t}_0 = 0$, $\tilde{g}(0) = 10.10$; $\tilde{t}_1 = 0.5$, $\tilde{g}(1) = 6.36$; $\tilde{t}_2 = 1.0$, $\tilde{g}(2) = 3.86$ ...... $\tilde{t}_{14} = 7.0$, $\tilde{g}(14) = 0.011$; $\tilde{t}_{15} = 7.5$ $\tilde{g}(15) = 0.006$; $\tilde{t}_{16} = 8$, $\tilde{g}(16) = 0.003$

From here, obtain m_0 using Simpson's rule as given in Eq (13.64) on p 431.

$$m_0 = \frac{0.5}{3}\left[10.10 + (4 \times 10.085) + 2(6.102) + 0.03\right]$$

$$m_0 = 10.44 \quad \Rightarrow \quad \boxed{\hat{K} = 10.44}$$

Next, obtain m_1 by setting $j = 1$ in Eq (13.63) on p. 431. (Keep in mind that time is now shifted).

$$m_1 = \frac{0.5}{3}\left\{0 + (4 \times 10.9265) + (2 \times 9.573) + 8 \times 0.003\right]$$

$$m_1 = 10.479$$

From here we obtain $\mu_1 = 10.479/10.44$

or $\mu_1 = 1.0037$

implying $\boxed{\hat{\tau} = 1.0037.}$

A plot of the theoretical impulse response of the process

whose transfer function is

$$g(s) = \frac{10.44e^{-2s}}{1.0037s + 1}$$

along with the given data is shown in FIG S13.4. The model fit appears adequate. (Other methods of estimation may be used; the results may be different from that obtained here, however.)

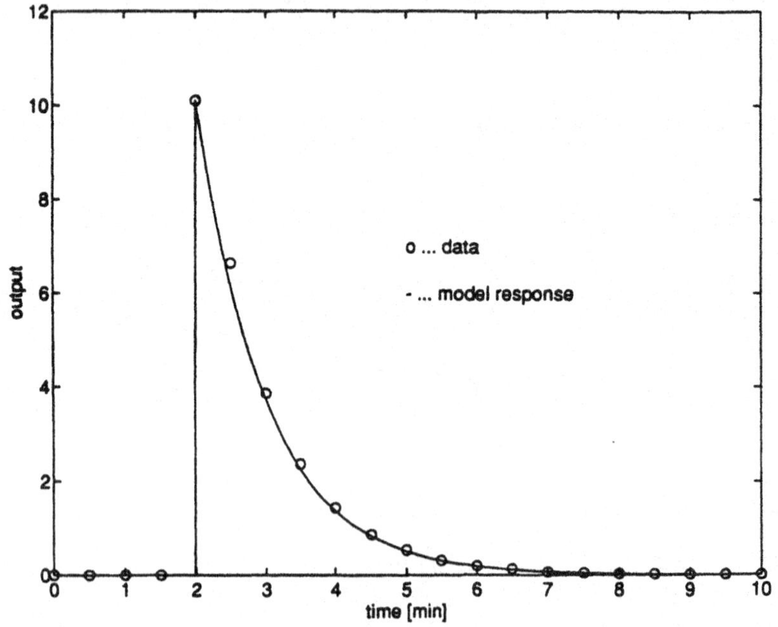

FIG S13.4

13.8 (a) A plot of the data is shown in FIG S13.5. A quick comparison of this plot with Fig 5.6 in the text, p.149 (or even FIG S13.4 above) shows a very crucial point: the data set fails to show the abrupt instantaneous rise that is characteristic of a first order system's impulse response.

Conversely, a comparison of the plot with Fig 6.11 on p198 of the text indicates a response that is more likely to be second order.

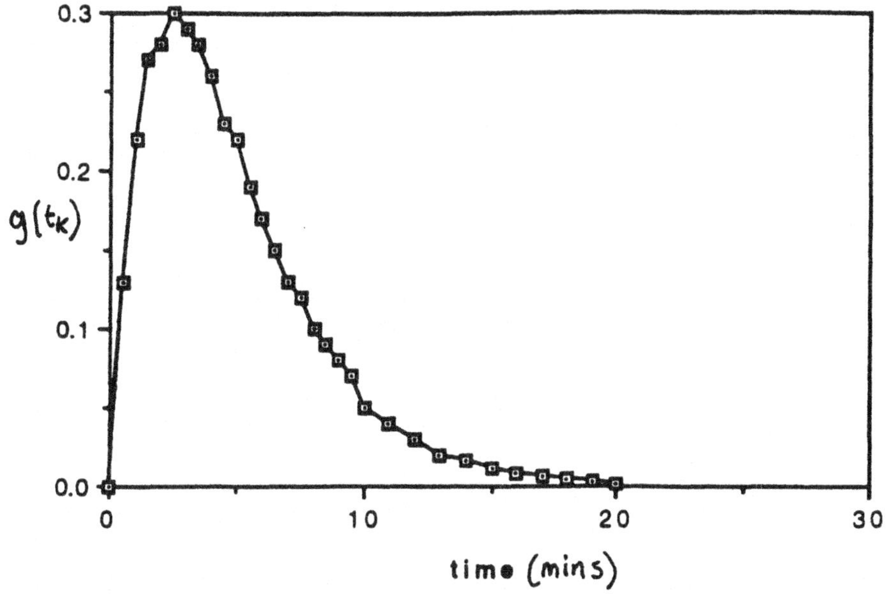

FIG S13.5

(b) From the data, and using Simpson's rule to evaluate the required integrals, obtain

$$m_0 = \int_0^{10} g(t)\,dt + \int_{10}^{20} g(t)\,dt \approx 1.978 \qquad (13.11)$$

$$m_1 = \int_0^{10} t\,g(t)\,dt + \int_{10}^{20} t\,g(t)\,dt \approx 9.673 \qquad (13.12)$$

$$m_2 = \int_0^{10} t^2 g(t)\,dt + \int_{10}^{20} t^2 g(t)\,dt \approx 69.965 \qquad (13.13)$$

It is important to separate the integrals as shown because Δt is different in each time span. (see the hint).

From here obtain

$$\mu_1 = 4.891 \qquad (13.14a)$$

$$\mu_2 = 35.378 \qquad (13.14b)$$

Observe first that were we to entertain a 1st order model, then (13.14a) and (13.14b) provide the following estimates of τ:

from (13.14a) $\mu_1 = \hat{\tau} = 4.891$

from (13.14b) $2\tau^2 = 35.378 \Rightarrow \hat{\tau} = 4.206$.

These estimates are different enough to confirm our initial assessment that a first order model will not represent the observed behavior adequately enough.

If a second order model is to be obtained, the parameter estimates are obtained using Eqn's (13.50) and (13.51) in the text; i.e.,

$$a_1 = \mu_1 = 4.891$$
$$a_2 = \mu_1^2 - \frac{\mu_2}{2} = 6.233$$

Thus an estimated second order transfer function is:

$$g(s) = \frac{1.978}{6.233s^2 + 4.891s + 1}$$

13.9 (a) A plot of the data (see FIG S13.6) indicates a high frequency asymptote for the AR plot with a slope which is close to -1; also the HFA for the ϕ plot appears to be close to $-90°$. These two aspects are enough to suggest a first order model.

Thus, the frequency response characteristics from which K and τ are to be estimated are as given in Eqns (9.34a) and (9.34b) in the text.

These equations may be rearranged to give:

$$\tan\phi = -\omega\tau \qquad (13.15)$$

$$AR = K\left[1+\omega^2\tau^2\right]^{-1/2} \qquad (13.16)$$

In principle, we <u>should</u> estimate τ and K simultaneously using the given data in conjunction with (13.15) and (13.16) above.

However, at the expense of "global optimality" of the parameter estimates, we may obtain an "independent" least squares estimate for τ from (13.15) alone; introducing this estimate into (13.16) then makes it possible to obtain the corresponding least squares estimate of K from (13.16). (This procedure, even though simpler, is not ideal: estimation errors associated with $\hat{\tau}$ are passed on to $\hat{K}$ estimates. Nevertheless it produces reasonable estimates in this case).

By following such a procedure, obtain

$$\hat{\tau} = 0.21$$

and $$\hat{K} = 0.71$$

(c) A plot of the theoretical frequency response for the system with the transfer function

$$g(s) = \frac{0.71}{0.21s + 1}$$

is shown in the solid lines in FIG S13.6; the data set is shown in circles. We may now observe that not only is a first order model adequate, the estimated parameters are also quite reasonable.

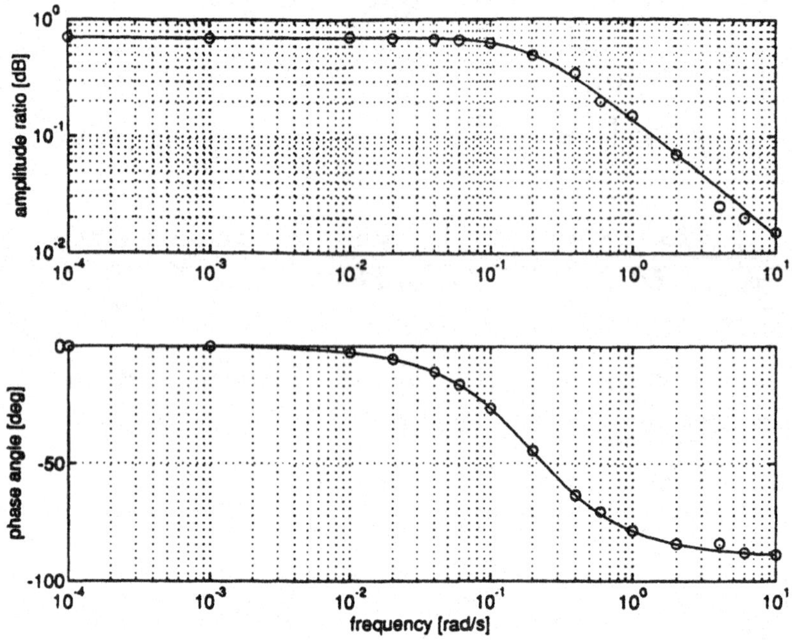

FIG S 13.6

13.10 (a) In the absence of any evidence to the contrary it is advisable to start by entertaining the simplest possible model and increase model complexity only if the simpler model proves inadequate.

In this case, a first-order-plus-time-delay model is a good starting point. The following transfer function was estimated from the data:

$$g(s) = \frac{27.5e^{-15s}}{28s + 1}$$

(b) FIG S 13.7 shows a simulation of the theoretical response of this system to the input data, F_T as given in Table P13.5. (Dashed line). The supplied output data, in solid line, agrees reasonably with the model.

(c) The indicated agreement between the data and the model appears good enough. This is real data from a real process which is not exactly linear, even in the operating regime. A more complicated linear model does not necessarily "do a better job." Some improvement may be obtained by introducing some non-linear terms, but this is outside the intended scope of this exercise.

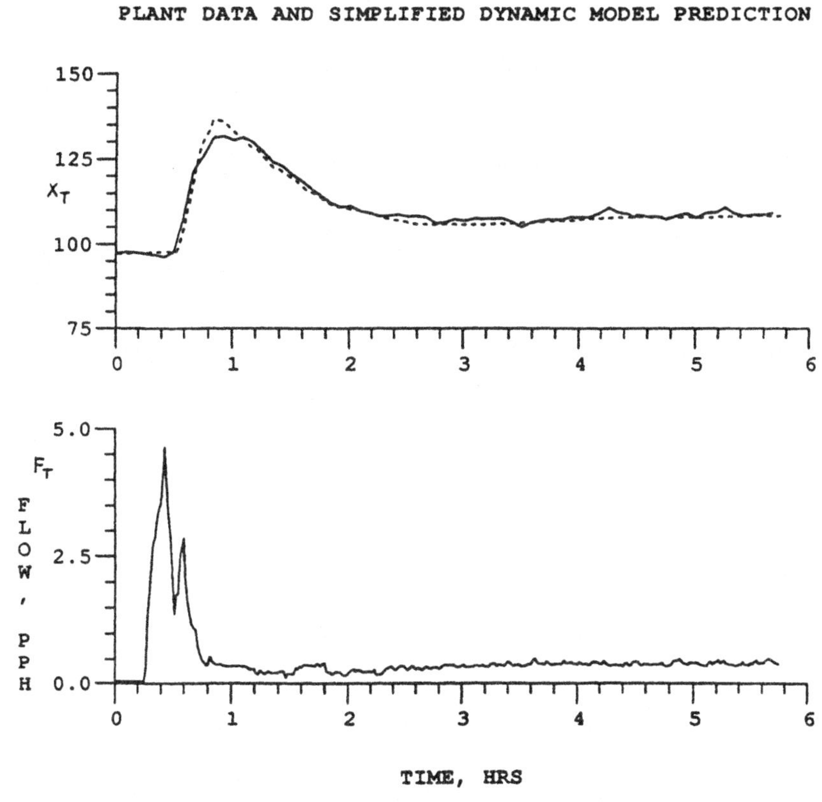

FIG S13.7

FIG S13.8 shows the X_T data and the theoretical model response on an expanded scale. The model fit may be better evaluated from this plot.

PLANT DATA AND SIMPLIFIED DYNAMIC MODEL PREDICTION

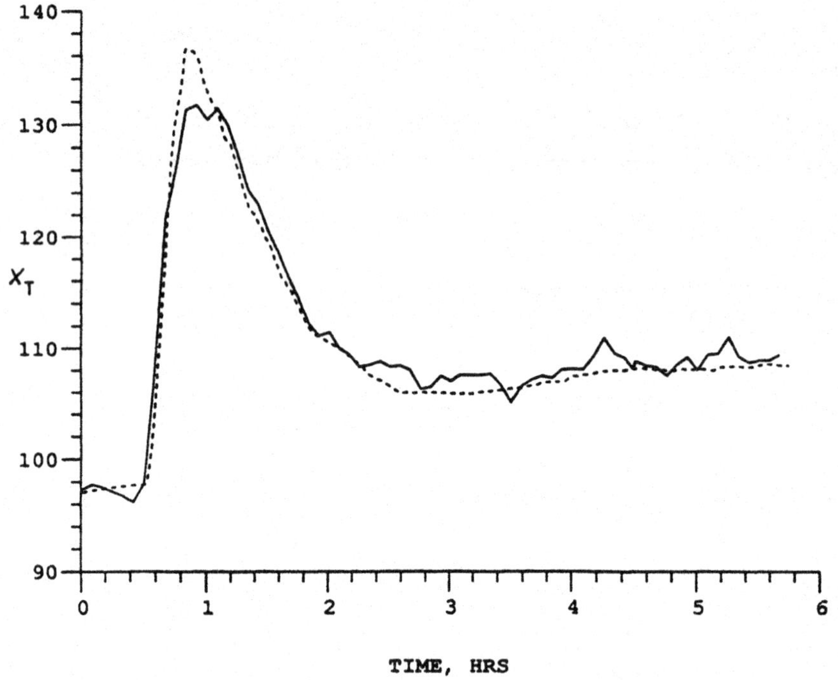

FIG S13.8

CHAPTER 14

14.1 (a) Define deviation variables

$$y = T - T^{*}\ ; \quad u = Q_F - Q_F^{*}\ ; \quad d_1 = T_i - T_i^{*}$$
$$d_2 = F - F^{*}$$

The required block diagram is shown below in FIG S14.1

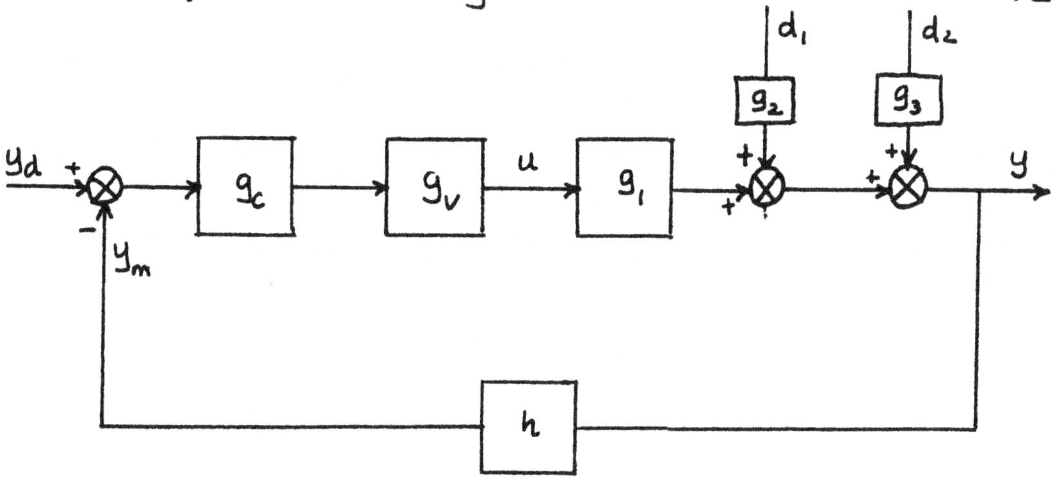

FIG S14·1

(b) Define $d_3 = P_F - P_F^{*}$; the block diagram is modified as below.

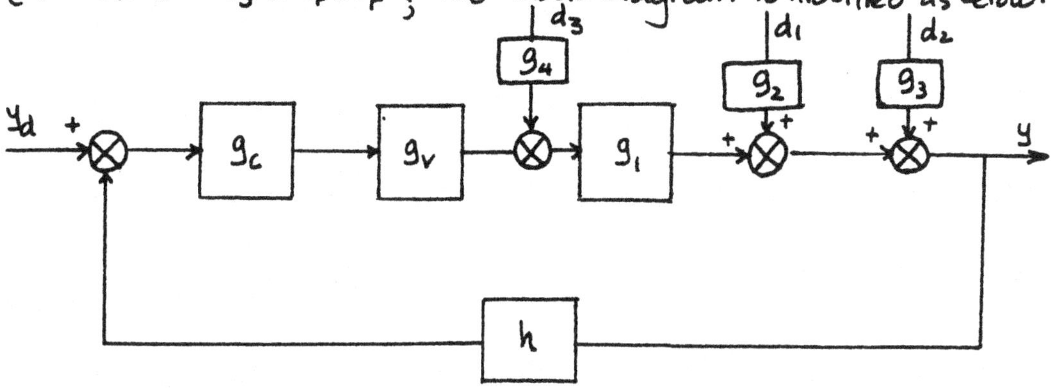

FIG S14·2

14.2 (a) Define deviation variables:

$$y = T - T^* \; ; \quad u = F_B - F_B^* \; ; \quad d = T_i - T_i^*$$

The block diagram for the process is shown below, with the indicated transfer functions given as follows:

$$g(s) = \frac{-(1-0.5e^{-10s})}{(40s+1)(15s+1)}$$

$$g_d(s) = 0.5e^{-10s}$$

$$g_c = -K_c$$

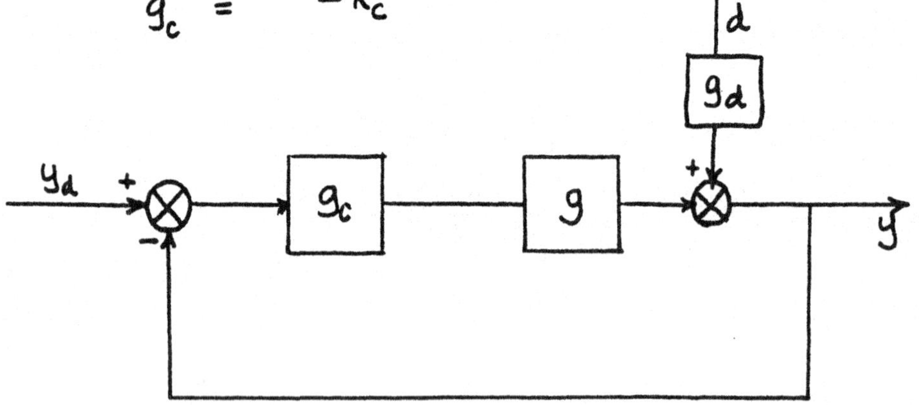

FIG S14.3

(b) The closed loop transfer function for this process is:

$$y = \frac{g g_c}{1 + g g_c} y_d + \frac{g_d}{1 + g g_c} d \qquad (14.1)$$

and the characteristic equation is

$$1 + g g_c = 0$$

which, in this specific case is

$$1 + \frac{K_c(1-0.5e^{-10s})}{(40s+1)(15s+1)} = 0$$

or

$$600s^2 + 55s + (1 + K_c) - 0.5K_c e^{-10s} = 0 \quad (14.2)$$

(c) Under the given circumstances,

$$d(s) = \frac{2}{s}$$

$$\text{Offset} = y_d - y_\infty$$

$$= 0 - \lim_{t \to \infty} y(t) = -\lim_{s \to 0} s y(s) \quad (14.3)$$

From Eq (14.1) above, with $y_d = 0$ and $d(s) = \frac{2}{s}$, we have

$$y(s) = \frac{g_d}{1 + gg_c} \cdot \frac{2}{s}$$

$$\therefore \lim_{s \to 0} s y(s) = \lim_{s \to 0} \left(\frac{2 g_d}{1 + gg_c} \right)$$

and upon introducing the expressions for the transfer function, we obtain

$$\lim_{s \to 0} = \frac{1}{1 + 0.5K_c}$$

Thus, the required offset, from Eq (14.3) is

$$\text{Offset} = -\frac{1}{1 + 0.5K_c}$$

14.3 (a) Mathematical model for the process may be obtained via a material balance.

$$\frac{d}{dt}(\rho A_c h) = F_C - (F_D + F_R) \quad (14.4)$$

where A_c = cross sectional area

$$= \frac{\pi (2 ft)^2}{4} = \pi ft^2$$

$$\rho = 30\ lb/ft^3 \quad \text{(given)}.$$

so that (14.4) becomes

$$(30\pi)\frac{dh}{dt} = F_C - (F_D + F_R) \qquad (14.5)$$

Define deviation variables: $y = h - h_s$; $u = F_D - F_{D_s}$;
$d_1 = F_C - F_{C_s}$; $d_2 = F_R - F_{R_s}$

and (14.5) becomes

$$\boxed{(30\pi)\frac{dy}{dt} = d - u} \qquad (14.6)$$

because F_R is constant. By taking Laplace transforms of Eq. (14.6) and rearranging, obtain

$$y(s) = \frac{-(1/30\pi)}{s}u(s) + \frac{(1/30\pi)}{s}d(s) \qquad (14.7)$$

as the required process transfer function model.
Note that (14.7) implies

$$g(s) = \frac{-(1/30\pi)}{s} \qquad (14.8a)$$

$$g_d(s) = \frac{(1/30\pi)}{s} \qquad (14.8b)$$

pure capacity system transfer functions.

(b) Define $c(s) = P_v - P_v^*$; the block diagram for the feedback control system is shown in FIG S14.4.

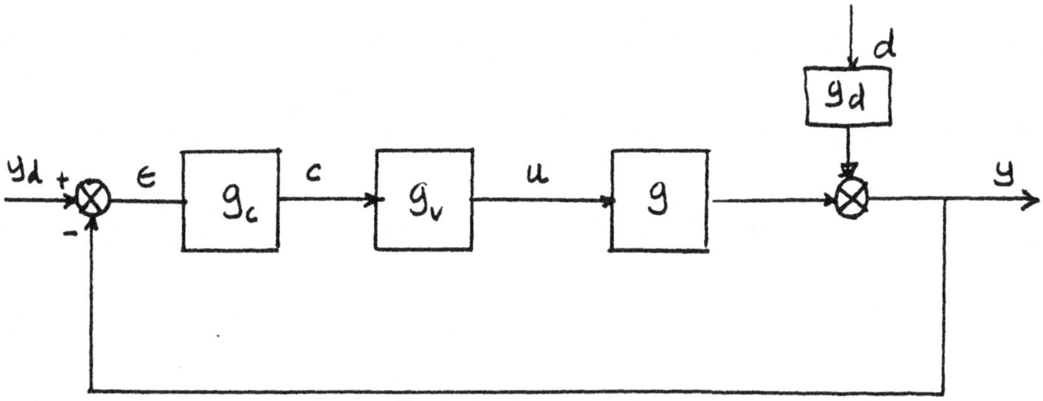

FIG S14.4

with $g_c = -K_c$; $g_v = \dfrac{0.02}{(1/300)s + 1}$

where the time constant for g_v has been converted to hours.

(c) The overall closed loop transfer function is:

$$y = \frac{g g_v g_c}{1 + g g_v g_c} y_d + \frac{g_d}{1 + g g_v g_c} d$$

For this specific case, $y_d = 0$; $d = \dfrac{2.5}{s}$; so that

$$y = \frac{g_d}{1 + g g_v g_c} \cdot \frac{2.5}{s}$$

Introducing appropriate expressions for the indicated transfer functions yields

$$y = \frac{\dfrac{(1/30\pi)}{s}}{1 + \left(\dfrac{1}{3\pi}\right)\left[\dfrac{0.02}{\left(\frac{1}{300}\right)s + 1}\right] \cdot \dfrac{1}{s}} \cdot \frac{2.5}{s}$$

which simplifies to:

$$y(s) = \frac{12.5\left[(1/300)s + 1\right]}{s(\tau_1 s + 1)(\tau_2 s + 1)} \qquad (14.9)$$

with $\tau_1 = 1/299.998$; $\tau_2 = 1/0.00212$

observe now that τ_1 is approximately equal to $\xi = 1/300$ in the numerator. Thus, there is an approximate pole-zero cancellation in (14.9) with the result

$$y(s) \approx \frac{12.5}{s(\tau_2 s + 1)}$$

from where Laplace inversion yields

$$y(t) = 12.5(1 - e^{-0.00212t})$$

(The complete, actual Laplace inversion of (14.9) gives

$$y(t) = 12.5(1 - e^{-0.00212t}) + 6.25 \times 10^{-10} e^{-299.98t}$$

and, as expected, the last term is essentially negligible).

The response is shown in FIG S14.5 — essentially a first order response with $\tau = 471$ hrs, a <u>very long</u> time constant, so long that, in "real time", the response will appear like that of a pure integrator.

Observe also the offset, 12.5 ft, attained after essentially 3000 hrs. The reason this pure capacity system, under proportional-only control, exhibits an offset is that the disturbance enters the process as a ramp, not as a bounded function. If the effect of the disturbance had been bounded, there would be no offset.

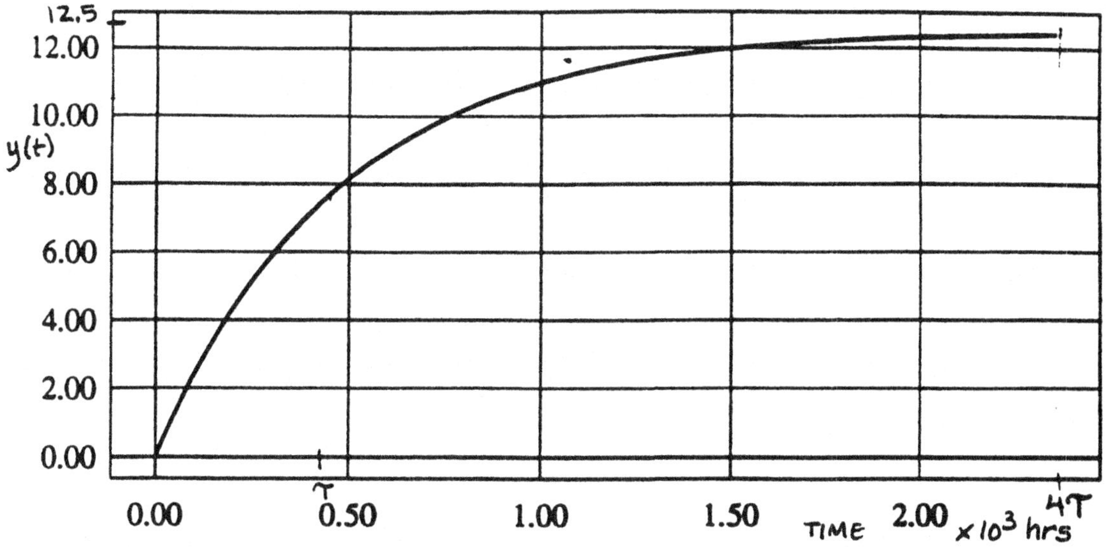

FIG S14·5

14.4 (a) consolidate inner loop first; then obtain overall closed loop transfer function as:

$$y = \psi y_d + \psi_{d_1} d_1 + \psi_{d_2} d_2$$

with

$$\psi = \frac{g_{c_1}\left(\frac{g_{c_2} g_2}{1 + g_{c_2} g_2 h_2}\right) g_1 g}{1 + g_{c_1}\left(\frac{g_{c_2} g_2}{1 + g_{c_2} g_2 h_2}\right) g_1 g h_3 h_1}$$

$$\psi_{d_1} = \frac{g_{d_1}}{1 + g g_1 \left(\frac{g_{c_2} g_2}{1 + g_2 g_{c_2} h_2}\right) g_{c_1} h_1 h_3}$$

$$\psi_{d_2} = \frac{\left(\dfrac{g_{d_2}}{1 + g_2 g_{c_2} h_2}\right)}{1 + g g_1 \left(\dfrac{g_2 g_{c_2}}{1 + g_2 g_{c_2} h_2}\right) g_{c_1} h_1 h_3}$$

which may be further simplified, if desired.

(b) Consolidate the inner loop first; then obtain overall closed loop transfer function as

$$\psi = \frac{\dfrac{g g_c}{1 + g_K g_c}}{1 + \dfrac{g g_c}{1 + g_K g_c}}$$

which simplifies to:

$$\psi = \frac{g^* e^{-\alpha s} g_c}{1 + (g + g_K) g_c} \qquad (14.10)$$

so that with $g_K = g^* - g^* e^{-\alpha s}$, ψ becomes

$$\psi = \frac{g^* g_c e^{-\alpha s}}{1 + g^* g_c}$$

and the characteristic equation,

$$1 + g^* g_c = 0$$

does not include the time delay element.

with $g_k = 0$, the characteristic equation is:

$$1 + g^* g_c e^{-\alpha s} = 0$$

which contains the time delay element.

14.5 If the outputs of each of the parallel processes are designated v_1 and v_2 then

$$y = g_d d + v_1 + v_2$$

with
$$v_1 = g_1 g_{c_1} (y_d - y)$$
$$v_2 = g_2 g_{c_2} (y_d - y)$$

so that

$$y = g_1 g_{c_1} (y_d - y) + g_2 g_{c_2} (y_d - y) + g_d d$$

This rearranges to give

$$y = \psi(s)\, y_d + \psi_d(s)\, d \qquad (14.11)$$

where

$$\psi = \frac{g_1 g_{c_1} + g_2 g_{c_2}}{1 + g_1 g_{c_1} + g_2 g_{c_2}} \qquad (14.12)$$

and

$$\psi_d = \frac{g_d}{1 + g_1 g_{c_1} + g_2 g_{c_2}} \qquad (14.13)$$

Now, for there to be no offset, we have the following conditions which ψ and ψ_d must satisfy.

$$\lim_{s \to 0} \psi(s) = 1 \qquad (14.14a)$$

and $$\lim_{s \to 0} \psi_d(s) = 0 \qquad (14.14b)$$

(these conditions may be obtained using the final value theorem of Laplace transforms.)

Under circumstances in which

$g_{c_1} = K_{c_1}$, a pure proportional controller

$g_{c_2} = \dfrac{K_{I_2}}{s}$, a pure integral controller

Eq (14.12) becomes

$$\psi(s) = \frac{s g_1 K_{c_1} + g_2 K_{I_2}}{s + s g_1 K_{c_1} + g_2 K_{I_2}}$$

so that $$\lim_{s \to 0} \psi(s) = \frac{g_2 K_{I_2}}{g_2 K_{I_2}} = 1$$

Similarly, Eq (14.13) becomes

$$\psi_d(s) = \frac{s g_d}{s + s g_1 K_{c_1} + g_2 K_{I_2}}$$

so that $$\lim_{s \to 0} \psi_d(s) = 0$$

thus establishing that, as required, there will be no offsets.

14.6 The characteristic equation for this human pupil servomechanism is:

$$1 + \frac{Ke^{-0.18s}}{(s+10)^3} = 0$$

or

$$(s+10)^3 + Ke^{-0.18s} = 0 \qquad (14.15)$$

(a) A Padé approximation (first-order) for the $e^{-0.18s}$ term in (14.15) yields the following approximation to the true characteristic equation:

$$(s+10)^3 + \frac{K(1-0.09s)}{(1+0.09s)} = 0$$

or

$$0.09s^4 + 3.7s^3 + 57s^2 + (390 - 0.09K)s + (1000 + K) = 0$$

On the verge of instability, with $K = K_u$, a pair of complex conjugate roots of this characteristic equation lie precisely on the imaginary axis. These roots are $s = \pm j\omega$ where ω is the frequency of sustained oscillation we seek.

Substituting $s = \pm j\omega$, collecting the real terms, as well as the imaginary terms, we obtain the two equations to be solved for ω and K_u as:

$$0.09\omega^4 - 57\omega^2 + 1000 + K_u = 0 \qquad \text{(Real part)} \; (14.16)$$

$$-3.7\omega^3 + (390 - 0.09K_u)\omega = 0 \qquad \text{(Imag. part)} \; (14.17)$$

with the following solutions:

$$\omega = 0 \; ; \qquad K_u = -1000 \qquad \text{(trivial solution)}$$

$$\boxed{\omega = 7.57498 \; ; \; K_u = 1974} \leftarrow \text{the required solution.}$$

$$\omega = 32.13632 \; ; \; K_u = -38124$$

(Alternatively, use the Routh array to obtain K_u from the characteristic equation; substitute into Eq. (14.17) and solve for ω.)

(b) By substituting $s = j\omega$ directly into the true characteristic equation in Eq. (14.15) we obtain

$$(j\omega)^3 + 30(j\omega)^2 + 300(j\omega) + 1000 + K_u e^{-0.18 j\omega} = 0$$

and since $e^{-0.18j\omega} = \cos 0.18\omega - j \sin 0.18\omega$, we may collect the imaginary part, and the real part, to obtain the following two equations to be solved for ω and K_u:

$$-30\omega^2 + K_u \cos 0.18\omega + 1000 = 0 \qquad (14.18)$$

$$-\omega^3 + 300\omega - K \sin 0.18\omega = 0 \qquad (14.19)$$

Apart from the trivial solution: $\omega = 0$, $K_u = -1000$, these two equations have an infinite number of solutions, the primary one being:

$$\boxed{\omega = 7.13 \; ; \qquad K_u = 1853}$$

Observe therefore, that, compared with the solution obtained in (a), the true values differ only slightly so that the approximation is in fact reasonable.

14.7 (a) Stability is determined for this system by investigating the roots of the characteristic equation:

$$1 + \frac{5e^{-8s} K_c(\cancel{1+3s})}{(15s+1)\cancel{(3s+1)}} = 0$$

which simplifies to

$$15s + 1 + 5K_c e^{-8s} = 0 \qquad (14.20)$$

Best to use the method of direct substitution. Let $s = j\omega$ in (14.20) above to obtain

$$15j\omega + 1 + 5K_c e^{-8j\omega} = 0$$

the solution of which requires that the real part as well as the imaginary part vanish simultaneously. i.e.,

$$0 = 1 + 5K_c \cos 8\omega \qquad \text{(real)} \qquad (14.21)$$

$$0 = 15\omega - 5K_c \sin 8\omega \qquad \text{(imaginary)} \qquad (14.22)$$

Apart from the trivial solution, $\omega = 0$, $K_c = -1/5$, the other meaningful solutions are obtained by solving these two equations simultaneously. From (14.21) obtain

$$K_c = \frac{-1}{5\cos 8\omega} \qquad (14.23)$$

substitute into (14.22) for K_c and rearrange to obtain

$$15\omega = -\tan(8\omega) \qquad (14.24)$$

Solving (14.24) for ω requires iteration. (A graphical method may also be employed). Of the infinite solutions to (14.24) only the following is of importance:

$$\omega = \pm 0.231411 \quad ;$$

Substituting this now into (14.23) yields

$$K_c = 0.7225$$

as the largest value of K_c for which the system is stable. Thus for stability $K_c < 0.7225$ must be satisfied.

(b) Under these new circumstances, the characteristic equation is easily rearranged to give

$$\tau^2\tau_I s^3 + 2\tau\tau_I\zeta s^2 + \tau_I(1+KK_c)s + KK_c = 0 \tag{14.25}$$

a cubic. The Routh array for the general cubic $a_0 s^3 + a_1 s^2 + a_2 s + a_3 = 0$, is:

1	a_0	a_2
2	a_1	a_3
3	b_1	0
4	a_3	

with $b_1 = \dfrac{a_1 a_2 - a_0 a_3}{a_1}$

Thus if a_0, a_1, a_2, a_3 are all positive, then the general cubic will have no roots in the RHP when

$$a_1 a_2 > a_0 a_3$$

(See Problem 11.8c)

In this case, with $a_0 = \tau^2\tau_I$; $a_1 = 2\tau\tau_I\zeta$; $a_2 = \tau_I(1+KK_c)$ and $a_3 = KK_c$

assuming $\tau_I > 0$; $K > 0$; $K_c > 0$ and $\zeta \geq 0$ (i.e. allowing for a purely undamped system) we have that for stability

$$(2\tau\tau_I\zeta)\tau_I(1+KK_c) > \tau\tau^2\tau_I KK_c$$

$$\boxed{\tau_I > \frac{\tau KK_c}{2\zeta(1+KK_c)}} \qquad (14.26)$$

in addition to

$$\tau_I(1+KK_c) > 0$$

which will always be satisfied for positive τ_I, and positive KK_c.

(In the special case of $\zeta = 0$, it can be shown using a technique similar to the one employed in Example 14·8 in the text that the system will be unstable for all values of $K_c > 0$.)

Thus the inequality shown above in Eq. (14.26) defines the range of values of K_c and τ_I for which the system will remain stable.

14.8 (a) The characteristic equation in this case is

$$1 + \frac{5K_c}{2s+1} = 0$$

or

$$2s + 1 + 5K_c = 0$$

with its single root located at

$$s = -\frac{(1+5K_c)}{2}$$

from where we observe that for all $K_c > 0$, this root will always lie in the LHP and the closed loop system will thus be stable.

(b) Introducing $h(s) = e^{-s}$ and in addition using the approximation

$$h(s) = e^{-s} \approx \frac{1 - \frac{1}{2}s}{1 + \frac{1}{2}s} \qquad (14.27)$$

the characteristic equation is modified to

$$1 + \frac{5K_c}{2s+1} \frac{(1 - \frac{1}{2}s)}{(1 + \frac{1}{2}s)}$$

or

$$s^2 + \left(\frac{5}{2} - \frac{5}{2}K_c\right)s + (1 + 5K_c) = 0$$

a quadratic for which stability now requires that

$$\frac{5}{2} - \frac{5}{2}K_c > 0$$

or

$$\boxed{K_c < 1.} \qquad (14.28)$$

Thus we see clearly that a system which was stable for all values of $K_c > 0$ has had its stability margin severely curtailed to $0 < K_c < 1$ as a result of the presence of the time delay (which has been approximated here as in (14.27) above). The destabilizing effect of the time delay is therefore obvious.

(c) With the introduction of the PD controller in place of the pure proportional controller, the new characteristic equation is:

$$1 + \frac{5}{2s+1} K_c (1 + 0.2s) \frac{(1 - \frac{1}{2}s)}{(1 + \frac{1}{2}s)} = 0$$

or

$$\left(1 - \tfrac{1}{2}K_c\right)s^2 + \left(\tfrac{5}{2} - \tfrac{3}{2}K_c\right) + (1 + 5K_c) = 0$$

and stability now requires that

(i) $1 - \frac{1}{2}K_c > 0 \Rightarrow K_c < 2$ (14.29)

(ii) $\frac{5}{2} - \frac{3}{2}K_c > 0 \Rightarrow K_c < \frac{5}{3}$ (14.30)

And since (14.29) violates (14.30), but (14.30) satisfies (14.29), the required condition for stability is

$$\boxed{0 < K_c < 5/3} \qquad (14.31)$$

Observe now that the maximum allowable K_c value has been increased by almost 67% as a result of introducing derivative action, clearly illustrating the stabilizing effect of this mode of control.

Now, for the general PD controller, with $g_c = K_c(1 + \tau_D s)$ the closed loop transfer function in this case is

$$\psi = \frac{g g_c g_v}{1 + g g_c g_v} = \frac{\frac{5e^{-s}}{2s+1} K_c(1+\tau_D s)}{1 + \frac{5}{2s+1} K_c(1+\tau_D s)e^{-s}}$$

or
$$\psi(s) = \frac{5K_c(1+\tau_D s)e^{-s}}{2s+1 + 5K_c(1+\tau_D s)e^{-s}}$$

For there to be no offset,

$$\lim_{s \to 0} \psi(s) = 1$$

But in this case,

$$\lim_{s \to 0} \psi(s) = \frac{5K_c}{1+5K_c} \neq 1 \quad \text{for all } K_c < \infty$$

thus establishing the required result: irrespective of τ_D or K_c, the closed loop system will suffer from steady state offset.

14.9 Problem 14.8 b revisited.

The true characteristic equation is

$$(2s+1) + 5K_c e^{-s} = 0$$

Use the method of direct substitution, and let $s = j\omega$; rearrange to obtain

$$(1 + 5K_c \cos\omega) + j(2\omega - 5K_c \sin\omega) = 0$$

The solution requires that both real and imaginary parts vanish simultaneously, i.e.

$$1 + 5K_c \cos\omega = 0 \qquad (14.32)$$

$$2\omega - 5K_c \sin\omega = 0 \qquad (14.33)$$

Ignoring the trivial solution ($\omega = 0$, $K_c = -1/5$), solve these equations simultaneously: first, from (14.32)

$$K_c = \frac{-1}{5\cos\omega} \qquad (14.34)$$

which, upon substitution into (14.33) yields

$$\omega = -\tfrac{1}{2}\tan\omega \qquad (14.35)$$

Of the infinite solutions to (14.35), only

$$\omega = \pm 1.836 \qquad (14.36)$$

is the primary solution (obtain numerically, or graphically); introduction of (14.36) into (14.34) now yields

$$K_c = 0.763$$

with the result that under the conditions introduced in Problem 14.8b, the range of K_c values for which the closed loop system is stable is actually

$$0 < K_c < 0.763$$

which is somewhat more restrictive than the range

$$0 < K_c < 1$$

obtained using a Padé approximation.

Problem 14.8c revisited

The true characteristic equation here is:

$$(2s+1) + 5K_c e^{-s} + K_c s e^{-s} = 0$$

Again, employ the method of direct substitution; let $s = j\omega$; rearrange to obtain

$$(2\omega + K_c\omega\cos\omega - 5K_c\sin\omega)j + [1 + K_c(\omega\sin\omega + 5\cos\omega)] = 0$$

with the solution requiring:

$$1 + K_c(\omega \sin\omega + 5\cos\omega) = 0 \quad (14.37)$$

$$2\omega - 5K_c \sin\omega + K_c\omega\cos\omega = 0 \quad (14.38)$$

Obtain from (14.37)

$$K_c = \frac{-1}{\omega\sin\omega + 5\cos\omega} \quad (14.39)$$

substitute into (14.38) and rearrange to obtain

$$\omega = -\frac{1}{2}\left[\frac{5\tan\omega - \omega}{\omega\tan\omega + 5}\right] \quad (14.40)$$

solve iteratively to obtain primary solution

$$\omega = 2.21$$

and upon introduction into (14.39) yields

$$K_c = 0.827$$

with the result that the actual range of K_c values required for stability is

$$0 < K_c < 0.827$$

which is better than $0 < K_c < 0.763$ obtained under pure P control. However, in Prob 14.8b, using the Padé approximation led to a range of K_c values of

$$0 < K_c < 5/3$$

indicating an upper acceptable limit more than double the

actual value of 0.827 obtained here.

14.10 From the given information, we may deduce the following about the elements of the closed loop:

$$g = \frac{1}{(s+1)(0.5s+1)}$$

$$h = \frac{1}{(\frac{1}{3}s+1)} \quad ; \quad g_c = K_c$$

The characteristic equation is therefore

$$1 + \frac{K_c}{(\frac{1}{3}s+1)(s+1)(0.5s+1)} = 0$$

or

$$\frac{1}{6}s^3 + s^2 + \frac{11}{6}s + (1+K_c) = 0 \qquad (14.41)$$

When on the verge of instability, $s = j\omega$ will satisfy this characteristic equation, with ω being the frequency of sustained oscillation. Thus, substituting $s = j\omega$ in (14.41) and rearranging, we obtain

$$(1+K_c) - \omega^2 + \left(\frac{11}{6}\omega - \frac{1}{6}\omega^3\right)j = 0$$

requiring

$$(1+K_c) - \omega^2 = 0 \qquad (14.42)$$

and

$$\left(\frac{11}{6}\omega - \frac{1}{6}\omega^3\right) = 0 \qquad (14.43)$$

with the solutions $\omega = 0$ (trivial) and $K_c = -1$, or

$$\omega = \pm\sqrt{11} \quad \text{and} \quad K_c = 10 \qquad (14.44)$$

Thus this system will oscillate with a frequency of $\omega = \sqrt{11}$ radians/sec. when on the verge of instability.

With the new device for which $h = 1$, the characteristic equation is modified to

$$0.5s^2 + 1.5s + (1 + K_c) = 0$$

and the system will never be unstable for any positive value of K_c. Thus the system will <u>never</u> be on the verge of instability.

14.11 Consolidate the inner loop first; then obtain the overall closed loop relationship as

$$y = \frac{gg_c}{1 + g'g_c + gg_c} y_d \qquad (14.45)$$

with a characteristic equation

$$1 + g'g_c + gg_c = 0 \qquad (14.46)$$

(a) with $g' = 0$, from (14.46) with $g' = 0$, or from the original block diagram, obtain the closed loop system characteristic equation as

$$1 + gg_c = 0$$

or

$$1 + \frac{K_c(1 - 5s)}{(5s+1)(3s+1)} = 0$$

or

$$15s^2 + (8 - 5K_c)s + (1 + K_c) = 0$$

a quadratic, which indicates that stability requires that

$$8 - 5K_c > 0 \Rightarrow K_c < 8/5$$

or $K_c < 1.6$ as required.

(b) with $g'(s)$ as given, the characteristic equation becomes

$$1 + K_c\left[\frac{10s}{(5s+1)(3s+1)} + \frac{(1-5s)}{(5s+1)(3s+1)}\right] = 0$$

$$1 + K_c\left[\frac{\cancel{(5s+1)}}{\cancel{(5s+1)}(3s+1)}\right] = 0$$

or

$$3s + (1 + K_c) = 0$$

with a single root located

$$s = -\frac{(1+K_c)}{3}$$

and for all $K_c > 0$, this root will always be in the LHP, so that the closed loop system will always be stable in this case. Observe therefore that the minor loop has extended the stability range considerably: it has converted a system with the restrictive stability range $0 < K_c < 1.6$ to one which is always stable!

14.12 The given $g(s)$ combined with the given $g_v(s)$ show 3 poles at

$$s = -0.05; \; -2.0; \text{ and } -1.0$$

Unaccounted for are:

1 pole at $s = 0$ (the origin)

2 zeros at $s = -0.5; \; -1.0.$

Clearly $g_c = K_c\left(1 + \frac{1}{\tau_I s} + \tau_D s\right)$, a PID controller

because g_c rearranges to

$$g_c = \frac{K_c(\tau_I s + \tau_I \tau_D s^2 + 1)}{\tau_I s} \qquad (14.47)$$

contributing a single pole at the origin ($s=0$) and 2 zeros at the roots of the indicated numerator quadratic. Having thus deduced the controller type, now note that from the given zeros, the numerator of g_c should be

$$\alpha(s+0.5)(s+1) = \alpha(s^2 + 1.5s + 0.5) \qquad (14.48)$$

↑ a scaling (multiplication) constant.

Comparing this with the numerator quadratic in (14.47) above, especially if (14.48) is rearranged as

$$0.5\alpha(2s^2 + 3s + 1)$$

observe that this implies

$$\tau_I = 3, \quad \tau_I \tau_D = 2$$

or $\quad \tau_D = 2/3$

Thus, the controller type is PID; $\tau_I = 3$; $\tau_D = 2/3$.

14.13 (a) The root locus diagram is shown in FIG S14.6 from where we observe that when $K_c = 2$, the roots are $0 \pm 2j$, so that

(i) the closed loop system is stable for $K_c > 2$;
(ii) when on the verge of instability, the system oscillates with a frequency $\omega = 2$ rad/sec.

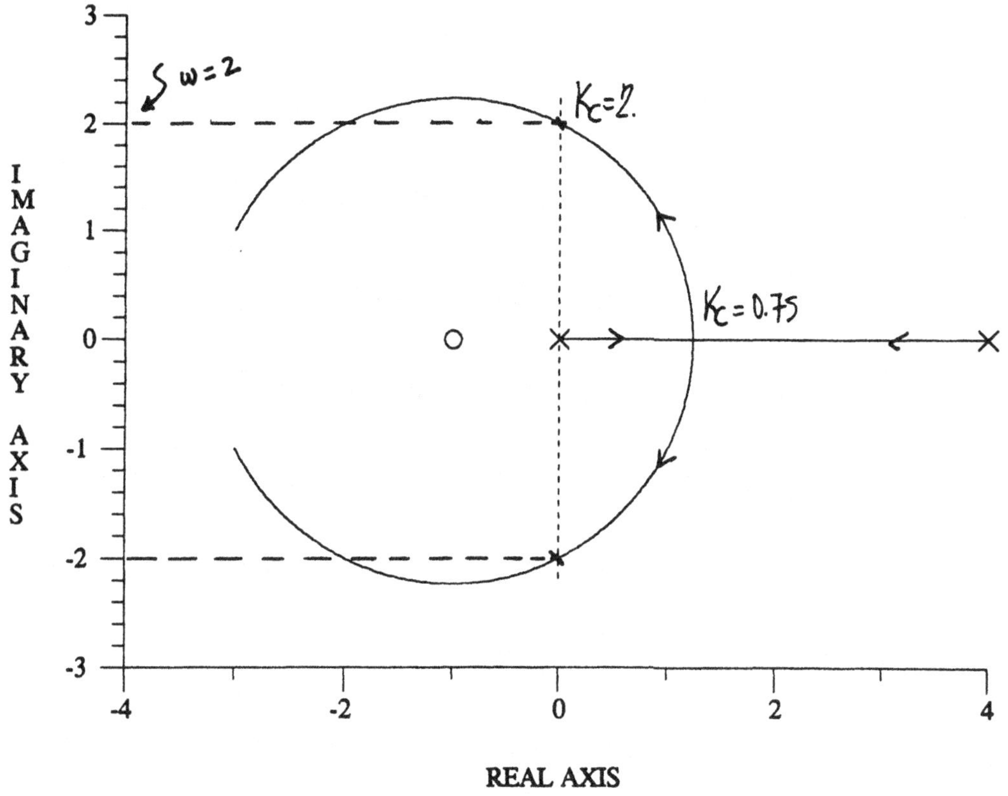

FIG S14.6

It is important to note that stability requires $K_c > 2$; for $K_c < 2$ observe that the roots are in the RHP.

(b) The characteristic equation is:

$$1 + K_c\left(1 + \frac{1}{s}\right)\left(\frac{2}{s-4}\right) = 0 \qquad (14.49)$$

or $$s^2 + (2K_c - 4)s + 2K_c = 0 \qquad (14.50)$$

a quadratic which indicates that for stability,

$$2K_c - 4 > 0$$

or $$2K_c > 4$$

or $$K_c > 2$$

The critical value $K_c = 2$ substituted into (14.50) gives

$$s^2 + 4 = 0$$

with the pair of purely imaginary roots

$$s = \pm 2j$$

so that the frequency of sustained oscillation is 2 rad/sec, confirming the results of part (a).

14.14 Solution not provided. Exercise requires anticipatory input from students which is later confirmed (or refuted!) by actual simulation.

CHAPTER 15

15.1 The closed loop transfer function relation for this control system is given by

$$y = \frac{\left(\frac{-0.025}{1.5s+1}\right) K_c \left(1 + \frac{1}{\tau_I s}\right)}{1 + \left(\frac{-0.025}{1.5s+1}\right) K_c \left(1 + \frac{1}{\tau_I s}\right)} \; y_d$$

or
$$y = \frac{-0.025 K_c (\tau_I s + 1).}{1.5\tau_I s^2 + (1 - 0.025 K_c)\tau_I s - 0.025 K_c} \; y_d. \quad (15.1)$$

(a) The closed loop poles are determined from the quadratic

$$1.5\tau_I s^2 + (1 - 0.025 K_c)\tau_I s - 0.025 K_c = 0$$

or, dividing through by $1.5\,\tau_I$ (provided $\tau_I \neq 0$),

$$s^2 + \frac{(1 - 0.025 K_c)}{1.5} s - \frac{0.025 K_c}{1.5\tau_I} = 0 \quad (15.2)$$

These poles are to be assigned to the same location as those of the roots of

$$s^2 + s + 1 = 0 \quad (P15.2)$$

By comparing Eq (15.2) with (P15.2), observe that appropriate pole assignment requires that

$$\frac{1 - 0.025K_c}{1.5} = 1 \quad \Rightarrow \quad K_c = -20 \,;$$

and

$$\frac{-0.025K_c}{1.5\,\tau_I} = 1 \quad \Rightarrow \quad 1/\tau_I = 3 \,, \text{ or } \tau_I = 1/3$$

With these parameter values, the closed loop zero is located at $s = -1/\tau_I$ or $s = -3$. (see Eq (15.1) above.) Note the negative value taken by K_c; this is because the process gain is also negative.

(b) The performance of this PI controller on the process, for a step change of -10 ppm in set point, is shown below in FIG S15.1

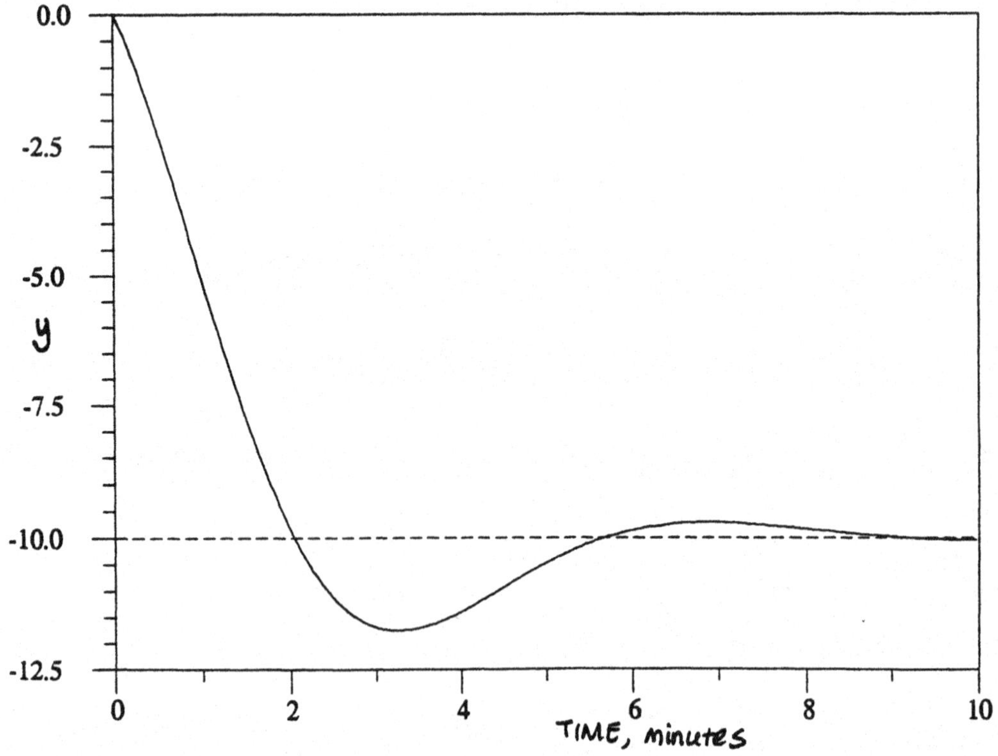

FIG S15.1

(c) A block diagram for the "real" process including the controller designed in (a) and implemented in (b) is shown below in FIG S15.2

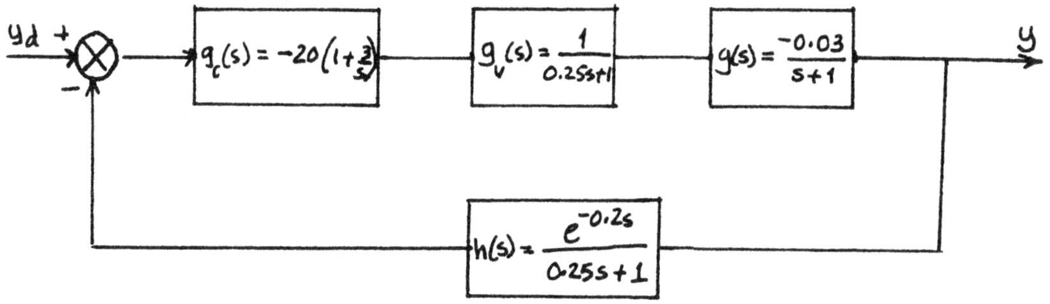

FIG S15.2

The closed loop response to the same −10 set-point change is shown below in FIG S15.3. The difference in the performance is significant enough to warrant a redesign: the "real" process response is <u>too</u> oscillatory.

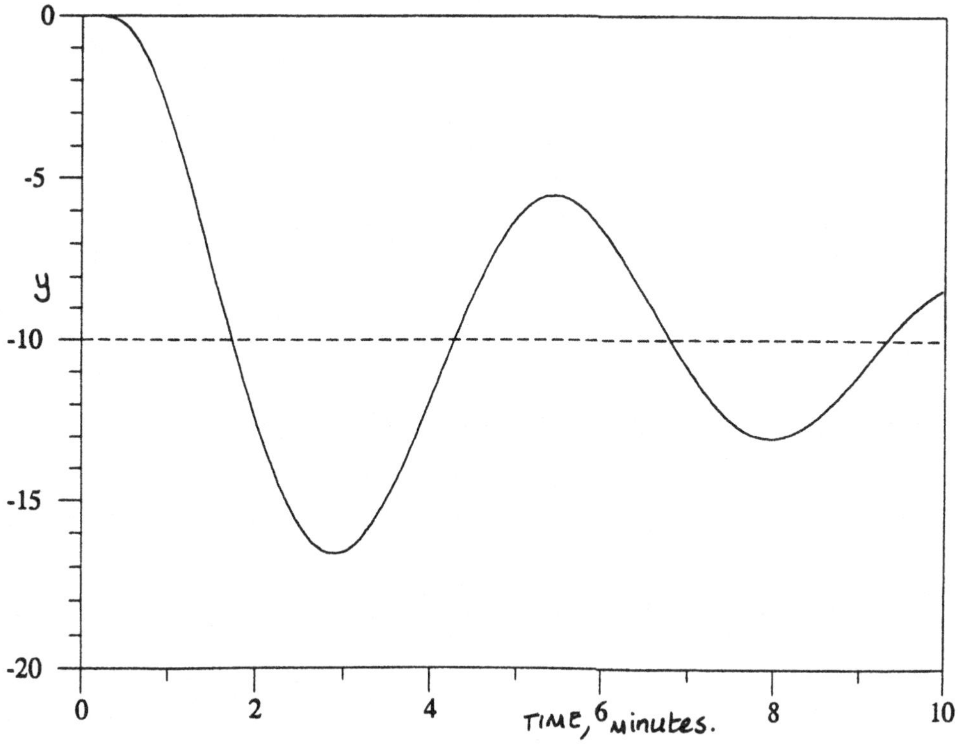

FIG S15.3

15.2 The new controller now has the form

$$g_c = -2\left(1 + \frac{3}{s}\right)$$

Implementation on the ideal process produces the response shown in FIG S15.4a below; it is definitely more sluggish than the response shown in FIG S15.1.

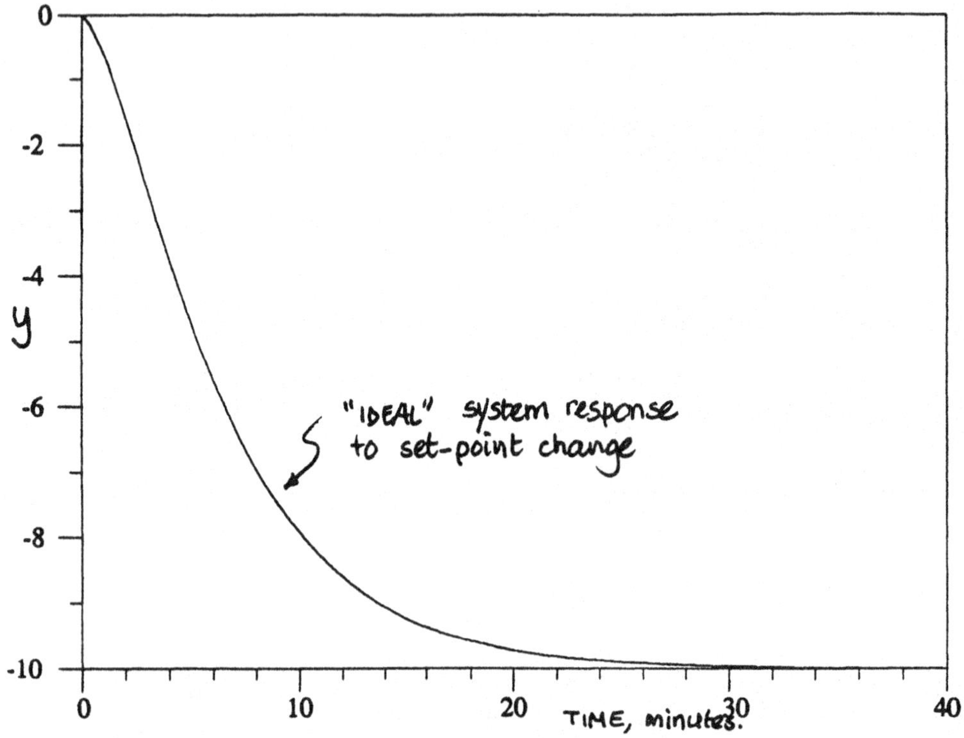

FIG S15.4a

When this controller is implemented on the real process (block diagram in FIG S15.2), the response to the same setpoint change is shown in FIG S15.4b, where we see immediately that there is <u>no</u> qualitative difference between this response and the one shown above in FIG S15.4a. These results should be compared with those in Prob 15.1 in which the FIG S15.3 response is <u>very different</u> from the "ideal" response of FIG S15.1.

Observe therefore that by using the smaller gain, the

resulting control system was made less sensitive to the effect of plant/model mismatch. This significantly lower sensitivity has been achieved however at the expense of the speed of response.

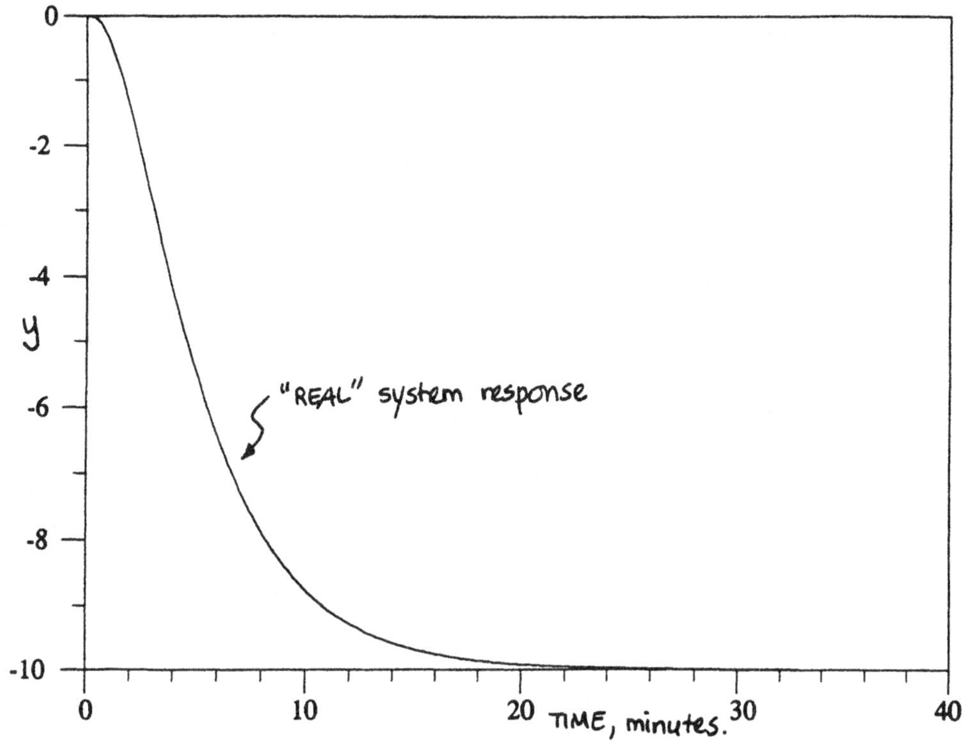

FIG S15.4b

15.3 (a) The "process reaction curve" obtained by implementing a unit step in $(F_B - F_B^*)$ on the process model is shown in FIG S15.5a.

By employing the techniques discussed in section 15.4.1 in the text, obtain an approximate first-order-plus-time-delay characterization

$$g(s) = \frac{-0.5e^{-15s}}{24.5s + 1} \tag{15.3}$$

A comparison of the "actual" PRC and the approximate characterization is shown in FIGS15.5b.

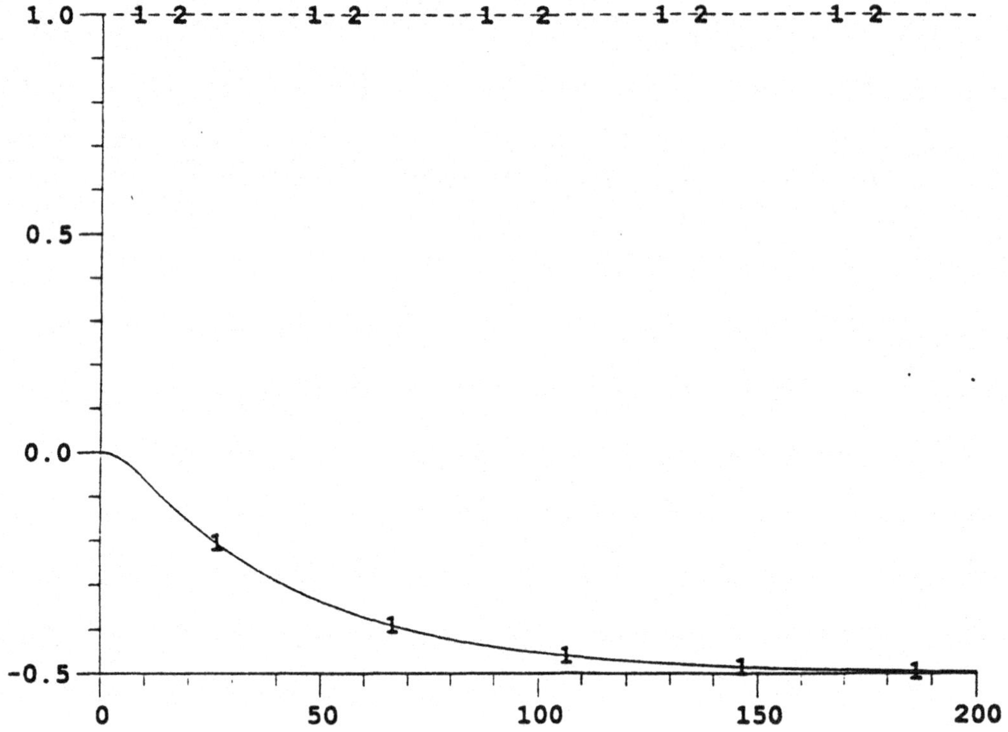

FIG S15.5a

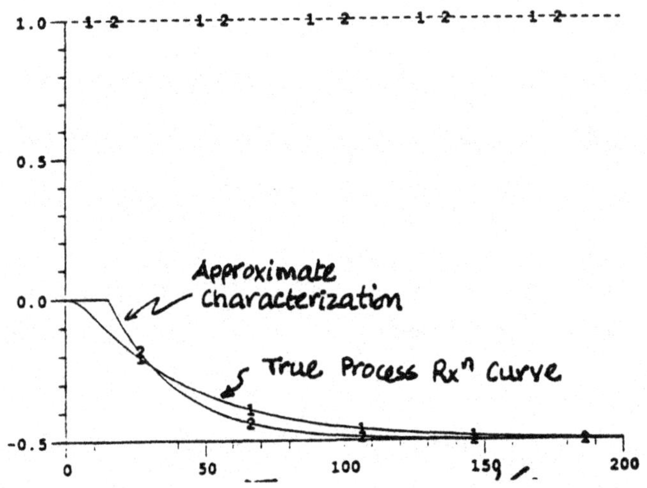

FIG S15.5b

The PI controllers designed on the basis of the three techniques, using Eq (15.3) above, are shown below.

TECHNIQUE \ CONTROLLER PARAMETERS	K_c	τ_I
ZIEGLER-NICHOLS	−2.94	49.95
COHEN-COON	−3.11	22.5
IMC	−2.13	32

(For $\lambda = 30$)
$\lambda > 1.7\alpha$ (= 25.5)

(b) The controller performaces are shown in FIGS 15.6 where we observe that, in general, the Z-N based design tends to produce a somewhat more sluggish controller; the C-C based design on the other hand tends to produce a more aggressive controller. The response obtained for the IMC controller (with $\lambda = 30$) seems to be just about right. Different values of λ will, of course, produce different responses.

If overshoot can be tolerated in the process, the C-C based controller may be desirable; if not, the IMC based controller (with $\lambda = 30$) is recommended.

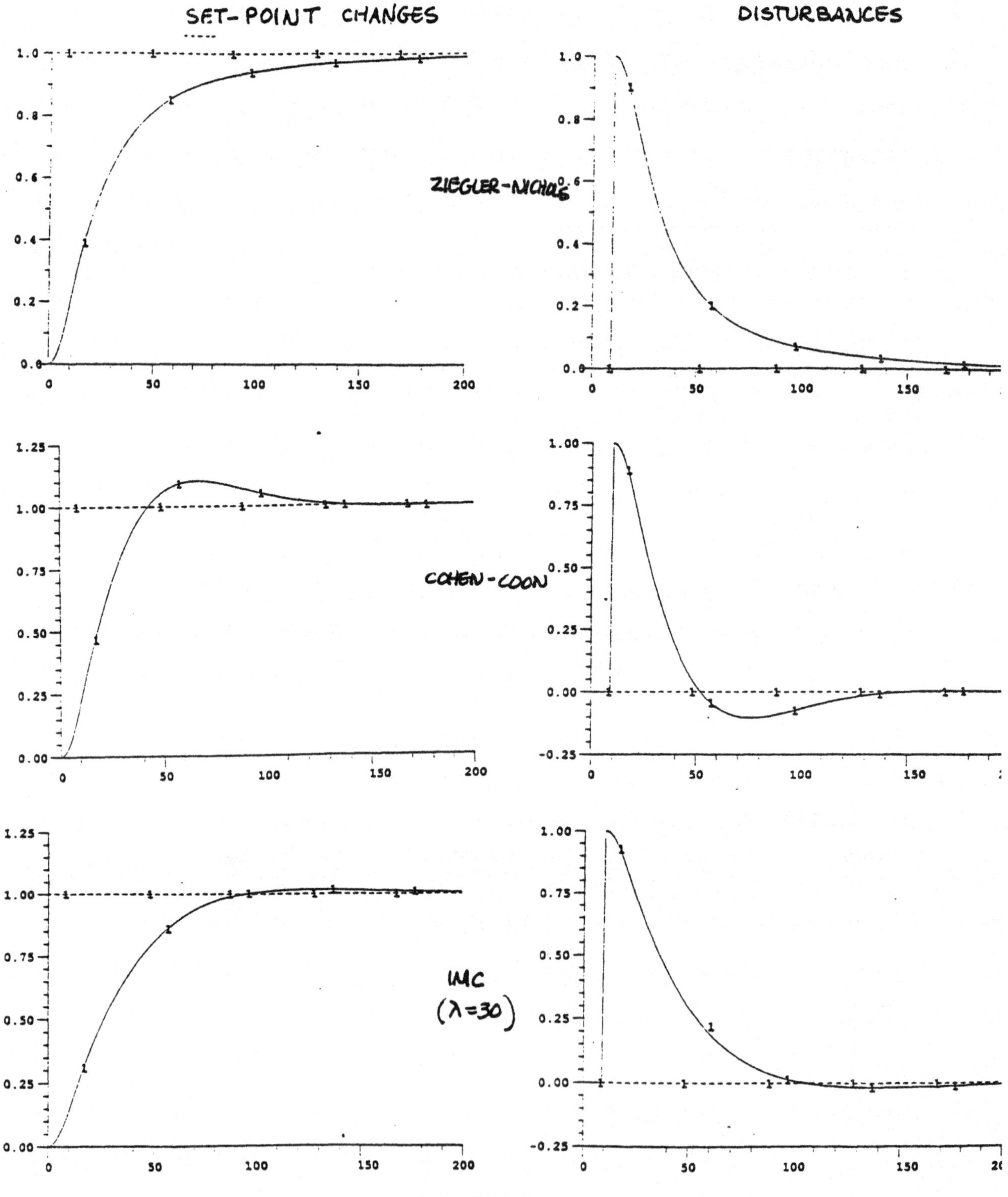

SET-POINT CHANGES
DISTURBANCES
ZIEGLER-NICHOLS
COHEN-COON
IMC
(λ=30)

FIG S15.6

15.4 (a)

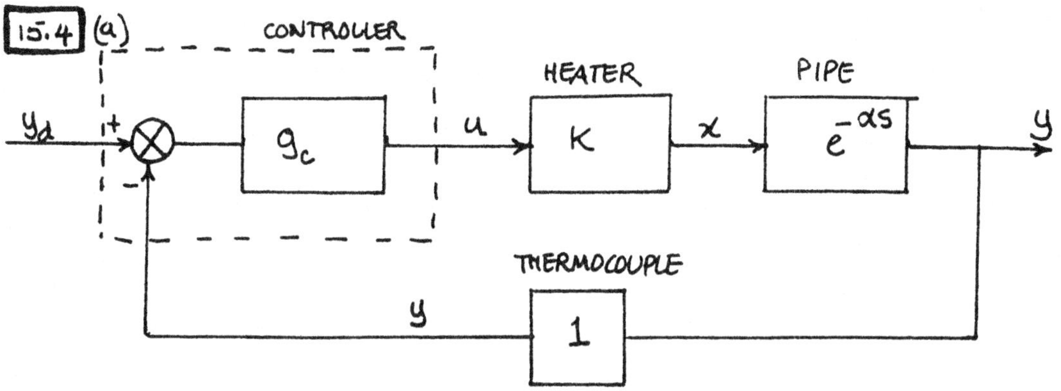

FIG S15.7

FIG S15.7 above shows the required block diagram.

(b) The process model is given by

$$y(s) = Ke^{-\alpha s}\, u(s) \qquad (15.4)$$

so that the process transfer function is:

$$g(s) = Ke^{-\alpha s} \qquad (15.5)$$

From the provided information, $K = 0.4$; α is obtained by noting that

$$\alpha = L/v$$

and

$$v = \frac{\text{Volumetric flow rate}}{\text{Cross sectional area}} = \frac{0.2\,m^3/min}{0.05\,m^2}$$

$$= 4\ m/min.$$

Thus

$$\alpha = \frac{1\,m}{4\,m/min} = 0.25\ min.$$

To obtain K_u, given $g(s) = Ke^{-\alpha s}$, a sketch of the Bode diagram is all that is required. The OLTF for this closed loop system is:

$$gg_c = KK_c e^{-\alpha s}$$

and, in this case

$$\frac{AR}{KK_c} = 1 \quad \text{at all frequencies,} \qquad (15.6)$$

$$\phi = -\alpha\omega \qquad (15.7)$$

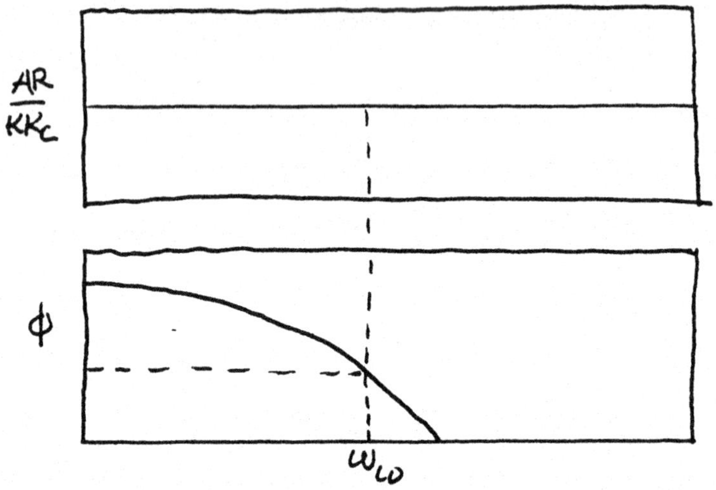

From here we see that at $\omega = \omega_{co}$ (as is indeed the case for all other frequencies)

$$\frac{AR}{KK_c} = 1$$

Thus, $K_u = \dfrac{1}{K}$

and in this specific case with $K = 0.4$,

$$K_u = 1/0.4 = 2.5$$

(c) We require P_u. This is obtained from Eq (15.7) above, and the fact that at the crossover frequency $\phi = -180°$ or $-\pi$ radians. Thus,

$$-\pi = -\alpha\omega_{co}$$

so that $$\frac{\pi}{\omega_{co}} = \alpha \qquad (15.8)$$

By definition, $P_u = \frac{2\pi}{\omega_{co}}$, so that from (15.8), we have

$$P_u = 2\alpha = 0.5 \text{ min}$$

And now, using the Z-N settings, with $K_u = 2.5$ and $P_u = 0.5$ we obtain the following recommended PID controller parameters:

$$K_c = 1.5 \; ; \; \tau_I = 0.25 \; ; \; \tau_D = 0.0625$$

The response of the closed loop system to a 1°C step change in the fluid temperature setpoint is shown in FIG S15.8 below.

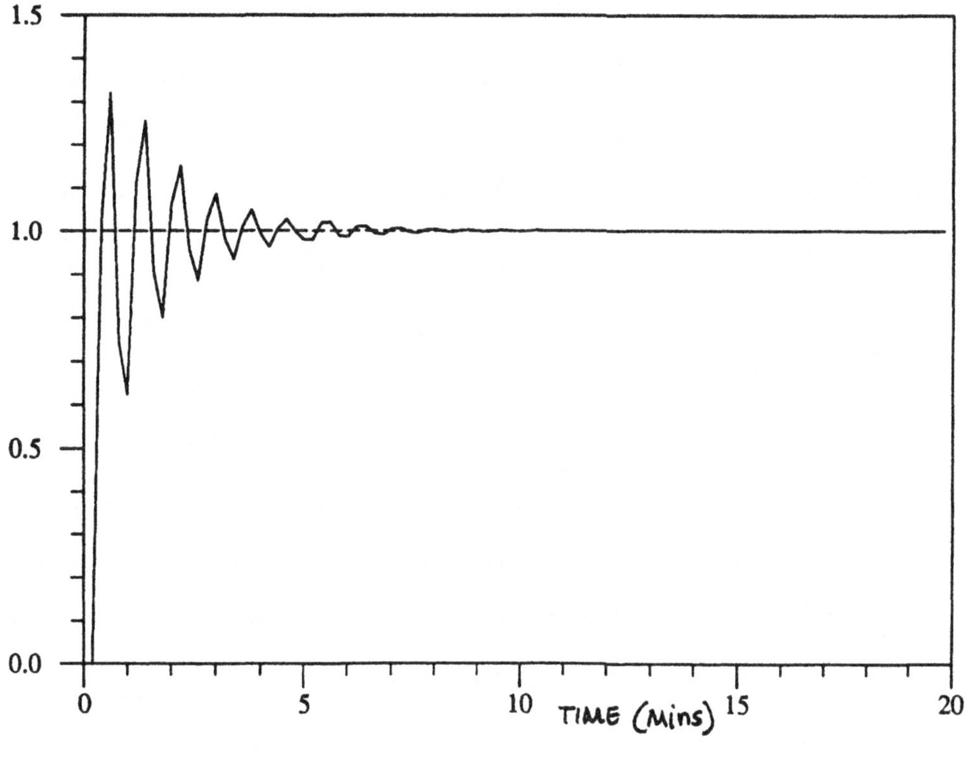

FIG S15.8

15.5 (a) A block diagram for the control system is shown below:

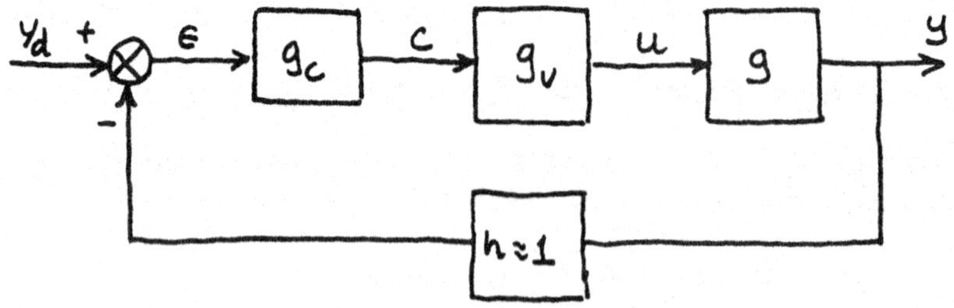

In this case, the OLTF g_L is given by

$$g_L = g_c g_v g = K_c\left(\frac{1.2}{0.8s+1}\right)\left(\frac{0.1(-2s+1)}{s(5s+1)}\right)$$

or

$$g_L = \frac{0.12K_c(-2s+1)}{4s^3+5.8s^2+s} \qquad (15.9)$$

A Bode plot for this g_L is shown in FIG S15.9. From here we obtain

$$\omega_c = 0.24 \; ; \; AR_c = 0.32$$

$$\Rightarrow K_{cu} = \frac{1}{AR_c} = 3.125$$

Using a gain margin of 1.7 gives

$$\boxed{K_c^{GM} = \frac{K_{cu}}{1.7} = 1.84} \qquad (15.10)$$

Returning to the Bode diagram, a phase margin of 30° gives $AR_{PM} = 0.67$ so that now

$$\boxed{K_c^{PM} = \frac{1}{0.67} = 1.5} \qquad (15.11)$$

Observe that the PM approach gives a more conservative controller.

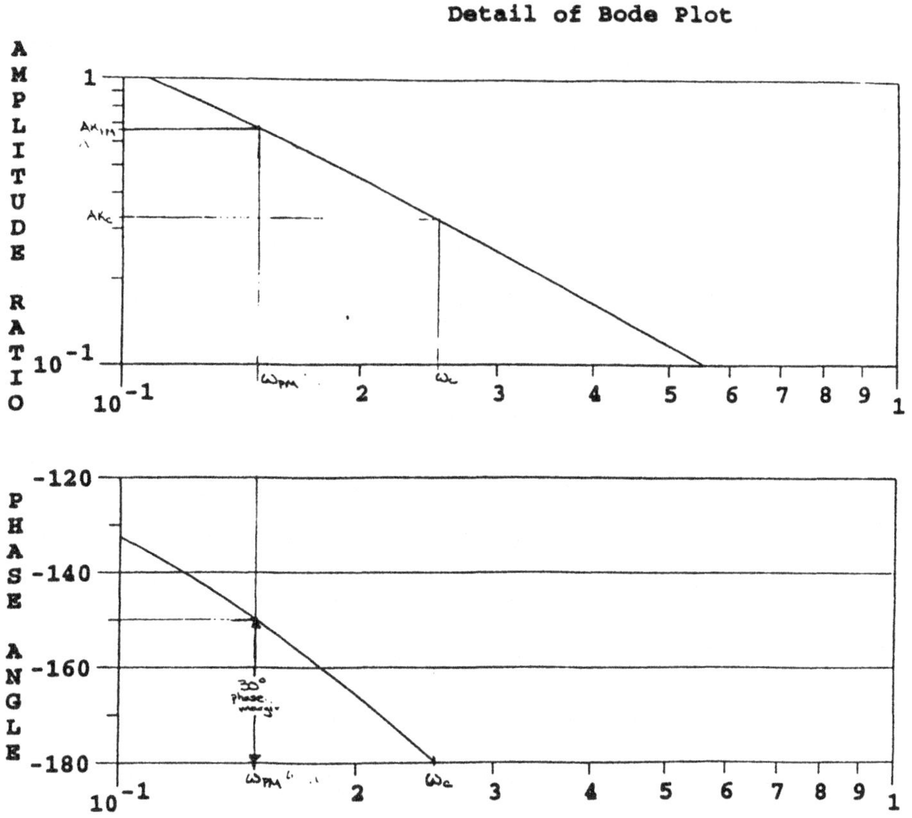

FIG S15.9

(b) The simulations showing the performance of each controller are in FIG S15.10a and FIG S15.10b. The controllers are essentially similar, the one based on GM considerations being somewhat more agressive. The overall performance obtained from the controllers are comparable. Note that there is no offset, even though we have implemented pure proportional controllers. This is because the process has an inherent integrator element. (see process transfer function.)

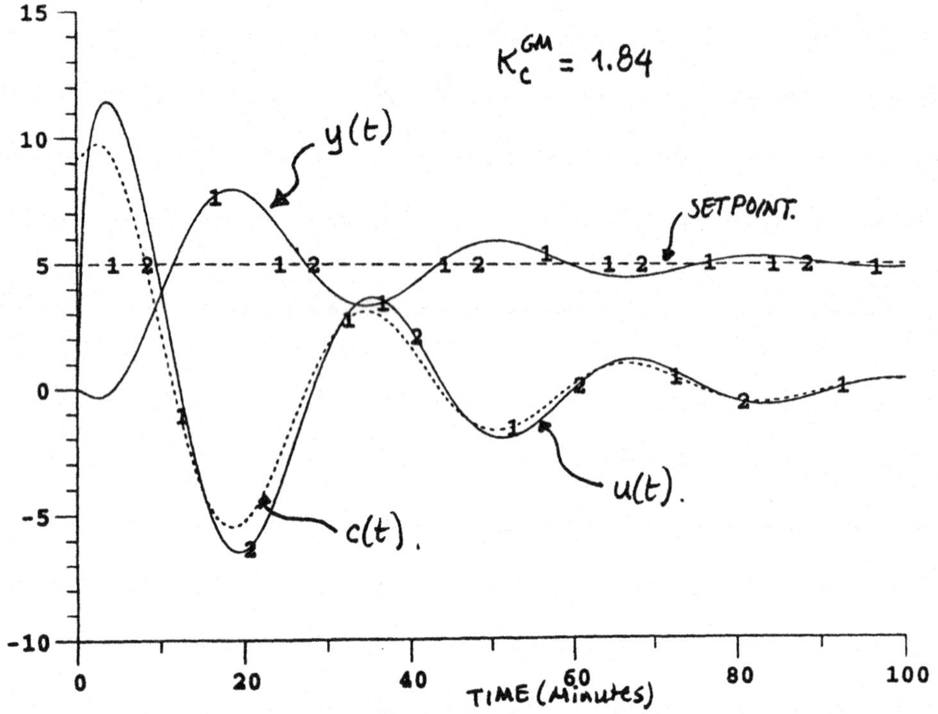

FIG S15.10a

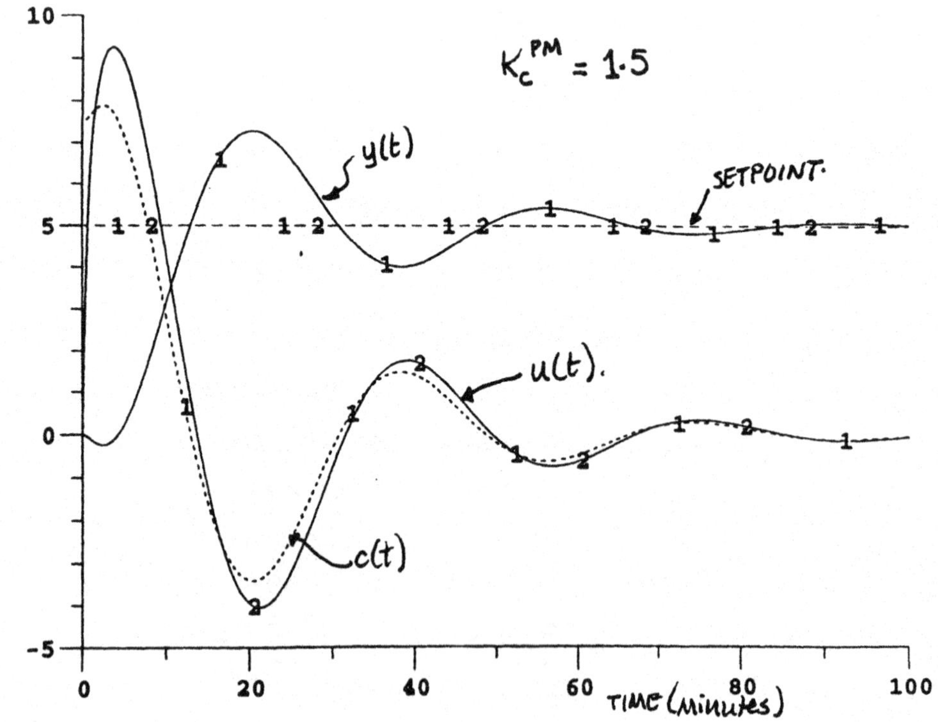

FIG S15.10b

15.6 (a) The "gain neutral" Bode plot for this control system is shown below in FIG S15.11 from where we obtain

$$\omega_{co} = 0.69 \; ; \quad AR_c = 0.14$$

given

$$K_{cu} = 1/AR_c = 7.14$$

$$P_u = \frac{2\pi}{\omega_{co}} = 9.11$$

The recommended PI controller tuning parameters from Table 15.1 are therefore

$$K_c = 3.21$$
$$\tau_I = 7.59$$

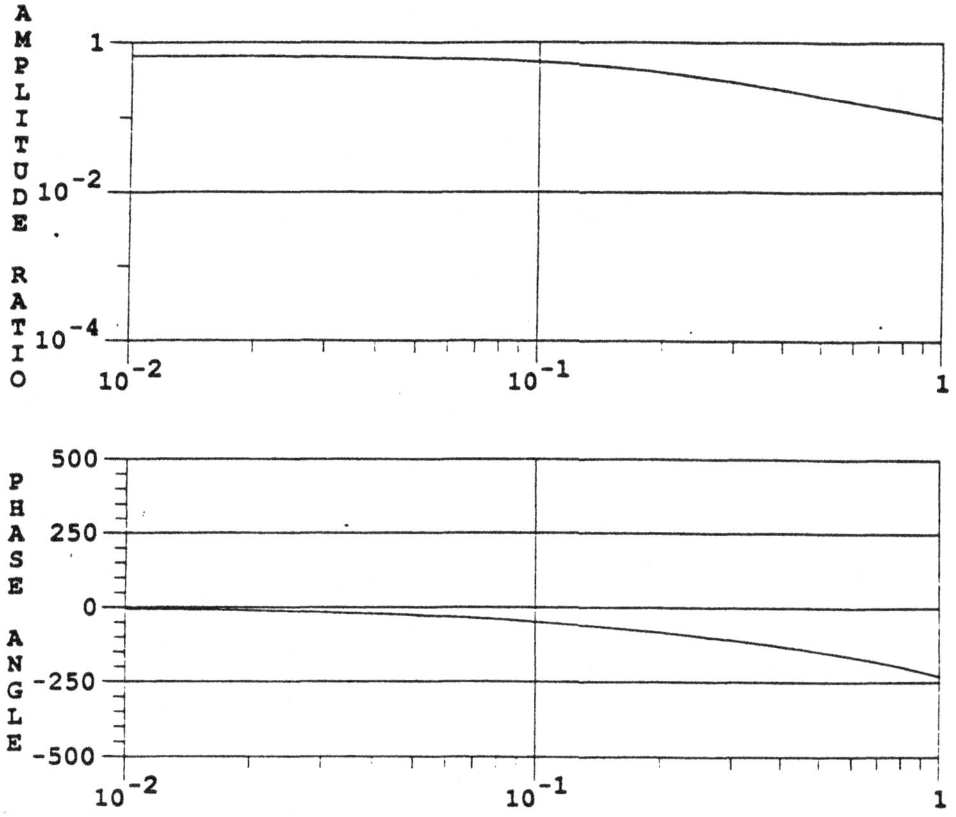

FIG S15.11

(b) Using the Fuentes technique, we have, in this case that

$$6.7\omega_c = \frac{1}{2} + \frac{\pi}{2}\left(\frac{6.7}{2.6}\right)$$

or

$$\omega_c = 0.679$$

(compare with $\omega_c = 0.69$ obtained earlier). From here we now employ Eq.(P15.10) to obtain AR_c, i.e.;

$$AR_c = \frac{K}{\sqrt{(1+\tau^2\omega_c^2}} = \frac{0.66}{\sqrt{1+(6.7\times 0.679)^2}}$$

$$AR_c = 0.1417$$

(compare with $AR_c = 0.14$ obtained earlier). These values for AR_c and ω_c now yield

$$K_{cu} = 1/AR_c = 7.06 \quad \text{and} \quad P_u = 2\pi/\omega_c = 9.254$$

with the corresponding Z-N settings

$$\boxed{\begin{aligned} K_c &= 3.177 \\ \tau_I &= 7.712 \end{aligned}}$$

which are in fact quite close to the values obtained earlier.

The Fuentes approach is not only simpler & more straightforward, it is less susceptible to the errors associated with reading numbers off graphs. It is quite useful in that it circumvents the need for obtaining a Bode diagram in the first instance. It is also clear (at least from this example) that the approximations involved are quite reasonable; and that these approximations do not lead to any significant loss of accuracy.

15.7 (a) We may use the analytical procedure for obtaining stability limits for this problem and from there use Table 15.1 for the Z-N design. The strategy is to assume a proportional controller with unspecified K_c, obtain a closed loop characteristic equation and then use the method of direct substitution to obtain ω_c and K_u.

For this process, the characteristic equation is therefore

$$1 + \frac{K_c(-2.5)}{(5s+1)^2(2.9s+1)} = 0$$

or

$$72.5s^3 + 54s^2 + 12.9s + (1-2.5K_c) = 0 \qquad (15.12)$$

substitute $s=j\omega$, rearrange, and equate both real and imaginary parts to zero to obtain.

$$-54\omega^2 + (1-2.5K_c) = 0$$

$$-72.5\omega^3 + 12.9\omega = 0$$

Solve simultaneously to obtain

$$K_u = -3.45$$

$$\omega_c = 0.422 \quad \Rightarrow \quad P_u = 2\pi/\omega_c = 14.89$$

From here, obtain the following controllers:

(i) P : $K_c = 0.5K_u \quad \Rightarrow \quad \boxed{K_c = -1.725}$

(ii) PI : $K_c = 0.45K_u$, $\tau_I = P_u/1.2 \quad \Rightarrow \quad \boxed{K_c = -1.552, \ \tau_I = 12.41}$

(iii) PID $K_c = 0.6K_u$ $\Rightarrow$ $K_c = -2.07$

$\tau_I = P_u/2$ $\tau_I = 7.445$

$\tau_D = P_u/8$ $\tau_D = 1.861$

(b) The performances of each of these 3 controllers are shown in FIG S15.12, S15.13, and S15.14 (for the P, PI and PID controllers respectively.

REMARKS

1. Both P and PI controllers give rise to feedback control systems that are too oscillatory; besides, the P controller leaves an unacceptable offset.
2. The PID controller is somewhat acceptable. It's performance is by far the best (with regards to oscillations). The oscillations could be further reduced by reducing integral action as well as the proportional gain..

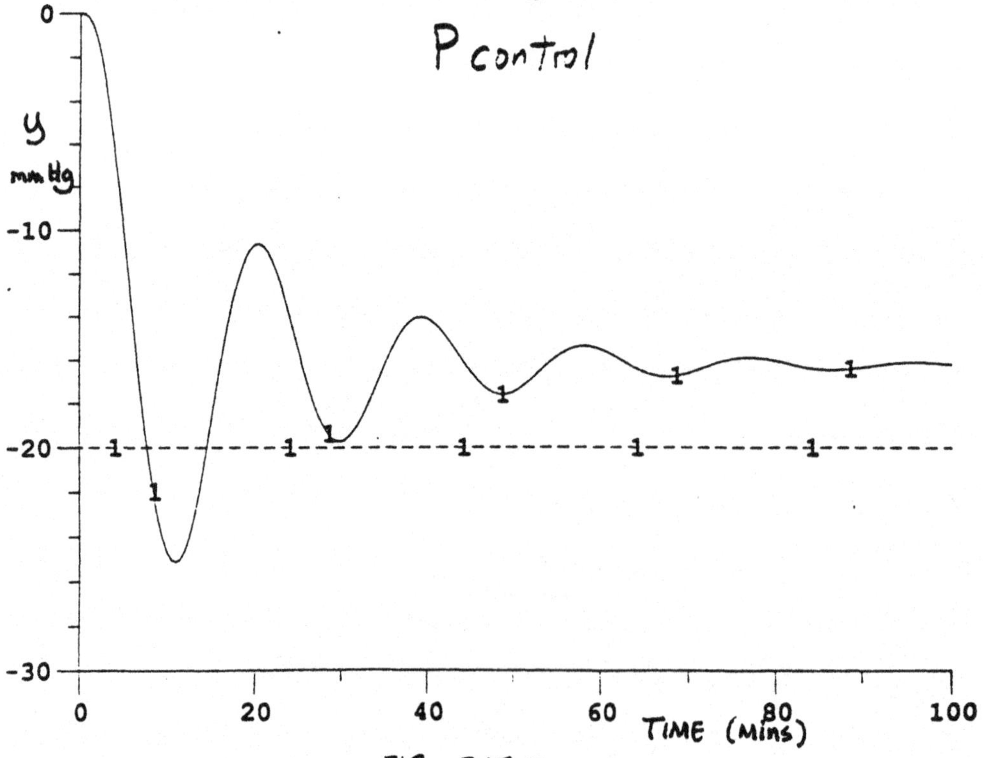

FIG S15.12

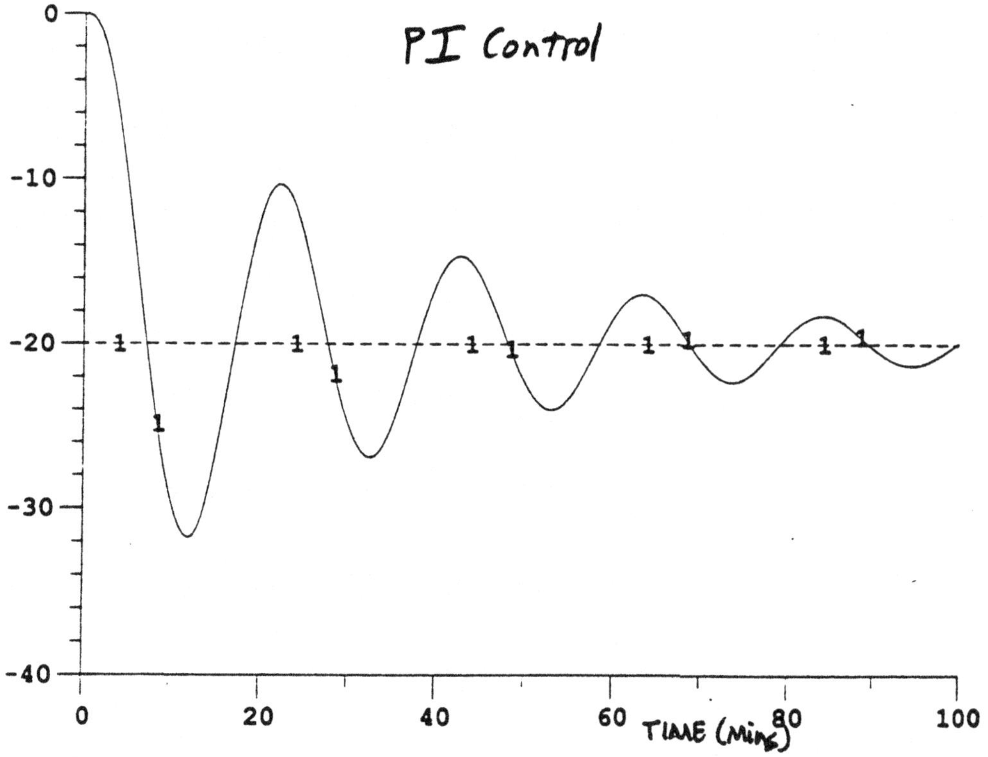

FIG S15.13

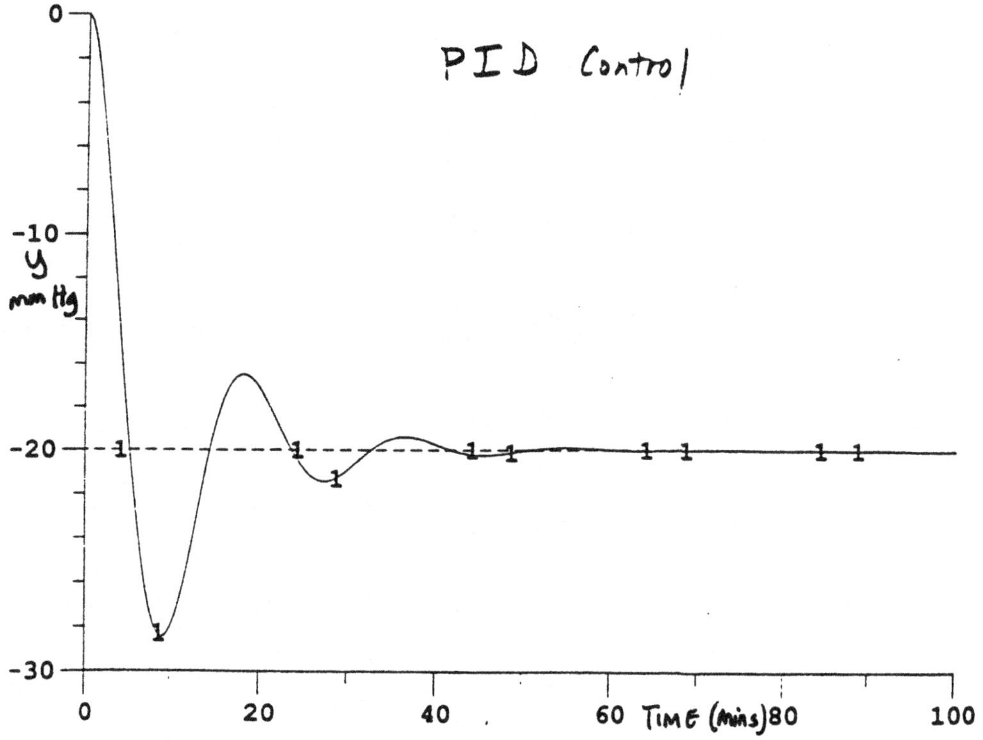

FIG S15.14

15.8 (a) Upon introducing the appropriate parameters and consolidating the process transfer function, we obtain for this process

$$g(s) = \frac{+1\times10^{-3}(8.35s+1)}{s(4.17s+1)} \qquad (15.13)$$

where the signs have been switched for transfusion.
The closed loop characteristic equation is now given by

$$1 + \frac{1\times10^{-3}K_c(8.35s+1)}{s(4.17s+1)} = 0$$

or

$$4.17s^2 + (1+8.35\times10^{-3}K_c)s + 1\times10^{-3}K_c = 0$$

$$\frac{+4.17}{1\times10^{-3}K_c}s^2 + \frac{(1+8.35\times10^{-3}K_c)}{+1\times10^{-3}K_c} + 1 = 0$$

To obtain a closed loop behavior with the characteristic equation

$$s^2 + \beta s + 1 = 0$$

requires that

$$\frac{+4.17}{1\times10^{-3}K_c} = 1$$

or

$$\boxed{K_c = +4{,}170.}$$

so that

$$\beta = 8.59$$

This controller will leave no offset because g(s) in (15.13) above includes an integrator element.

(b) The closed loop response is shown below in FIG S15.15.

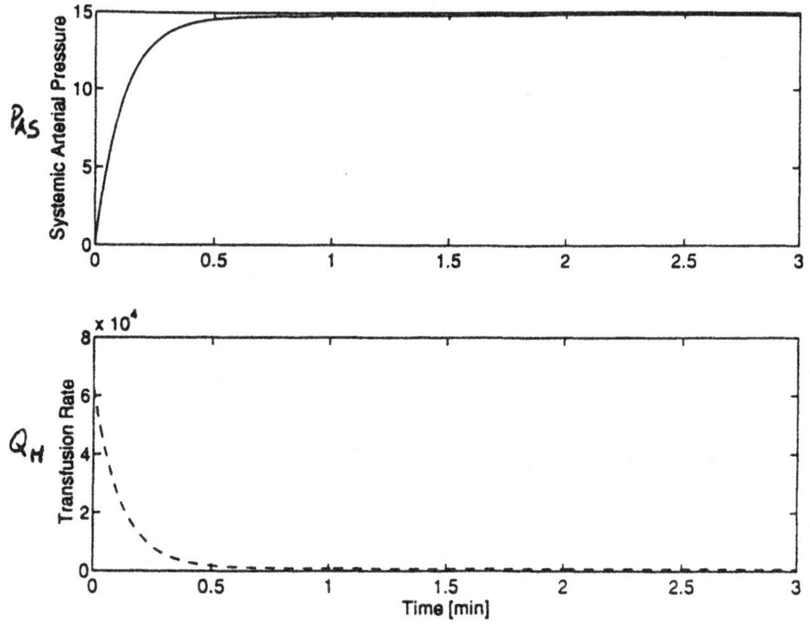

FIG S15.15

CHAPTER 16

16.1 (a) The disturbance enters the process via the non-manipulated stream; thus for feedforward control, measure either at A or B. Now, the objective of feedforward control is to compensate for the effect of disturbances before the process is affected. The manipulated stream is subject to a 3 second delay; thus for effective FF control, disturbance measurement must be obtained with enough time to allow the delayed manipulated stream to compensate adequately; i.e. the disturbance must be measured at least 3 secs before it enters the process. Hence measure at A.

(b) The required diagrams are shown below in FIG S16.1 a, b, c, d respectively for (i) Feedforward; (ii) Feedback + Cascade; (iii) Feedforward + Feedback; (iv) FF, Feedback + Cascade.

Each diagram shows the appropriate connection between transmitters and controllers. The redundant transmitters are left unconnected in each case.

i) Feed Forward

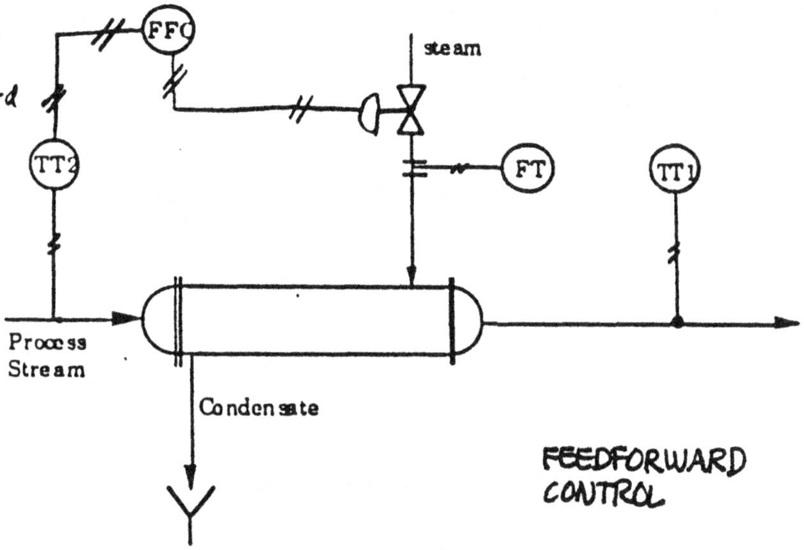

FIG S16.1 (a)

ii) Feed back / Cascade

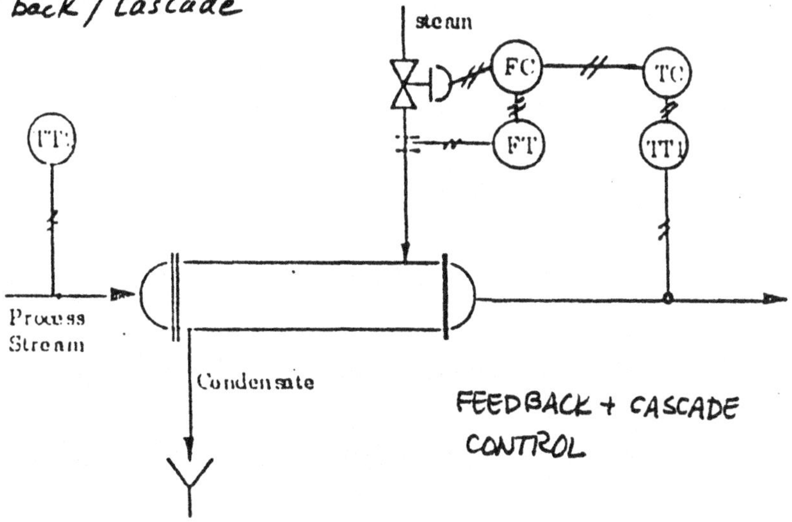

FIG S16.1 (b)

iii) Feed Forward/Feed Back

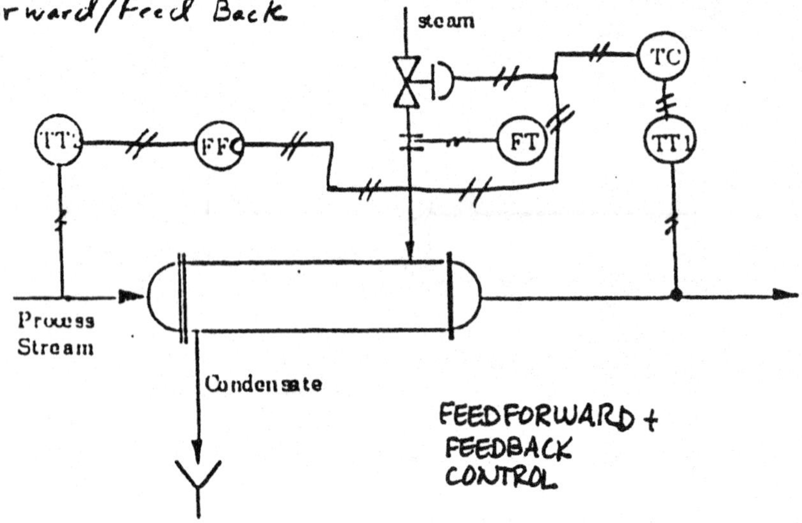

FIG S16.1 (c)

iv) Feed Forward/Feed Back/Cascade

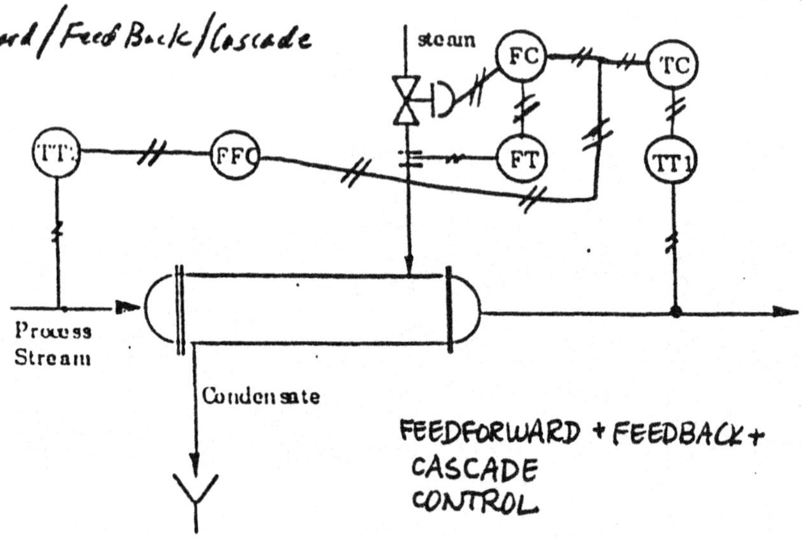

FIG S16.1(d)

16.2 (a) From Eq (16.7) in the text, upon introducing the elements of the transfer function model given for the polymer process, we obtain the following expression for the <u>feedforward only</u> controller:

$$u(s) = \underbrace{\left(\frac{11.0s+1}{-550.0}\right)}_{g_{st}} y_d - \underbrace{\left[\frac{4.5(11.0s+1)}{550(5s+1)(25s+1)}\right]}_{g_{ff}} d(s) \tag{16.1}$$

The implementation block diagram is shown below in FIG S16.2.

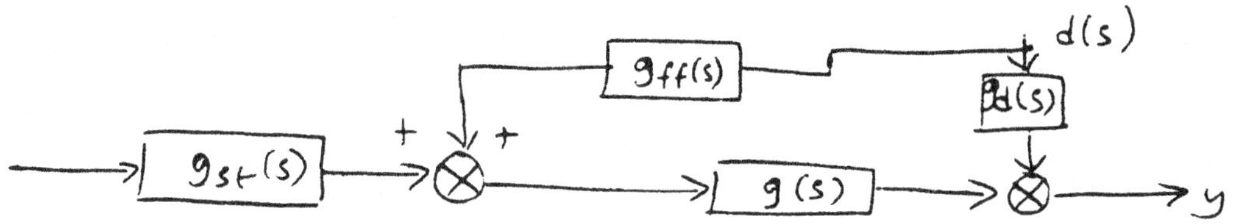

FIG S16.2

The indicated controllers are implementable because there are no predictive elements, neither are there any unstable elements. g_{st} is implementable as a PD controller; g_{ff} as a second order system with a single zero.

(b) The response is shown in FIG S16.3 (with the symbol).

(c) The response is shown in FIG S16.3 (with the symbol +) and the resulting offset is evident. (Note that in practice, however, such an offset of about 20 in the number average molecular weight will not be noticeable.) The offset is clearly a direct result of the model error.

(d) The response is shown in FIG S16.3 (with symbol □)

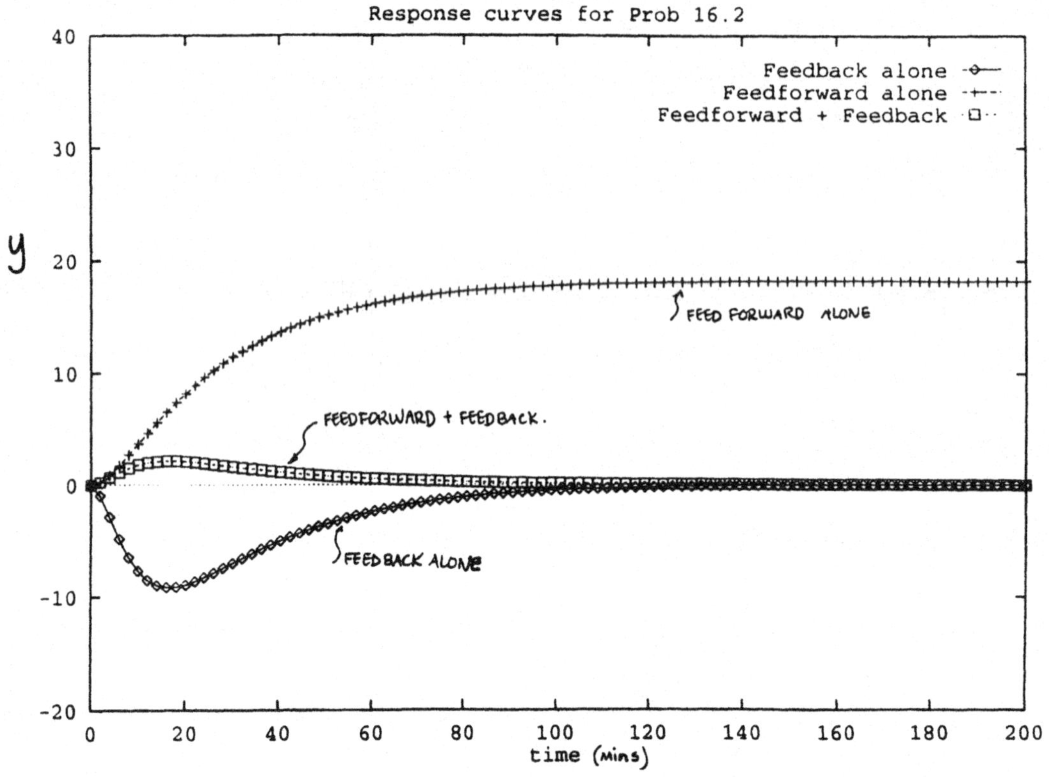

FIG S16.3

As illustrated here, clearly feedforward alone is quite susceptible to the effect of model errors. The amount of offset will depend on the amount of model error.

Feedback alone will leave no offset (if there is integral action) but observe that the process upset lasted for close to 100 minutes: feedback control is unable to correct for upsets until the process is affected.

When the feedforward scheme is augmented with feedback, we see the anticipatory nature of feedforward cutting down the maximum amount of deviation from set-point; we also see the feedback eliminating the offset. This combination yielded the best response.

16.3 (a)

$y \rightarrow$ Tray Temperature $(T - T^*)$
$u \rightarrow$ Reflux Flow rate $(R - R^*)$
$d \rightarrow$ Feed flow rate $(F - F^*)$

The process model becomes

$$y = \underbrace{\left(\frac{20e^{-15s}}{30s+1}\right)}_{g_d} d + \underbrace{\left(\frac{-5e^{-2s}}{6s+1}\right)}_{g} u \qquad (16.2)$$

(b) The transfer function for the feedforward controller, g_{ff} is given by

$$g_{ff} = \frac{-g_d}{g} = \frac{4e^{-13s}(6s+1)}{(30s+1)}$$

$$\boxed{g_{ff} = \frac{4(6s+1)e^{-13s}}{30s+1}} \qquad (16.3)$$

This is realizable because it contains no unstable elements; neither does it contain any predictive terms.

In physical terms, the delay associated with the effect of reflux flow rate changes on the tray temperature (2 minutes) is less than the delay associated with the effect of feed flow rate fluctuations on tray temperature. Thus a well designed feedforward controller has time to compensate for feed flow rate disturbances.

The explicit expression for changes in u as a result of observed changes in d is:

$$u(s) = \frac{4(6s+1)e^{-13s}}{30s+1} d(s) \qquad (16.4)$$

(c) The required response is obtained from

$$u(s) = g_{ff}\,\frac{2}{s}$$

$$= \frac{4(6s+1)e^{-13s}}{(30s+1)} \cdot \frac{2}{s}$$

By partial fraction expansion, followed by Laplace inversion, or by recalling the response of the lead/lag system given in chapter 5, obtain

$$u(t) = \begin{cases} 0 & t < 13 \\ 8 - 6.4e^{-(t-13)/30}; & t \geq 13 \end{cases}$$

The response is plotted below in FIG S16.4

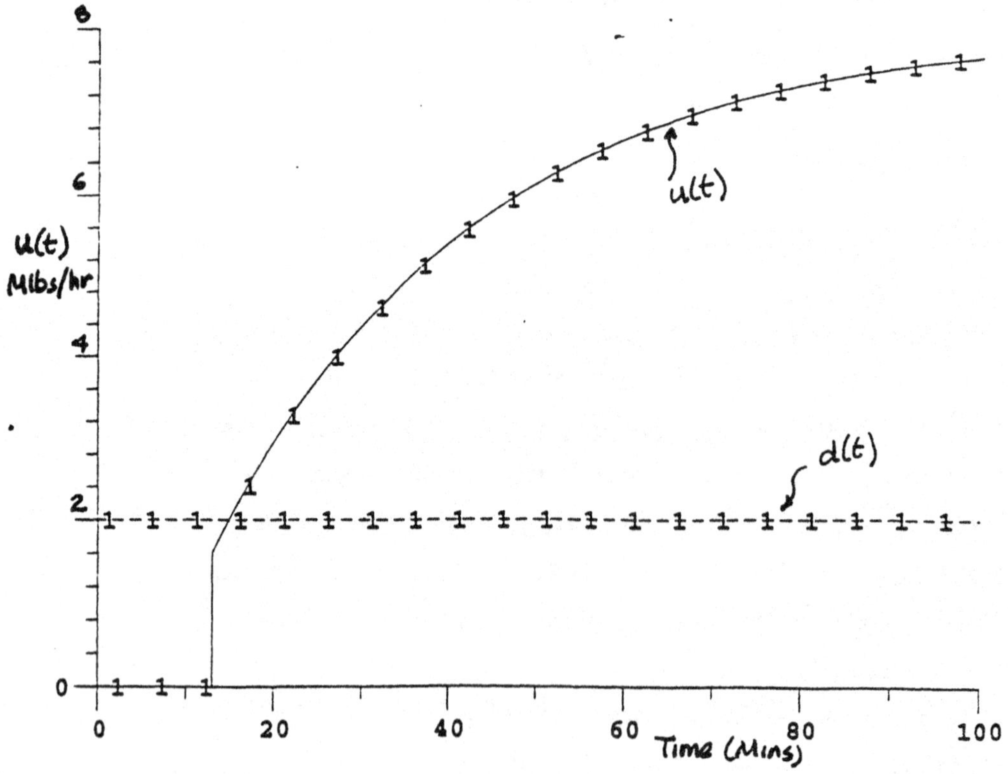

FIG S16.4

16.4 (a)

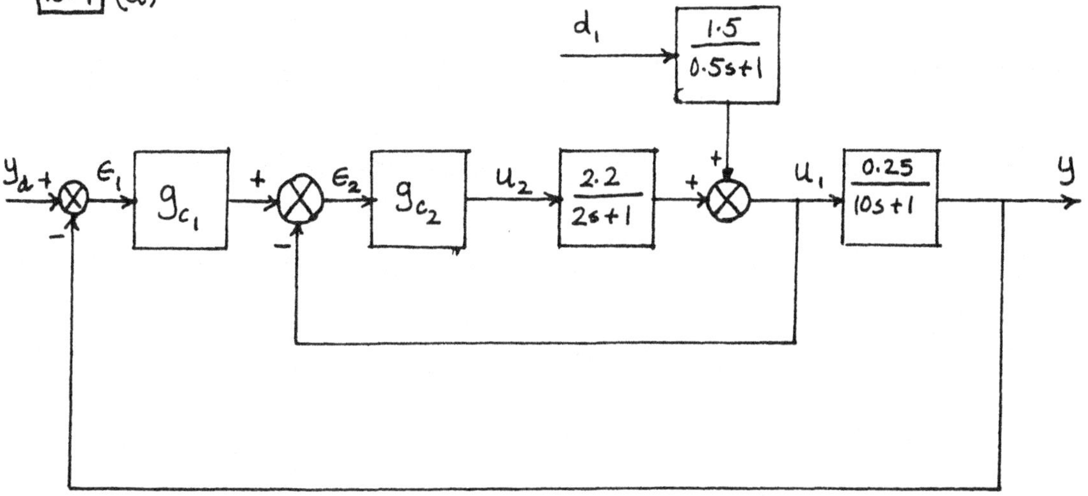

FIG S16.5

(b)

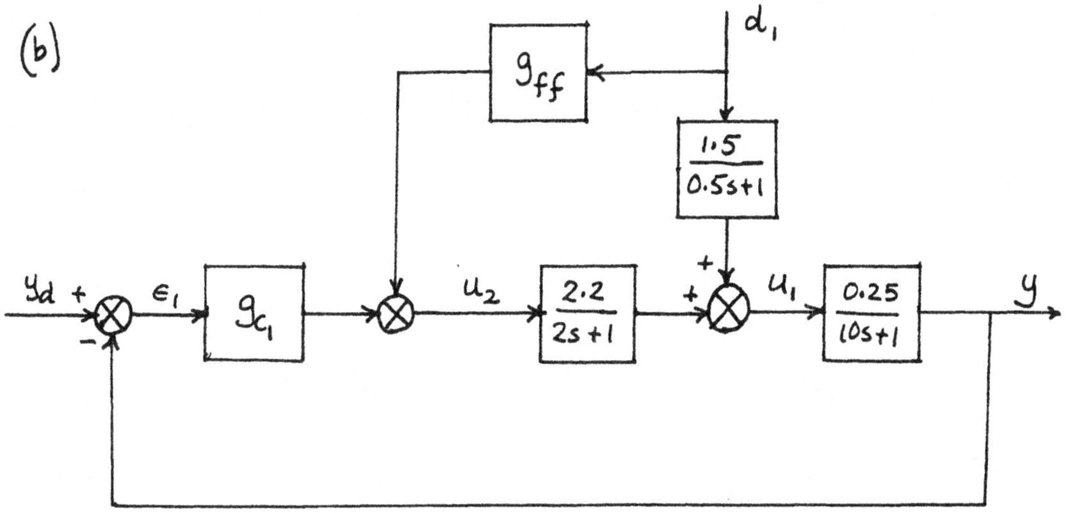

FIG S16.6

where

$$g_{ff} = -\frac{\frac{1.5}{0.5s+1}}{\frac{2.2}{2s+1}}$$

or

$$\boxed{g_{ff} = \frac{-0.6818\,(2s+1)}{(0.5s+1)}} \qquad (16.5)$$

when the process model is extremely well known, the feedforward control strategy is able to perform near perfect disturbance rejection using an easily implementable lead/lag element indicated in Eq (16.5) above. Feedback control will correct for any residual error due to imperfect modeling.

On the other hand, the cascade scheme relies on pure feedback in the inner loop, a faster loop to be sure, but a feedback loop all the same. This is bound to be less effective, when the process model is very well known.

When the process models are too approximate to be considered as "very accurate" process representations, the cascade scheme may very well provide an advantage over the feedforward scheme.

All the discussion above is based on the fact that d_1 is measurable, of course.

16.5 (a)

y = CSTR temperature (controlled variable);

y_d = CSTR temperature set-point;

g_{c_1} = Master Temperature controller;

u_{p_1} = Desired set-point for T_f, temperature of reactor feed ex-preheater;

g_{c_2} = Slave Controller: Preheater temperature controller;

g_v = Preheater valve transfer function;

g_2 = Transfer function representing preheater dynamics;

d_1 = Reactant temperature T_i (disturbance)

g_{d_1} = Transfer function relating effect of changes in T_i on T_f, actual reactor feed temp;

u = Preheater valve opening;

g = Transfer function for CSTR; relating T_f to T.

(b) Consolidate inner loop first to obtain

$$u_p = \underbrace{\left(\frac{g_2 g_v g_{c_2}}{1 + g_2 g_v g_{c_2}}\right)}_{g_1^*} u_{p_1} + \underbrace{\left(\frac{g_{d_1}}{1 + g_2 g_v g_{c_2}}\right)}_{g_{d_1}^*} d_1 \qquad (16.6)$$

where u_p is the input signal to the g block (i.e. u_p is $(T_f - T_f^*)$ where T_f is the temperature of the reactor feed ex-preheater).

Block diagram now becomes

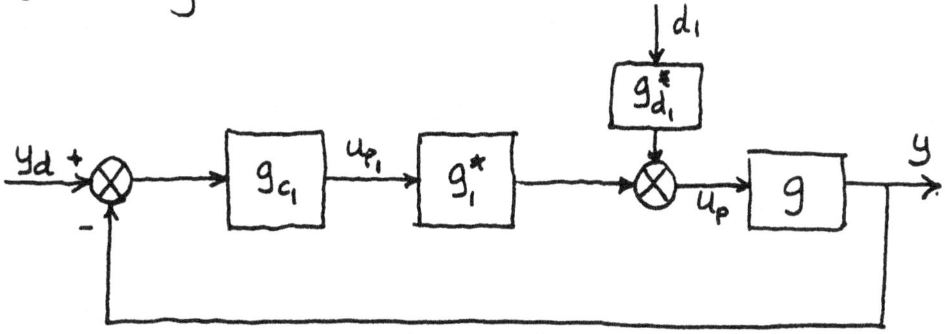

FIG S16.7

So that the overall closed loop transfer function relations are

$$y = \frac{g g_1^* g_{c_1}}{1 + g g_1^* g_{c_1}} y_d + \frac{g_{d_1}^*}{1 + g g_1^* g_{c_1}} d_1 \qquad (16.7)$$

with g_1^* and $g_{d_1}^*$ as indicated above in Eq (16.6).

The characteristic equation is

$$1 + g g_1^* g_{c_1} = 0$$

or, upon introducing the explicit expression for g_1^*,

$$1 + g_2 g_v g_{c_2}(1 + g g_{c_1}) = 0 \qquad (16.8)$$

(c) With $g_{c_1} = K_{c_1}$; $g_{c_2} = K_{c_2} = 2$

$$g(s) = \frac{2e^{-4s}}{3s+1} \approx \frac{2(1-2s)}{(1+2s)(3s+1)}$$

$$g_2 g_v = \frac{2.5}{(2s+1)}$$

The characteristic equation becomes.

$$1 + \frac{5}{2s+1}\left[1 + \frac{2K_{c_1}(1-2s)}{(2s+1)(3s+1)}\right] = 0$$

Simplifying to

$$12s^3 + 46s^2 + (32 - 20K_{c_1})s + (6 + 10K_{c_1}) = 0$$

stability now requires (for $K_{c_1} > 0$)

$$(32 - 20K_{c_1}) > 0 \quad \Rightarrow K_c < 1.6 \qquad (16.9)$$

and:

$$a_1 a_2 > a_0 a_3$$

in this case:

$$46(32 - 20K_{c_1}) > 12(6 + 10K_{c_1})$$

or $1040\, K_{c_1} < 1400$

$$\boxed{K_{c_1} < 1.346} \qquad (16.10)$$

so that the required condition for stability is as in Eqn (16.10) above.

(d) Under these simpler conditions, the block diagram is as shown below in FIG S16.8

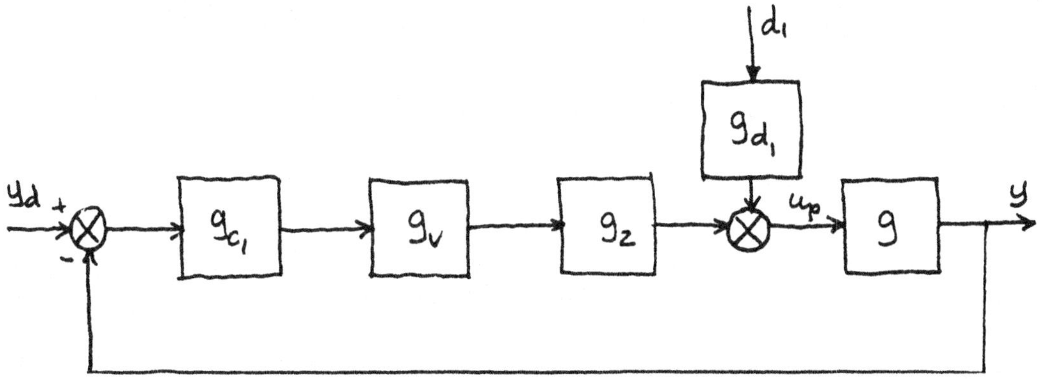

FIG S16.8

with a new characteristic equation:

$$1 + g g_2 g_v g_{c_1} = 0$$

With the given process information, this becomes

$$1 + \frac{5K_{c_1}(1-2s)}{(2s+1)^2(3s+1)} = 0$$

or

$$12s^3 + 16s^2 + (7 - 10K_{c_1})s + (1 + 5K_{c_1}) = 0 \qquad (16.11)$$

Stability now requires: (For $K_{c_1} > 0$)

$$(7 - 10K_{c_1}) > 0 \Rightarrow K_{c_1} < 0.7 \qquad (16.12)$$

and

$$16\,(7 - 10K_{c_1}) > 12(1 + 5K_{c_1})$$

or

$$220K_{c_1} < 100$$

$$\Rightarrow \boxed{K_{c_1} < 0.4545} \qquad (16.13)$$

And now by comparing Eqn (16.13) with the inequality in (16.10) we observe that the stability range for conventional feedback control is, in this case, almost 3 times narrower than the stability range for the cascade control system.

16.6 (a) The block diagram is shown below in FIG S16.9

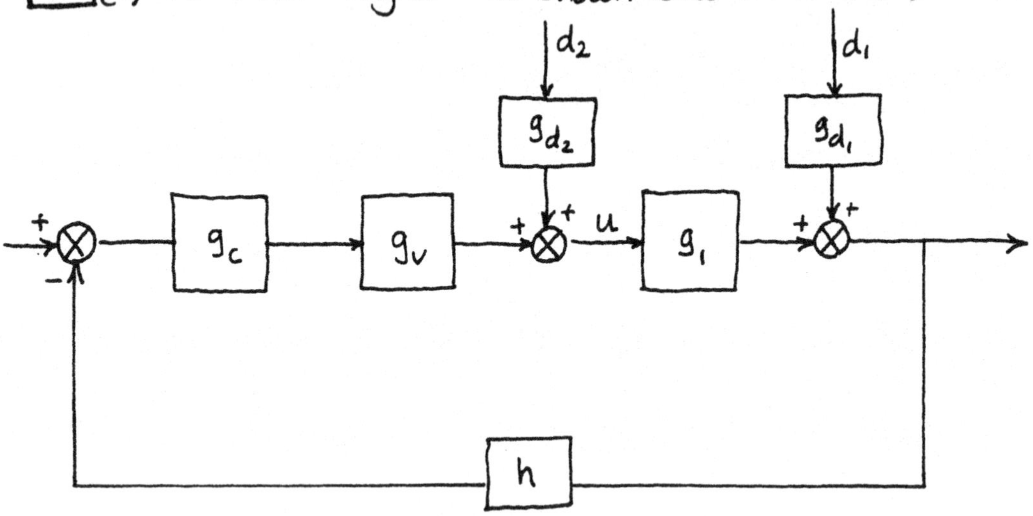

FIG S16.9

The closed loop transfer function equation is.

$$y = \frac{g_1 g_v g_c}{1 + g_1 g_v g_c h} y_d + \frac{g_1 g_{d_2}}{1 + g_1 g_v g_c h} d_2 + \frac{g_{d_1}}{1 + g_1 g_v g_c h} d_1 \qquad (16.14)$$

(b) The characteristic equation:

$$1 + g_1 g_v g_c h = 0$$

under the given circumstances becomes

$$1 + \left(\frac{0.5}{5s+1}\right)\left(\frac{2}{0.5s+1}\right) K_c \left(1 + \frac{1}{0.2s}\right)\left(\frac{1}{0.2s+1}\right) = 0$$

or

$$1 + \frac{K_c\,(0.2s+1)}{(5s+1)(0.5s+1)(0.2s)} \cdot \frac{1}{(0.2s+1)} = 0$$

or

$$0.2s\,(5s+1)(0.5s+1) + K_c = 0$$

$$0.5s^3 + 1.1s^2 + 0.2s + K_c = 0$$

Stability now requires

(i) $K_c > 0$, and

(ii) $a_1 a_2 > a_0 a_3$, in this case

$$1.1 \times 0.2 > 0.5K_c$$

$$\Rightarrow \boxed{K_c < 0.44} \qquad (16.15)$$

(c) The new controller should be configured to accept pressure measurements and manipulate the steam valve opening. The block diagram is shown in FIG S16.10 below.

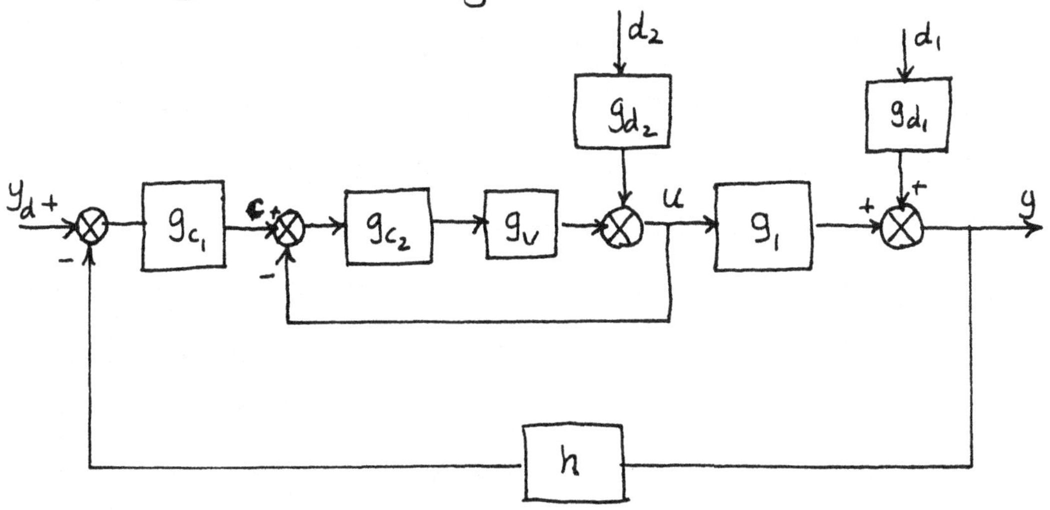

FIG S16.10

Consolidating the inner loop, we obtain

$$u = \underbrace{\left(\frac{g_v K_{c_2}}{1+g_v K_{c_2}}\right)}_{g_1^*} c + \underbrace{\left(\frac{g_{d_2}}{1+g_v K_{c_2}}\right)}_{g_{d_2}^*} d_2 \qquad (16.16)$$

and the overall closed loop characteristic equation becomes

$$1 + g_1 g_1^* g_{c_1} h = 0$$

with g_1^* as indicated in Eq (16.16) above. This equation rearranges to give

$$1 + g_v K_{c_2}(1 + g_1 g_{c_1} h) = 0$$

Upon introducing the given transfer functions, we obtain

$$1 + \frac{10}{0.5s+1} + \left(\frac{0.5}{5s+1}\right)\left(\frac{10}{0.5s+1}\right) K_c \left(\frac{0.2s+1}{0.2s}\right)\left(\frac{1}{0.2s+1}\right) = 0$$

which simplifies to

$$0.5s^3 + 11.1s^2 + 2.2s + 5K_c = 0$$

And now stability requires

(i) $K_c > 0$

(ii) $11.1 \times 2.2 > 0.5 \times 5K_c$

or $\boxed{K_c < 9.678}$ (16.17)

Comparing Eq (16.17) with Eq (16.15) shows how the cascade control loop has enlarged the stability range more than 20 times.

16.7 (a) With $K_{c_2} = 5$, upon consolidating the inner loop, the block diagram becomes what is shown below in FIG S16.11.

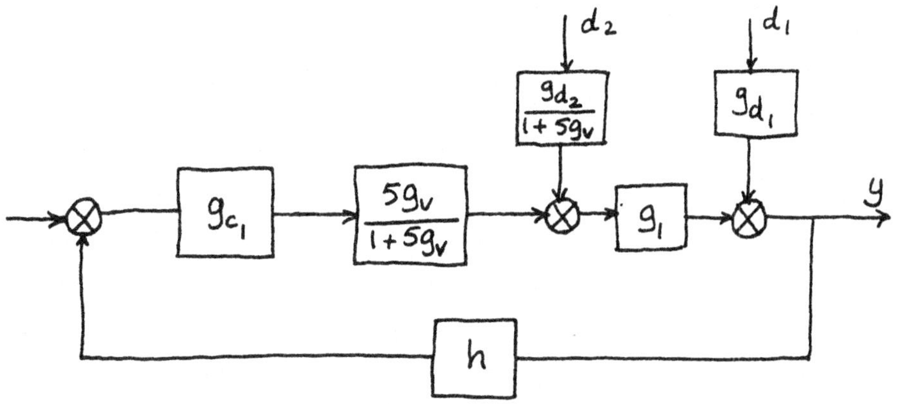

FIG S16.11

To employ the Ziegler-Nichols stability margin design method, we may plot the "gain neutral" Bode diagram for this systems OLTF. In this case

$$g_L = g_1\left(\frac{5g_v}{1+5g_v}\right)g_{c_1}h \qquad (\text{with } g_{c_1} = K_{c_1})$$

$$g_L = \left(\frac{0.5}{5s+1}\right)\left(\frac{10}{0.5s+11}\right)\left(\frac{K_{c_1}}{0.2s+1}\right)$$

$$g_L = \frac{5/11\, K_{c_1}}{(5s+1)\left(\frac{0.5}{11}s+1\right)(0.2s+1)} \qquad (16.18)$$

The Bode diagram for this g_L is shown in FIG S16.12 and from here we deduce the following:

$$K_u = 311.7$$

$$P_u = \frac{2\pi}{\omega_{co}} = \frac{2\pi}{10.72} = 0.586$$

The Z-N recommended settings for the PID controller are:

$K_c = 187.02$; $\tau_I = 0.195$; $\tau_D = 0.07325$

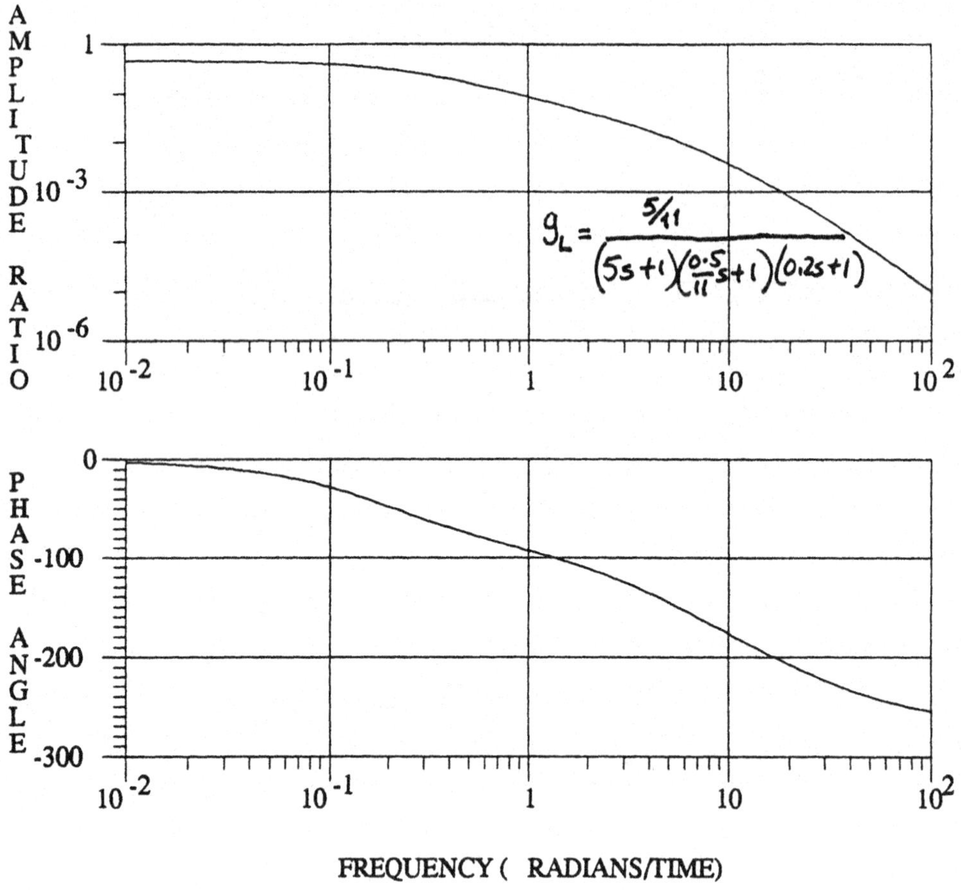

FIG S16.12

(b) The response shown in FIG S16.13 in actual fact is the overall system response to a simultaneous unit step change in d_1 and a step change of magnitude 5000 in d_2! The overall control system is so effective that disturbances in pressure (d_2) no longer affect the process seriously. Further illustrating this point is the "extra" response shown in FIG S16.14: this is the overall system response to a <u>unit</u> step change in d_2. Note that the scale on y is $\times 10^{-5}$.

Note that without control, a unit change in d_1 will cause a 1° change in y; a unit change in d_2 will cause a 0.15° change in y.

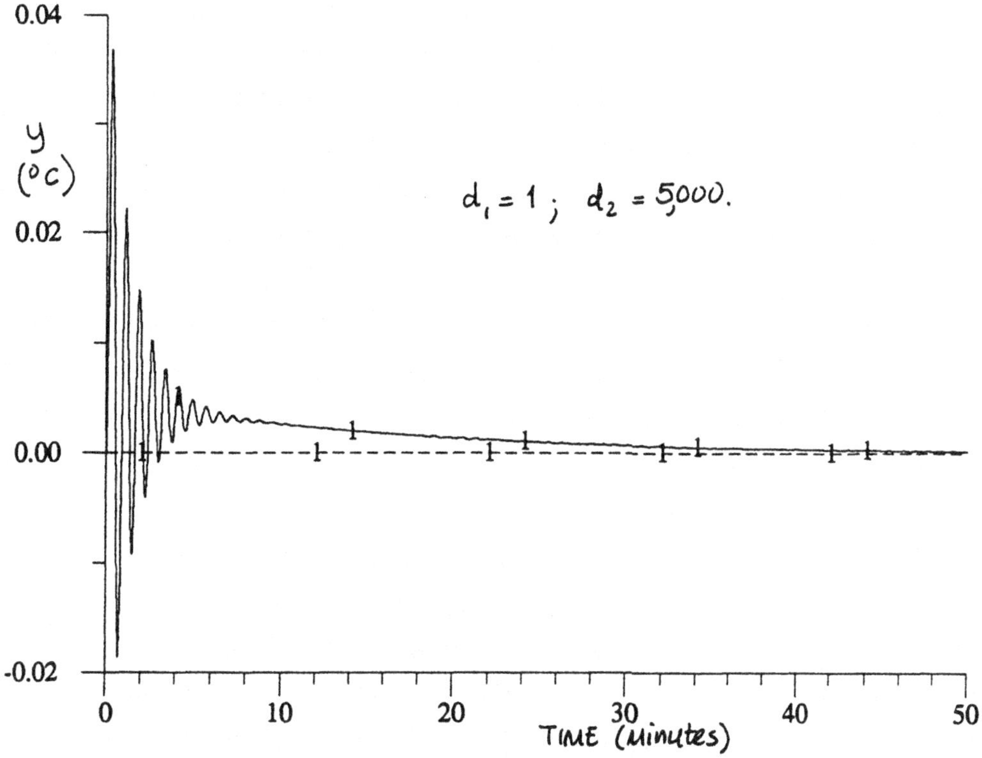

FIG S16.13

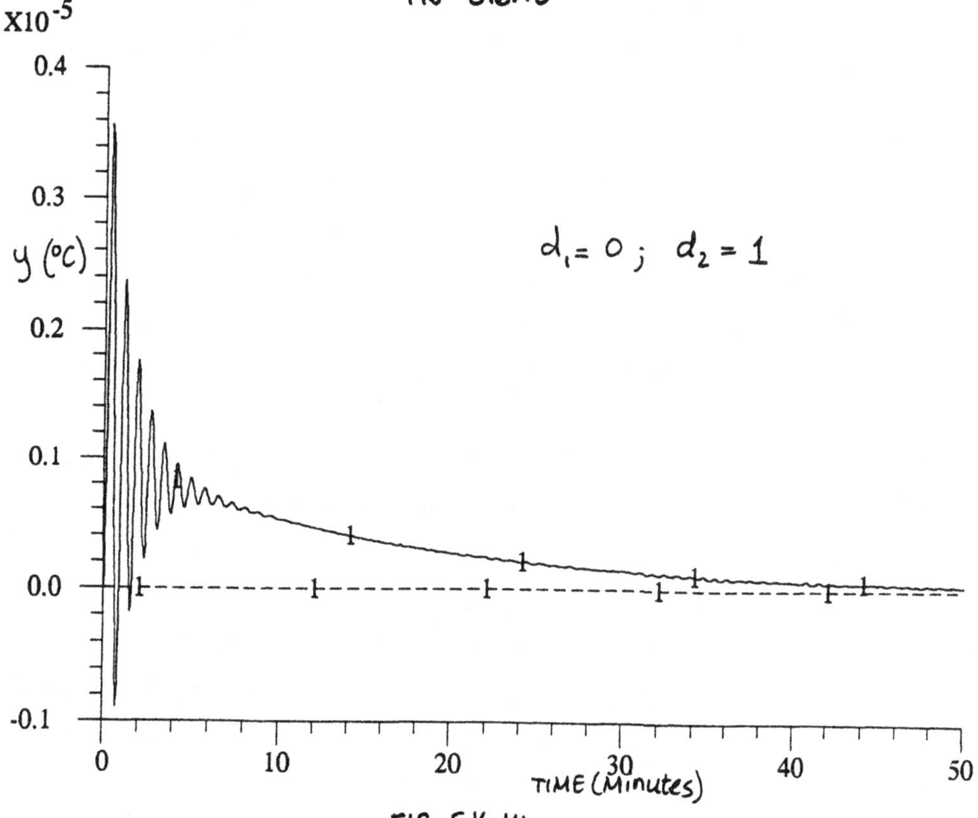

FIG S16.14

16.8 (a) The block diagram is shown below in FIG S16.15.

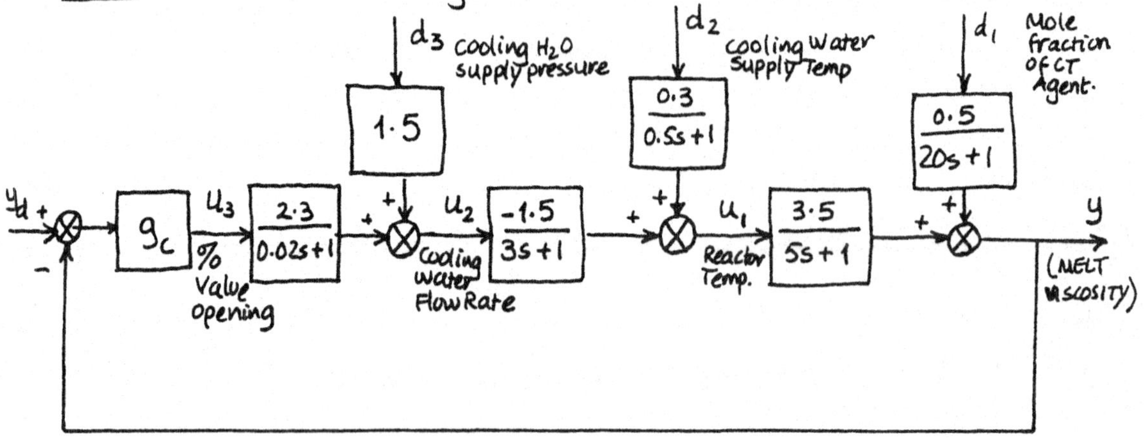

FIG S16.15 Block diagram for polymer process.

(b) For this process,

$$g_L = \frac{K_c\,(2.3)(-1.5)(3.5)}{(0.02s+1)(3s+1)(5s+1)} \qquad (16.19)$$

(K_c must be negative to "neutralize" this negative gain element in the closed loop.)

and the "gain neutral" Bode diagram is shown in FIG S16.16 From where we obtain the following:

$$\frac{AR}{|K_{cu}|} = 0.0298 \quad \Rightarrow \quad |K_{cu}| = 33.5$$

$$P_u = \frac{2\pi}{5.17} \quad \Rightarrow \quad P_u = 1.215$$

The recommended Z-N settings are now obtained as:

$$K_c = 0.45 K_{cu} = -15$$

$$\tau_I = \frac{P_u}{1.2} = 1.01$$

Upon implementing this controller on the process as indicated, the closed loop system goes UNSTABLE! The Z-N settings for PI control in general tend to be too aggressive; in this case they give rise to an unstable feedback system. The controller must now be detuned. This may be done several different ways. We choose to reduce τ_I from 1.01 to 0.1. The resulting closed loop system response is shown in FIG S16.17; it is still quite oscillatory, but now it is stable.

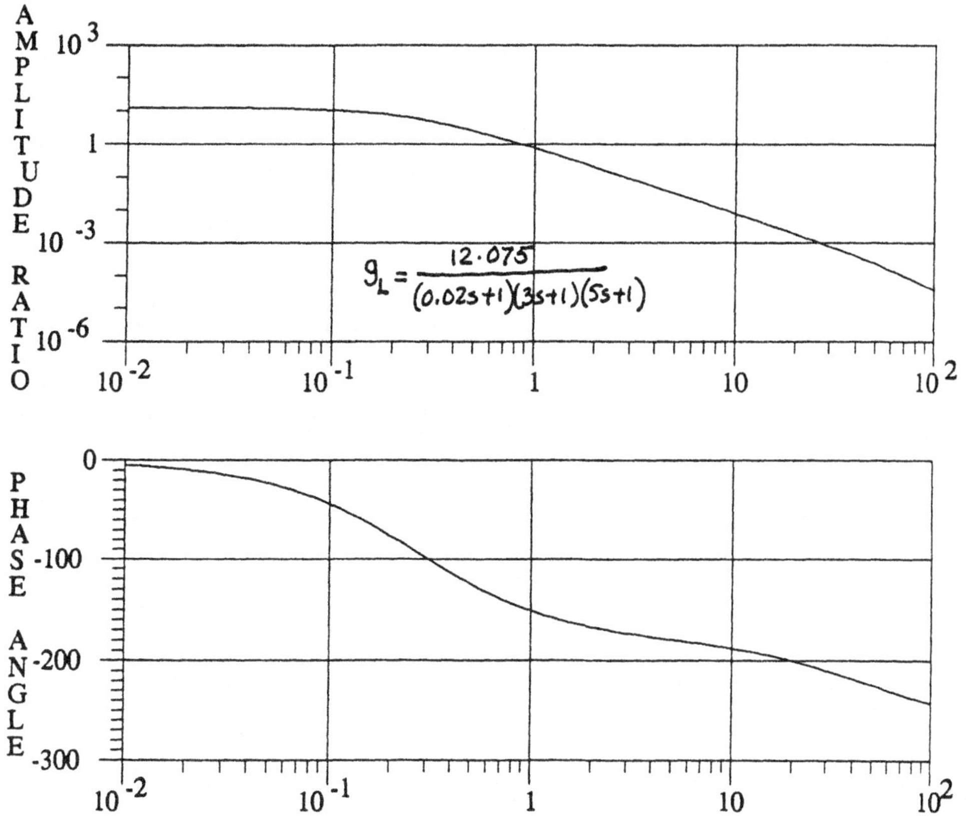

FIG S16.16

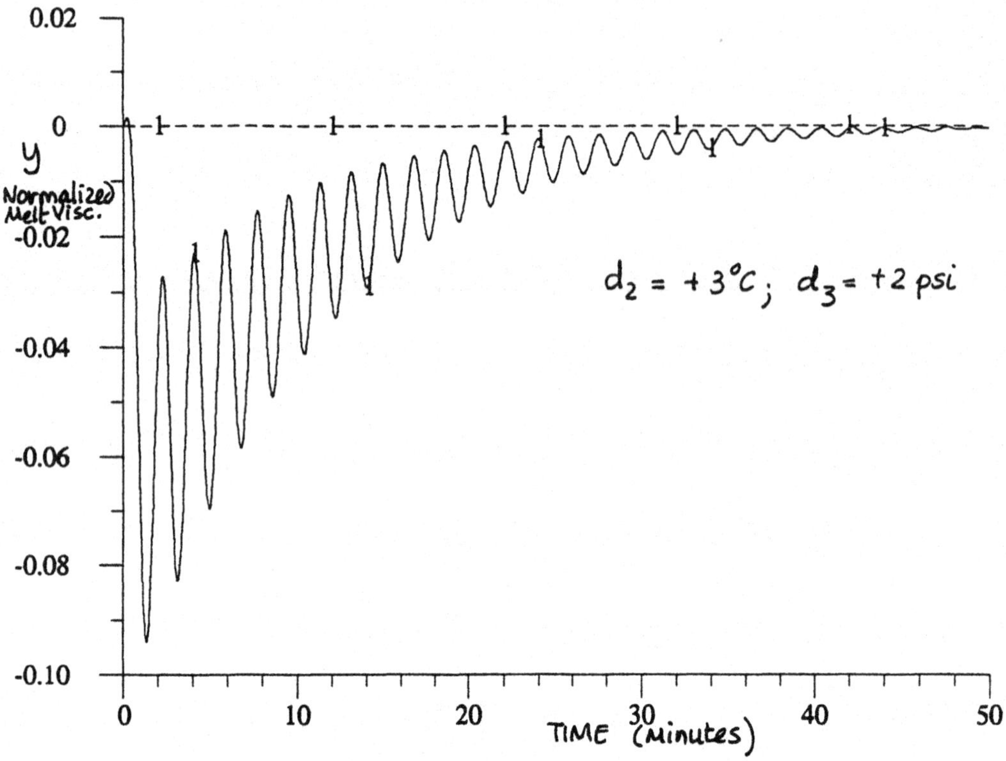

FIG S16.17

(c) Under the indicated cascade control scheme, the block diagram is modified as shown in FIG S16.18.

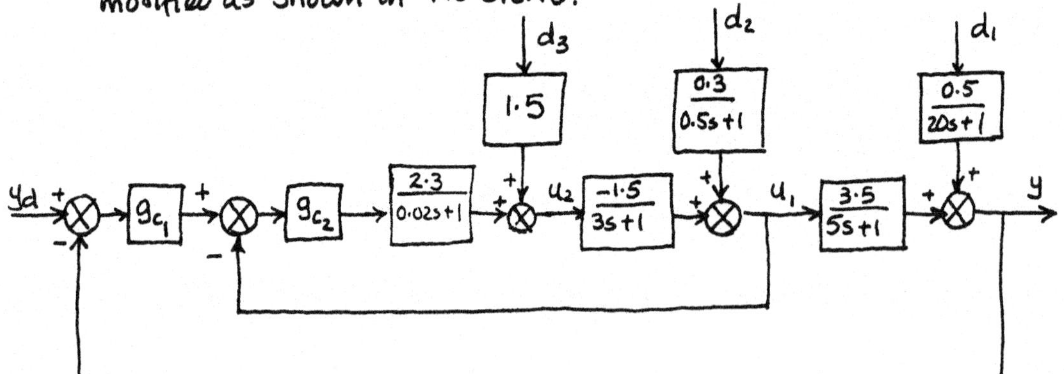

FIG S16.18 Block diagram for Polymer Process Under Cascade Control.

If now $g_{c_2} = K_{c_2} = -4.0$ we must now note that this controller, with its negative value has made the inner loop into a system with a POSITIVE gain. This implies that the controller in the outer loop (the original viscosity controller) MUST NOW HAVE A POSITIVE GAIN:

Thus, with $K_{c_1} = +15.0$; $\tau_I = 0.1$ (detuned from 1.01) for the VC (same as in part (b) except for the sign change!) the process response to the same inputs investigated in (b) is shown below in FIG S16.19.

Observe that not only is the maximum deviation from set-point significantly reduced (notice the scale for y is $\times 10^{-1}$) the overall behavior is better: the oscillations are gone. Thus the addition of this cascade loop has significantly improved process response.

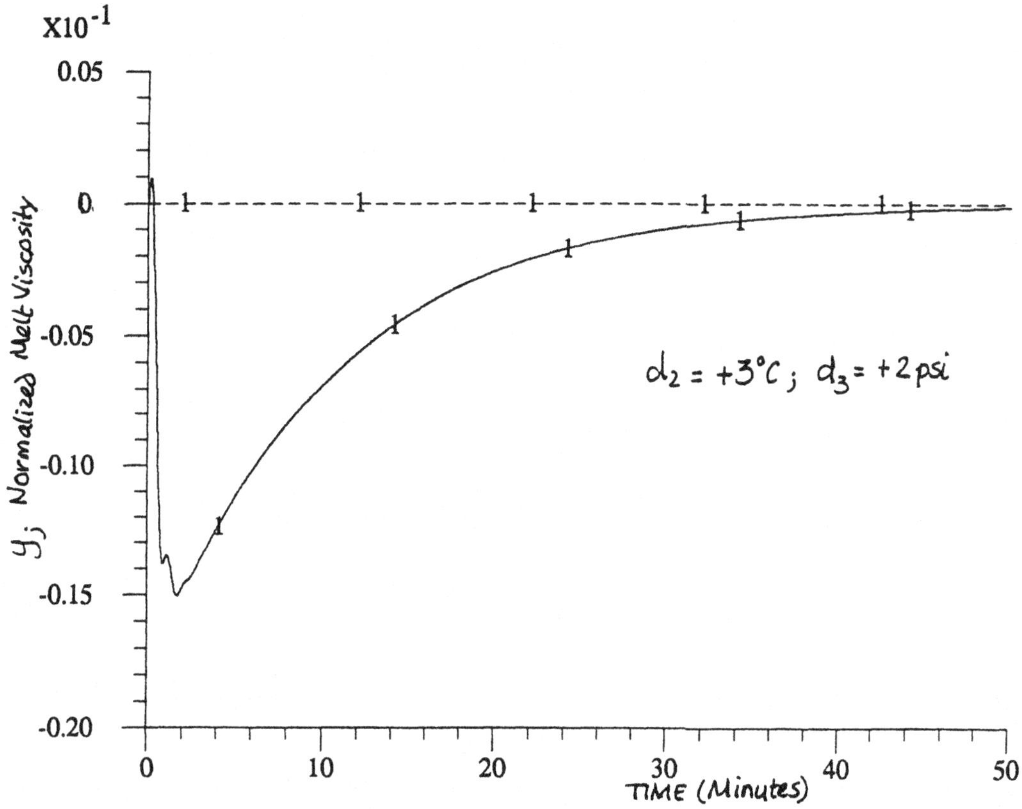

FIG S16.19

16.9 (a) The block diagram is shown below in FIG S16.20.

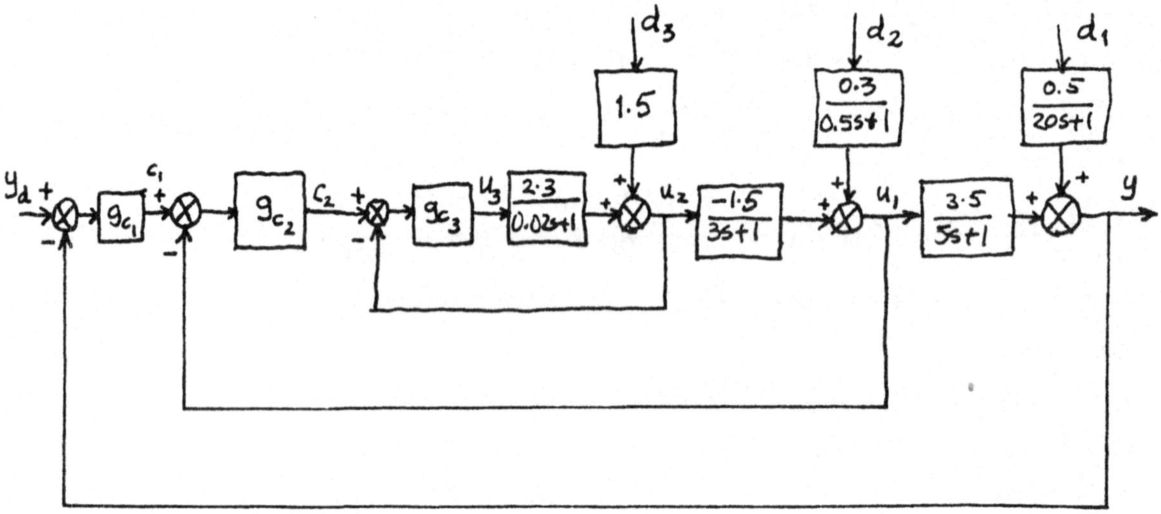

FIG S16.20 Block diagram for Polymer process under multilayered cascade control.

Having introduced c_1 and c_2 respectively for the control signals from g_{c_1} and g_{c_2}, we consolidate the internal loops one at a time, starting with the g_{c_3} loop to obtain:

$$u_2 = \frac{\dfrac{2.3 g_{c_3}}{0.02s+1}}{1 + \dfrac{2.3 g_{c_3}}{0.02s+1}} c_2 + \frac{1.5}{1 + \dfrac{2.3 g_{c_3}}{0.02s+1}} d_3$$

or

$$u_2 = \left(\frac{2.3 g_{c_3}}{0.02s + 1 + 2.3 g_{c_3}}\right) c_2 + \left[\frac{1.5(0.02s+1)}{0.02s + 1 + 2.3 g_{c_3}}\right] d_3 \qquad (16.20a)$$

or

$$u_2 = g_3^* c_2 + g_{d_3}^* d_3 \qquad (16.20b)$$

With this consolidation, the block diagram is now simplified to what is shown below in FIG S16.21

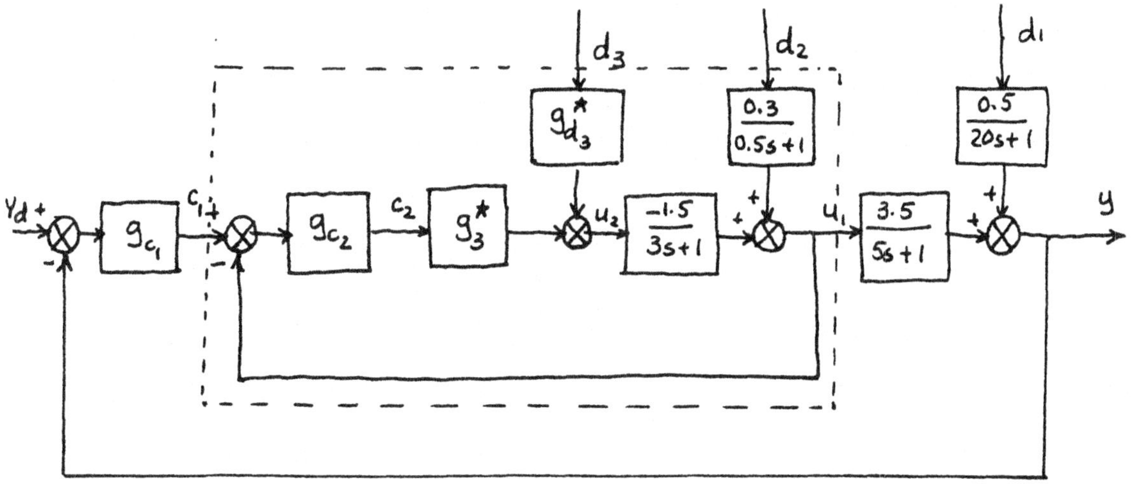

FIG S16.21 Polymer process block diagram of FIG S16.20 after consolidating first inner loop.

Next consolidate the newly consolidated single inner loop shown here in FIG S16.21, and obtain:

$$u_1 = \left[\frac{-1.5 g_3^* g_{c2}}{3s+1+(-1.5)g_3^* g_{c_2}}\right] c_1 + \left[\frac{-1.5 g_{d_3}^*}{3s+1+(-1.5)g_3^* g_{c_2}}\right] d_3 + \left[\frac{0.3(3s+1)}{(0.5s+1)(3s+1-1.5 g_3^* g_{c_2})}\right] d_2 \qquad (16.21a)$$

or

$$u_1 = g_1^* c_1 + g_{d_3}^{**} d_3 + g_{d_2}^* d_2 \qquad (16.21b)$$

The newly introduced transfer functions are as indicated above. With this the block diagram finally becomes:

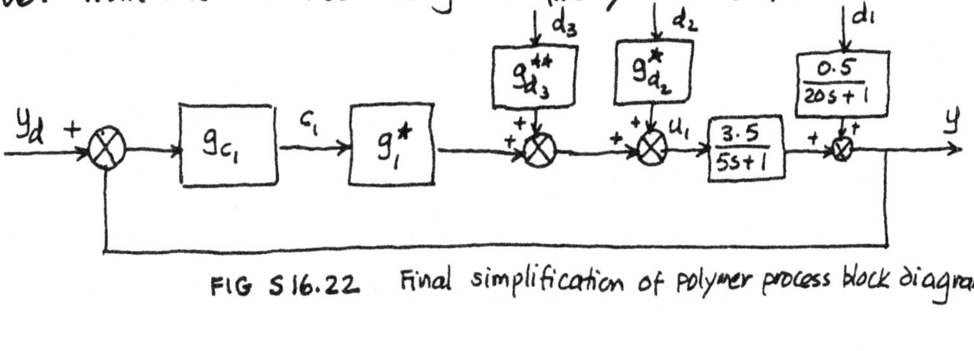

FIG S16.22 Final simplification of Polymer process block diagram.

The overall closed loop transfer function expression is now obtained as:

$$y = \left[\frac{3.5 g_1^* g_{c_1}}{5s + 1 + 3.5 g_1^* g_{c_1}}\right] y_d + \left[\frac{3.5 g_{d_3}^{**}}{5s + 1 + 3.5 g_1^* g_{c_1}}\right] d_3$$

$$+ \left[\frac{3.5 g_{d_2}^*}{5s + 1 + 3.5 g_1^* g_{c_1}}\right] d_2 + \left[\frac{0.5(5s+1)}{(20s+1)(5s+1+3.5 g_1^* g_{c_1})}\right] d_1 \qquad (16.22a)$$

or

$$y = \psi y_d + \psi_{d_1} d_1 + \psi_{d_2} d_2 + \psi_{d_3} d_3 \qquad (16.22b)$$

with the transfer functions ψ, ψ_{d_1}, ψ_{d_2}, ψ_{d_3} obvious from 16.22a and b.

(b) With $g_{c_3} = K_{c_3} = 0.5$ and retaining the two controllers of Problem 16.8, the response to the indicated disturbances is shown in FIG S16.23.

When compared with the response in FIG S16.17 (conventional feedback control) the current control scheme is seen to be clearly superior. When compared with FIG S16.19 (single cascade loop) we observe that the most obvious advantage of the double cascade control scheme is in the amount of time it takes to reject the disturbances — almost a 20 min advantage. The maximum deviation allowed is, however, somewhat in favor of the standard cascade control scheme (note that the y-axis scale in FIG S16.19 is $\times 10^{-1}$).

There is a complicated interaction between the flow control loop (with g_{c_3}) and the temperature control loop (with g_{c_2}) which is absent in standard cascade control. However, the double cascade scheme will always result in quicker response. (see FIG S16.24).

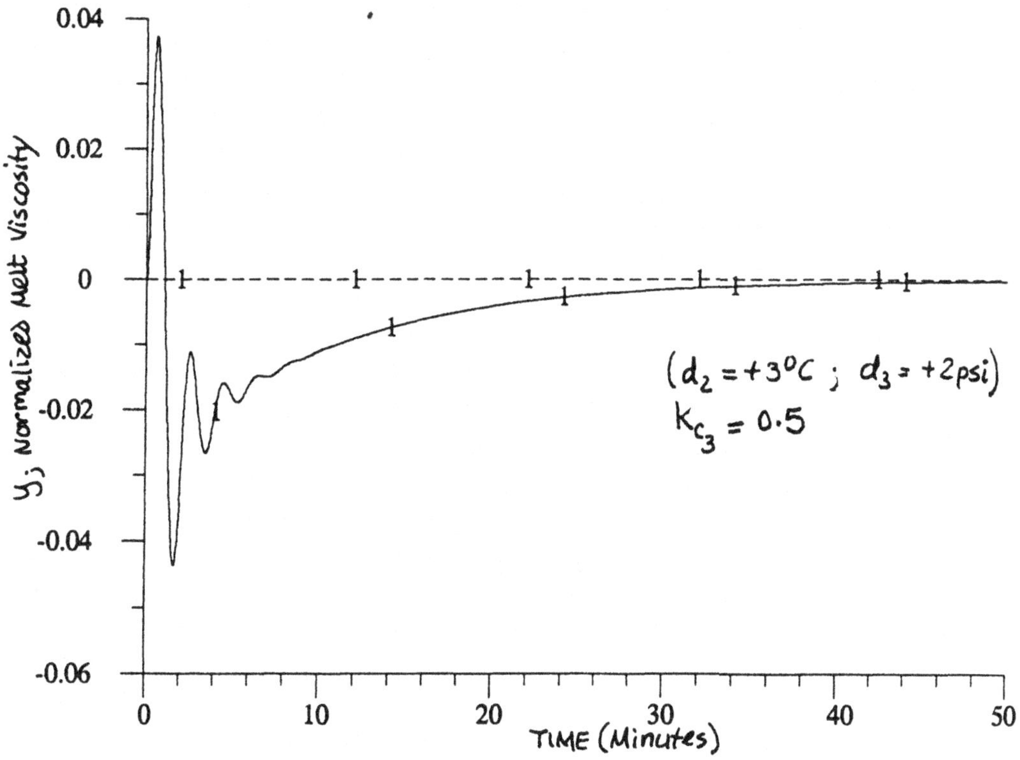

FIG S16.23

(c) The basic underlying philosophy of cascade control is what is coming into play here: with a single fast inner loop (for Temperature Control) disturbances in the cooling water supply temperature in particular can be rejected quicker before the polymer viscosity can be affected too drastically. But even the "inner" Temperature controller itself could benefit from its own cascade loop. In this case, disturbances in the water supply pressure can be contained by another fast flow controller so that the Temperature Controller can provide tighter temperature control.

As a further illustration of how much faster the double cascade scheme is able to reject disturbances by increasing the flow controller gain to $K_{c_3} = 10$ and retaining both the TC as well as the VC precisely as originally designed in Problem 16.8, the response to the same disturbances ($d_2 = +3°C$; $d_3 = +2$ psi) is shown below in FIG S16.24. The faster FC permits the observed much quicker rejection of the disturbances.

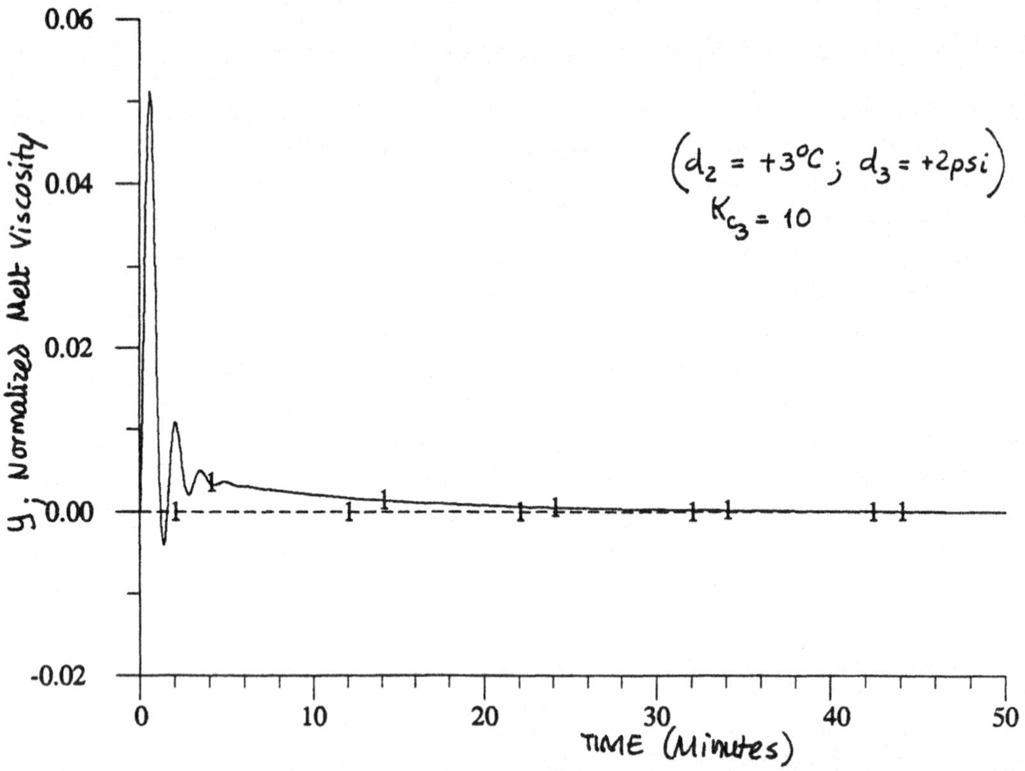

FIG S16.24

16.10 Because it is used to control production rate, the Vanadium catalyst flow rate is designated as the "wild stream." Clearly the co-catalyst flow is to be adjusted so that the desired fixed ratio, say r, is maintained.

The ratio control scheme may be designed as follows, using the configuration illustrated in Figure 16.12(b) in the text (the more common, and less problematic configuration).

(i) Measure Vanadium Catalyst Flow;
(ii) Multiply by the desired ratio, at a ratio station;

(iii) Use the result thus produces as the set-point for the Aluminum cocatalyst flow rate

(iv) Design an appropriate flow controller to receive this Setpoint and manipulate the valve on the cocatalyst line.

The process diagram for such a scheme is shown below in FIG S16.25.

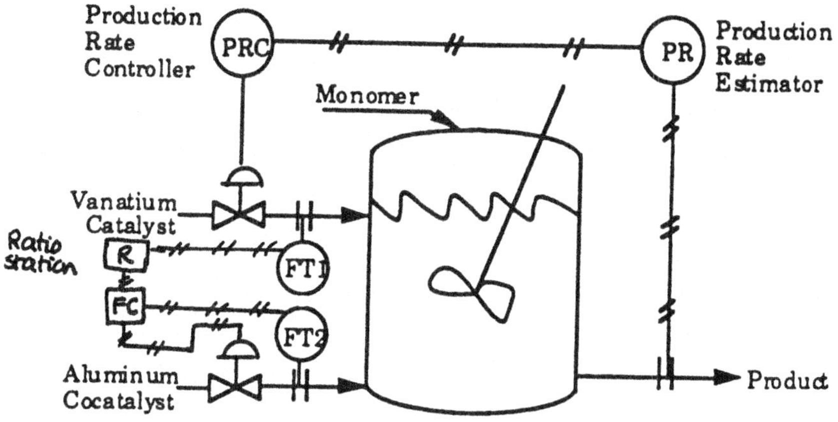

FIG S16.25 Polymerization process of Figure P16.8 modified to indicate Ratio control scheme.

16.11 (a)

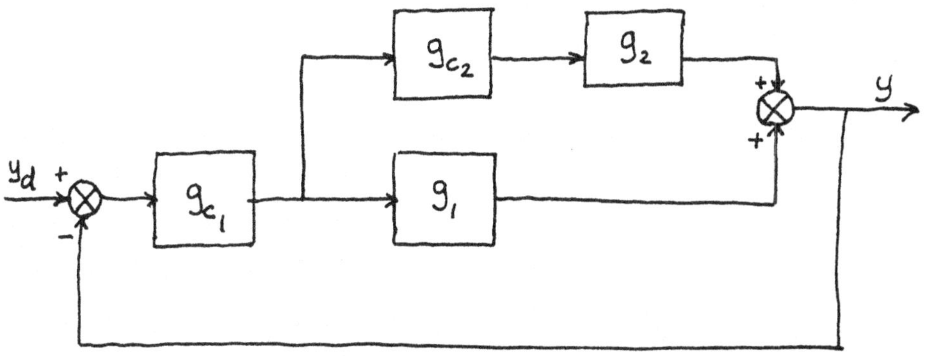

FIG S16.26 Block diagram for the Heat Exchanger control system.

(b) (i) The disturbance causes T (the mixed stream temperature) to drop; TC1, (a proportional-only controller) responds by closing off the bypass valve by an amount proportional to the observed error in T;

(ii) The TC1 control action causes a rise in T which is almost instantaneous; this same action is communicated to TC2

(iii) TC2 responds by increasing steam flow to the heat exchanger, which will cause T to increase but on a slower time scale.

(iv) As the TC2 actions begin to take effect, TC1 responds by opening up the bypass valve again until ultimately the bypass value is restored to its original starting value of 50%.

ULTIMATE EFFECT: TC1 provides instantaneous response, passes this information on to TC2 and readjusts as TC2 takes ultimate responsibility for the process regulation.

This parallel scheme will be more effective because it permits almost "perfect", almost instantaneous disturbance rejection; a scheme based on TC2 alone is limited by the slow dynamics associated with the effect of steam flow rate changes on T.

16.12 (a) With the model transfer function as

$$y_1(s) = \frac{-4e^{-0.1s}}{15s+1} \qquad (16.23)$$

we obtain the recommended Z-N PI settings from Table 15.2 as

$$K_c = \frac{0.9}{K}\left(\frac{\tau}{\alpha}\right) = \frac{0.9}{-4}\left(\frac{15}{0.1}\right) = -33.75 \; ; \quad \tau_I = 0.33$$

The control system response to a 1°F step change in feed temperature is shown below in FIG S16.27.

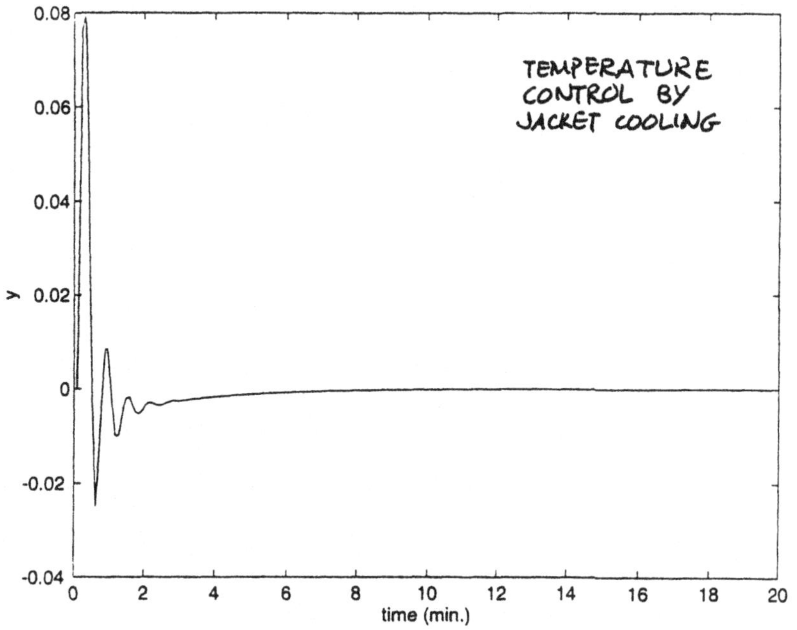

FIG S16.27

(b) Using the condenser unit to achieve temperature control implies that the appropriate model transfer function is

$$g_2 = \frac{-0.5e^{-.1s}}{2s+1} \qquad (16.24)$$

and the Z-N recommended settings are obtained as

$$K_c = \frac{0.9}{-0.5}\left(\frac{2}{0.1}\right) = -36.0$$

$$\tau_I = 0.333$$

In this case, the closed loop system response to the same 1°F step change in feed temperature, using the overhead condenser for temperature control, is shown in FIG S16.28.

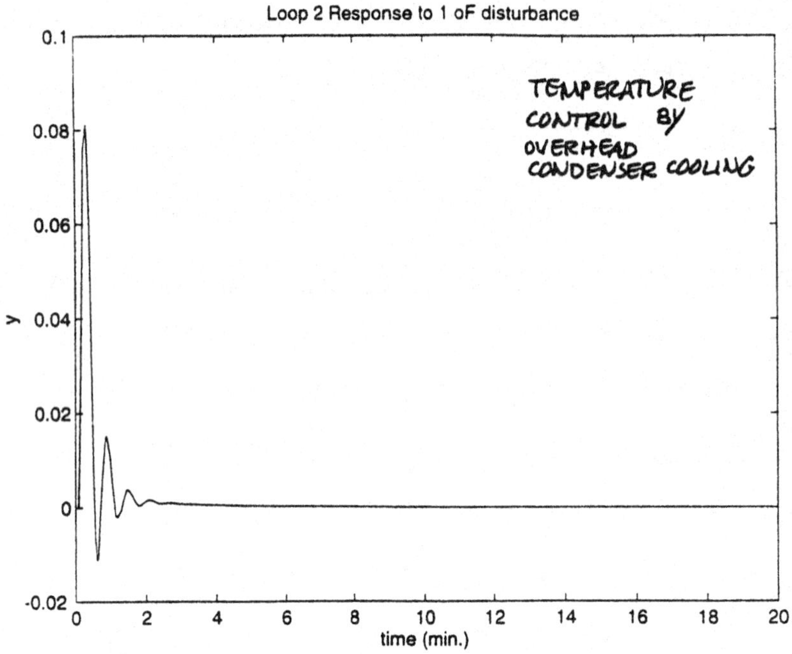

FIG S16.28

(c) when the two controllers are combined and used simultaneously, the overall process block diagram is as shown below in FIG S16.29

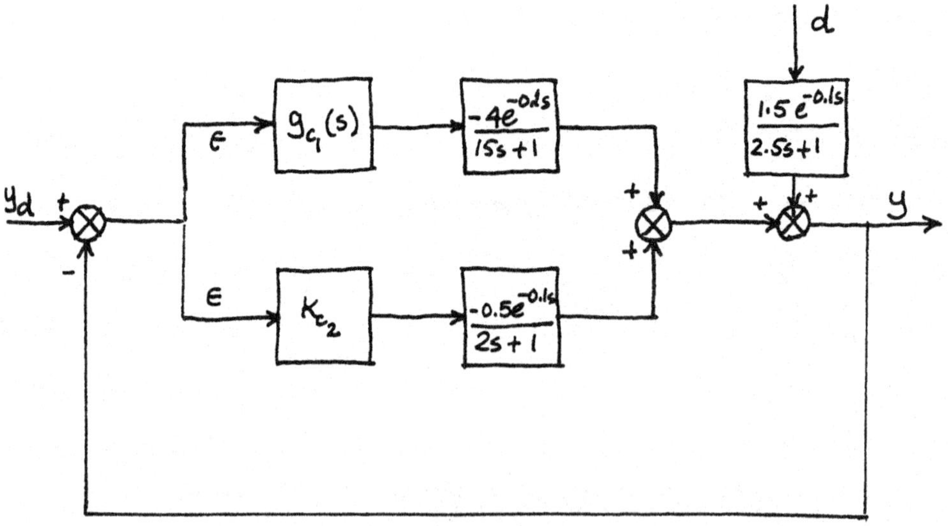

FIG S16.29 Exothermic reactor temperature control under "parallel control"

Now, with g_{c_1} as in part (a), i.e. $K_{c_1} = -33.75$; $\tau_{I_1} = 1/3$, <u>and</u> with g_{c_2} as the proportional part of the controller in part (b), i.e. $g_{c_2} = K_{c_2} = -36.0$, the closed loop system response under this simultaneous condenser <u>and</u> jacket cooling control is shown in FIG S16.30 where we see clearly that the system is unstable.

Intuitively, we should expect that if two individually aggressive controllers are to be combined, especially as in FIG S16.29 with the overall action being cumulative, this might lead to instability.

Mathematically, with g_1 and g_2 as defined in (16.23) and (16.24), the closed loop characteristic equation for the system under parallel control is:

$$1 + g_1 g_{c_1} + g_2 g_{c_2} = 0 \qquad (16.25)$$

and one should expect that the g_{c_1} which guarantees that

$$1 + g_1 g_{c_1} = 0$$

has stable roots (for jacket cooling control <u>by itself</u>), and the g_{c_2} which guarantees that

$$1 + g_2 g_{c_2} = 0$$

has stable roots (for condenser cooling <u>by itself</u>) will in fact not necessarily guarantee that (16.25) will have no unstable roots. FIG S16.30 shows that these controller parameters in fact make the system unstable.

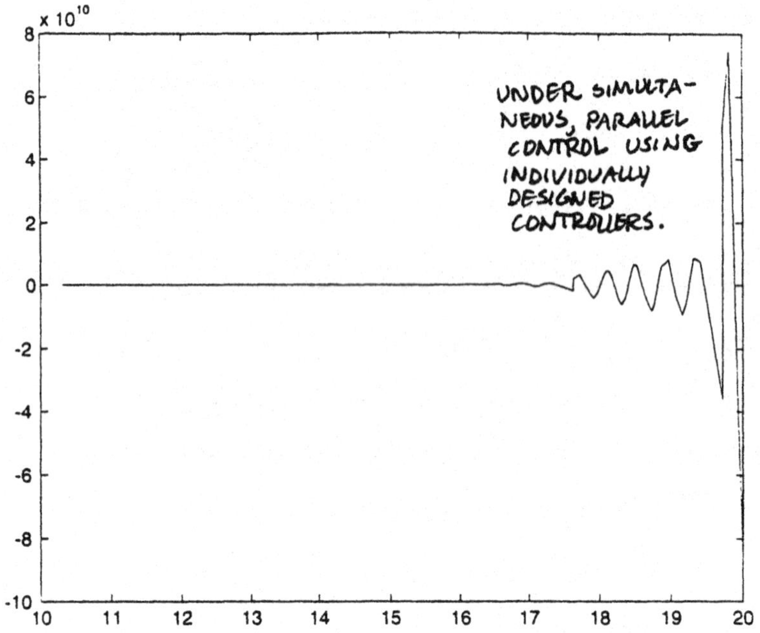

FIG S16.30

The appropriate procedure is to choose K_{c_1}, τ_{I_1} and K_{c_2} using (16.25) (or by any other means which takes the parallel and cumulative effect of the two controllers into consideration).

A straightforward way of detuning the controllers is simply to cut the gains back. For example, by cutting the gains in half, ($K_{c_1} = -16.875$; $K_{c_2} = -18.0$) the overall closed loop system response becomes stable and the actual performance is shown in FIG S16.31.

The corresponding control actions for u_1 (the jacket cooling water rate) and u_2 (condenser cooling water rate) are shown in FIGS S16.32 and S16.33 respectively. Note carefully that u_2 returns to zero while u_1 settles to a value of $(1.5/4)$ the ratio of the gains associated with g_d and g_1.

The overall response can be speeded up by increasing K_2.

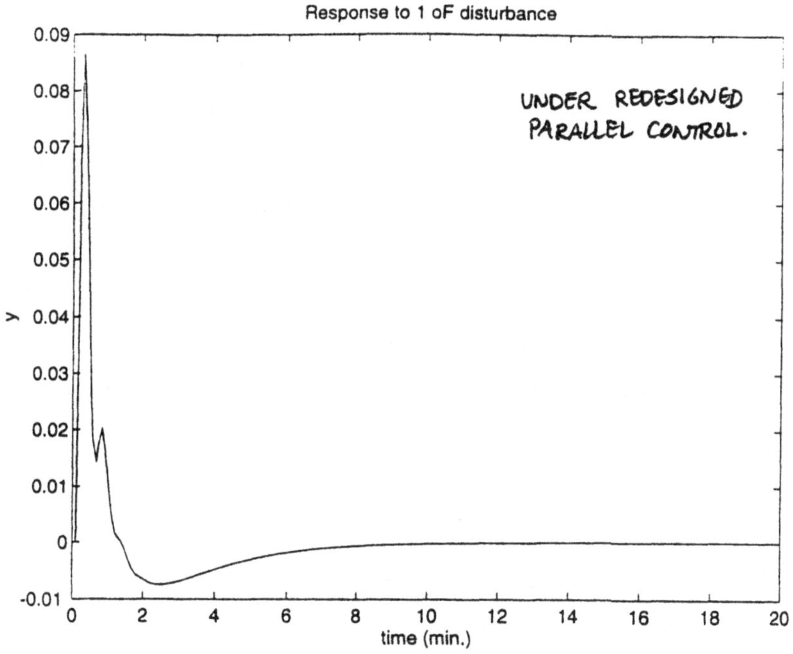

FIG S16.31

The control action taken by each controller is shown below.

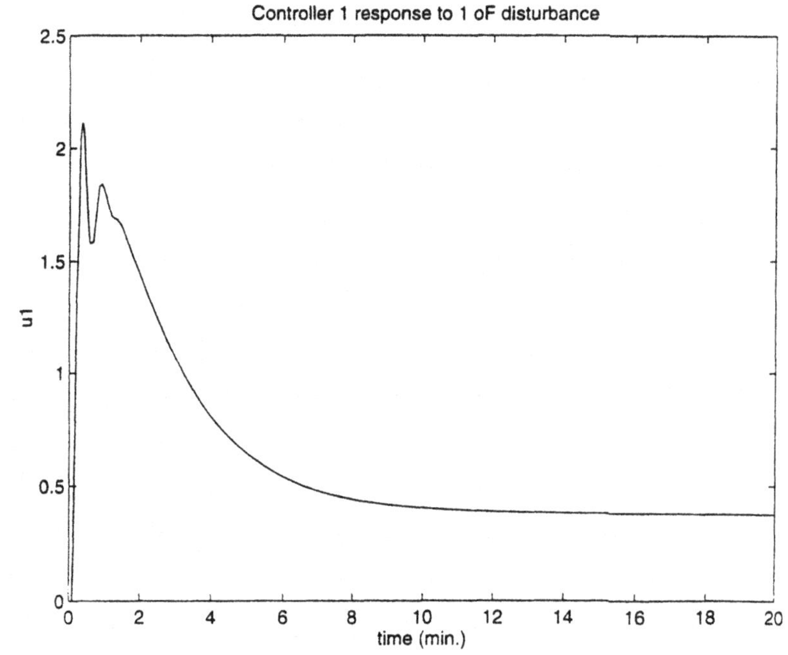

FIG S16.32

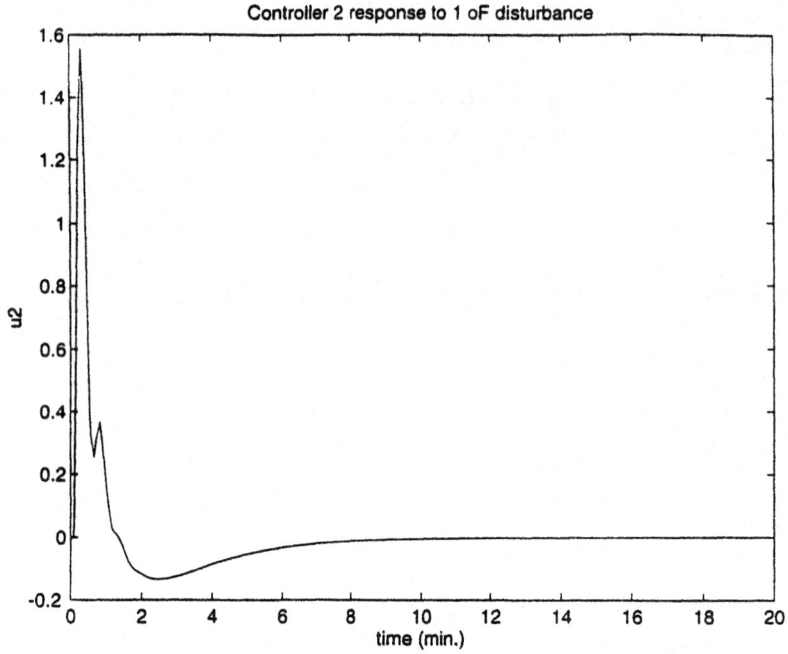

Controller 2 response to 1 oF disturbance
u2
time (min.)
1.6
1.4
1.2
1
0.8
0.6
0.4
0.2
0
-0.2
0
2
4
6
8
10
12
14
16
18
20

FIG S16.33

CHAPTER 17

17.1 The given process transfer function

$$g(s) = \frac{5.0e^{-10.0s}}{15.0s+1} \qquad (17.1)$$

implies $K = 5.0$; $\tau = 15.0$; $\alpha = 10$

(a) PI controller parameters (Cohen-Coon) obtained from Table 15.3 as:

$$K_c = 0.2867$$
$$\tau_I = 14.328$$

PID controller parameters (Cohen-Coon) obtained from Table 15.3, last set of entries, as:

$$K_c = 0.45; \quad \tau_I = 19.636; \quad \tau_D = 3.243$$

The closed loop system response to a unit step change in set-point is shown in FIG S17.1 for the PI controller; the corresponding response for the PID controller is shown in FIG S17.2.

The PID controller results in quicker settling, but at the expense of slightly increased initial overshoot (Note the different y-axis scale). The PI controller results in "smoother" system response, however. The overall performances are quite comparable.

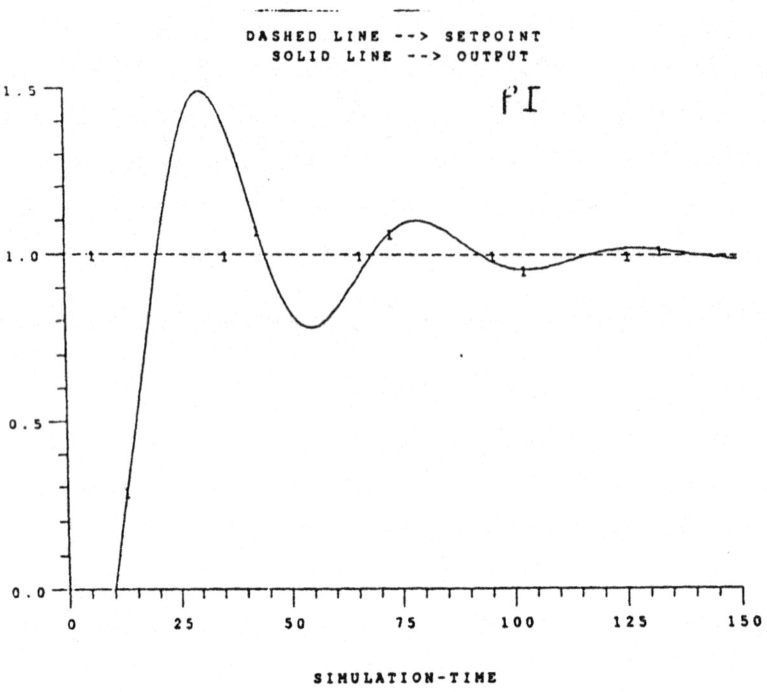

FIG S17.1 Closed Loop Response PI: (Cohen-Coon)

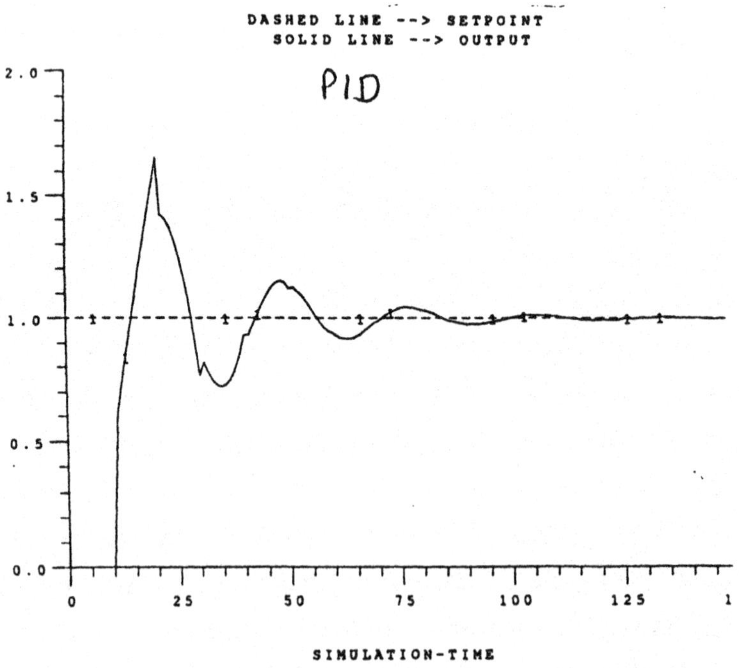

FIG S17.2 Closed Loop Response PID: (Cohen-Coon)

(b) Obtain the required ITAE tuning rules from Table 15.4, for set-point changes.
For PI control,

$$K_c = 0.1699 \; ; \quad \tau_I = 16.304$$

For PID control,

$$K_c = 0.273 \; ; \quad \tau_I = 21.49 \; ; \quad \tau_D = 3.17$$

The closed loop responses for each of these controllers are shown in FIGS S17.3 and S17.4 respectively for PI and PID. Both responses are comparable, with the PID response exhibiting no overshoot, while the PI response is "smoother".

Overall, compared to the Cohen-Coon controllers of part (a), the ITAE controllers give rise to far less system oscillation. This is consistent with the ITAE optimization objective.

17.2 (a) From Fig 17.4 in the text, the transfer function to be implemented in the Smith predictor minor loop is obtained as:

$$g^*(1 - e^{-10s})$$

where, in this case

$$g^* = \frac{5}{15s+1}$$

so that

$$g_{minor\,loop} = \left(\frac{5}{15s+1}\right)\left(1 - e^{-10s}\right) \qquad (17.2)$$

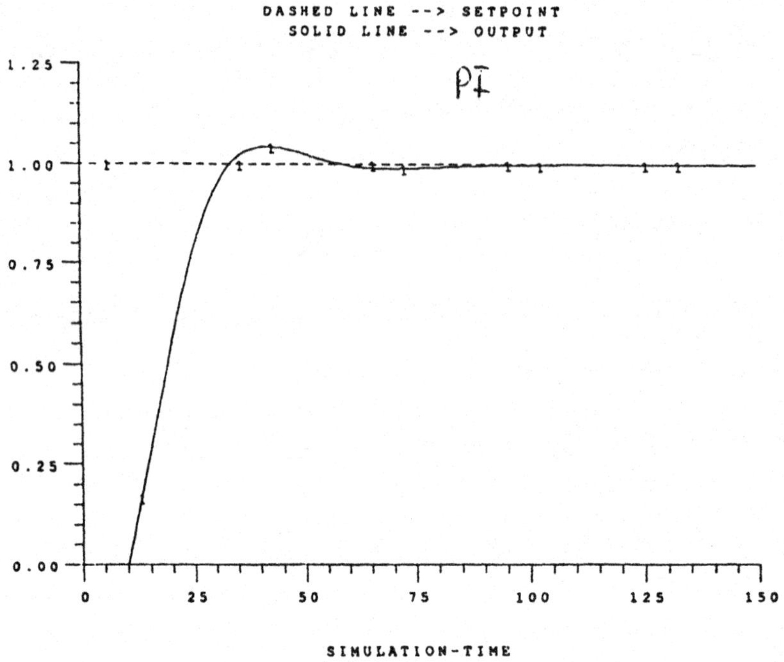

FIG S17.3 Closed Loop Response PI (ITAE)

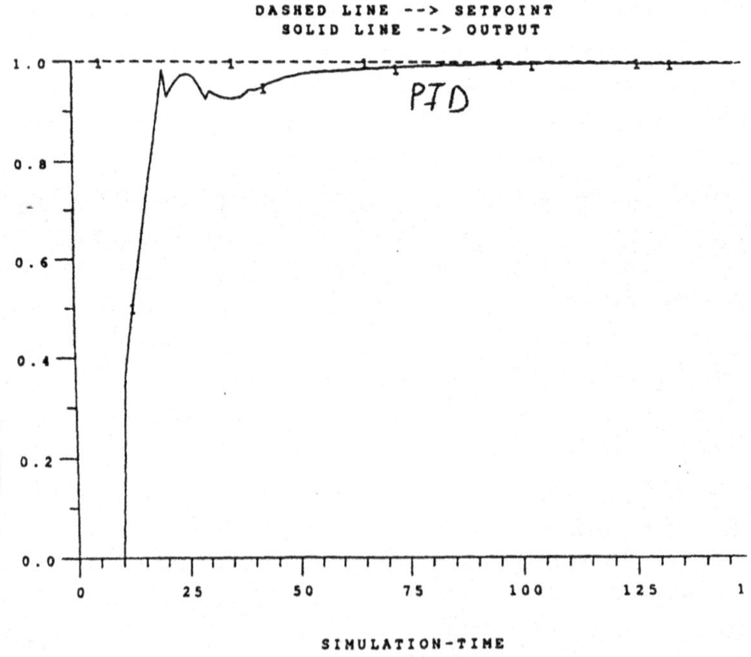

FIG S17.4 Closed Loop Response PID (ITAE)

(b) Using a PI controller with $K_c = 0.3$, $\tau_I = 15$, in conjunction with the $g_{minor\ loop}$ given above in the Smith predictor's minor loop, the overall closed loop system response to a unit step change in the set-point, is shown in FIG S 17.5 below.

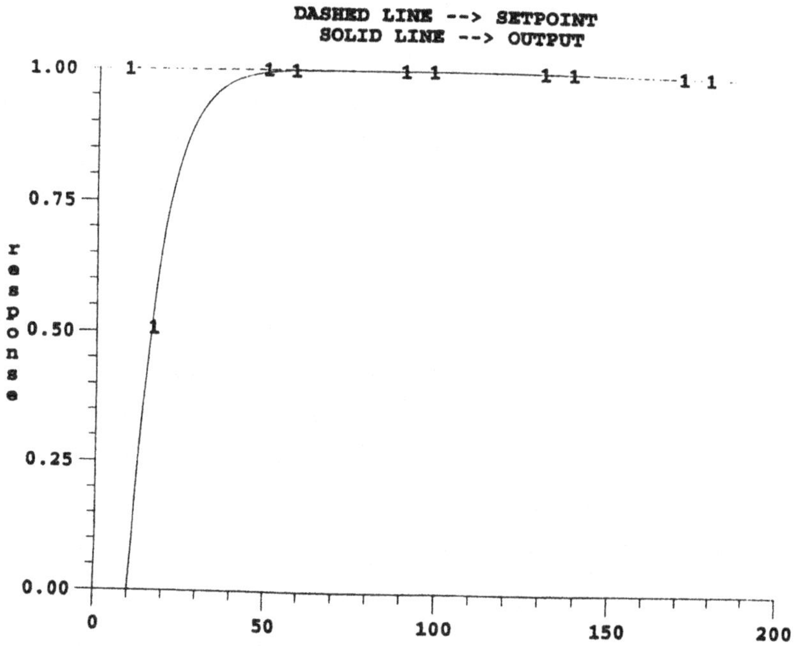

FIG S 17.5 Closed loop response: Smith Predictor Perfect Model.

(c) Given a "true" process transfer function

$$g(s) = \frac{5.5e^{-10s}}{15.0s + 1}$$

using the same controller and Smith predictor of part (b), the closed loop response to a unit step change in set-point is shown in FIG S17.6
This closed loop response is qualitatively still similar to that obtained in part (b) above.

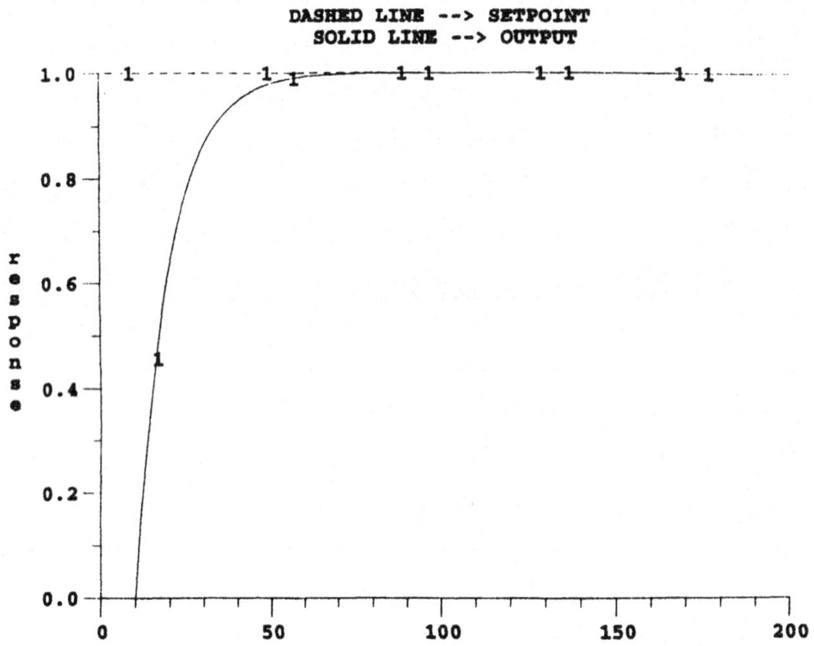

FIG S17.6 Smith Predictor; Process Gain error

(d) With a "true" transfer function

$$g(s) = \frac{5.0e^{-11.0s}}{15.0s + 1}$$

the required closed loop response is shown in FIG S17.7; Apart from the obvious increase of 1 min in the initial delay, this closed loop response is also qualitatively similar to that obtained in part (b).

(e) With the "true" transfer function now as

$$g(s) = \frac{5.0e^{-10.0s}}{16.5s + 1}$$

the closed loop response is shown in FIG S17.8. There is a slight overshoot indicated in this response; otherwise, it, too is essentially similar to all the other responses.

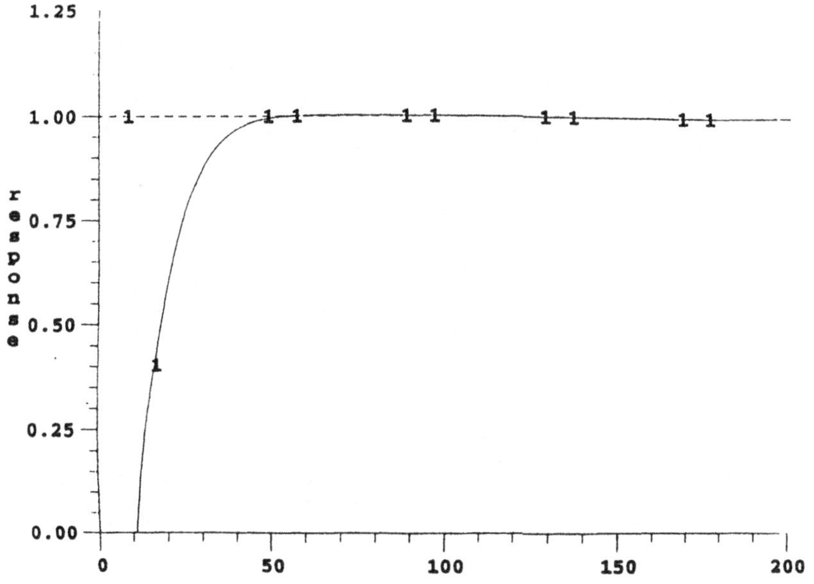

FIG S 17.7 Smith Predictor; Time delay error.

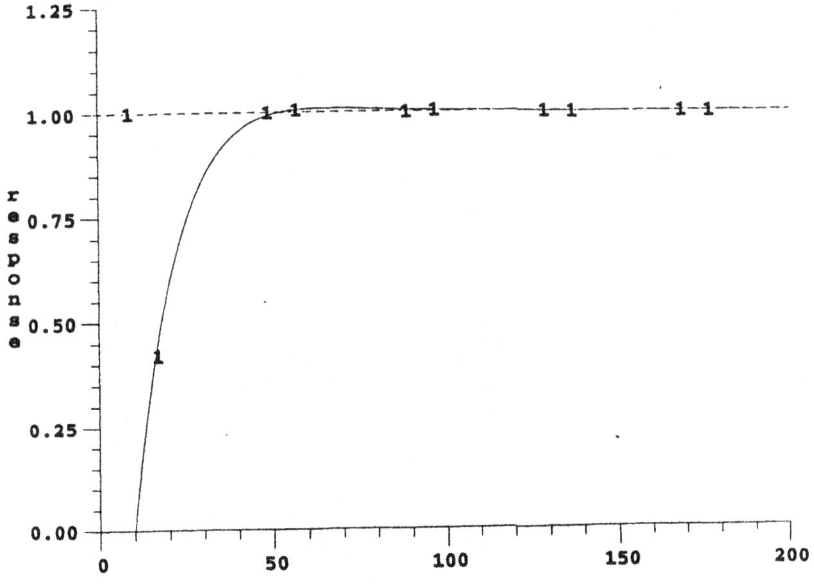

FIG S 17.8 Smith Predictor; Time constant error.

These responses indicate that if the controller used in the Smith predictor is not too aggressive, such relatively minor errors in process parameters do not cause significant performance degradation.

17.3 (a) For this specific process, the characteristic equation,

$$1 + gg_c = 0$$

becomes

$$1 + \frac{2K_c(1-4s)}{(2s+1)(5s+1)} = 0 \qquad (17.3)$$

which rearranges to:

$$10s^2 + s(7-8K_c) + (2K_c+1) = 0 \qquad (17.4)$$

The two roots of this quadratic equation will both have negative real parts if

$$(7-8K_c) > 0 \quad \Rightarrow \quad K_c < 7/8$$

and

$$(2K_c+1) > 0 \quad \Rightarrow \quad K_c > -1/2$$

The range of K_c values required for stability is now obtained as

$$\boxed{-1/2 < K_c < 7/8} \qquad (17.5)$$

(b) With $K_c = 0.5$, the closed loop transfer function relation:

$$y = \frac{gg_c}{1+gg_c} y_d$$

becomes

$$y = \frac{0.5 \times 2(1-4s)/(2s+1)(5s+1)}{1 + 0.5\times2(1-4s)/(2s+1)(5s+1)} \, y_d$$

or.

$$y = \frac{(1-4s)}{(2s+1)(5s+1)+(1-4s)} \cdot \frac{1}{s}$$

upon introducing $1/s$ for $y_d(s)$ in order to evaluate the step response. The final value is obtained (using the Final Value Theorem of Laplace Transform) as.

$$\lim_{t \to \infty} y(t) = \lim_{s \to 0} s \left\{ \frac{(1-4s)}{(2s+1)(5s+1)+(1-4s)} \cdot \frac{1}{s} \right\}$$

$$= \frac{1}{2}$$

Since the desired set-point is 1, the steady-state offset will be: $1 - 1/2 = 0.5$

(c) From the block diagram in Fig 17.10 in the text, with $g_c = K_c$, and $g'(s)$ as given in Eq. (P17.6) the "effective controller" transfer function, say g_c^*, is given by

$$g_c^* = \frac{K_c}{1 + K_c g'}$$

$$g_c^* = \frac{(2s+1)(5s+1)K_c}{(2s+1)(5s+1) + 10K_c s} \qquad (17.6)$$

And the overall closed-loop transfer function is given by

$$CLTF = \frac{g g_c^*}{1 + g g_c^*}$$

From where the characteristic equation is obtained as:

$$1 + gg_c^* = 0$$

$$1 + \frac{2(1-4s)}{\cancel{(2s+1)(5s+1)}} \cdot \frac{\cancel{(2s+1)(5s+1)}K_c}{(2s+1)(5s+1) + 10K_c s} = 0$$

or:

$$10s^2 + (7 + 10K_c - 8K_c) + 1 + 2K_c = 0$$

For stability, all the roots of this quadratic equation must have negative real parts, which will be true provided all the coefficients are positive; i.e.

$$7 + 2K_c > 0 \quad \Rightarrow \quad K_c > -3.5$$

and

$$1 + 2K_c > 0 \quad \Rightarrow \quad K_c > -0.5$$

Thus, <u>any</u> positive value of K_c will result in a stable closed-loop system.

(d) The apparent controller is shown in Eq. (17.6) above, from where we observe that

$$\lim_{K_c \to \infty} g_c^* = \frac{(2s+1)(5s+1)\cancel{K_c}}{10\cancel{K_c}s}$$

$$= \frac{10s^2 + 7s + 1}{10s}$$

$$= 0.7\left[1 + \frac{1}{7s} + \frac{1}{0.7}s\right] \qquad (17.7)$$

a PID controller with $K_c = 0.7$, $\tau_I = 7$, $\tau_D = 1/0.7$.

(e) The required closed-loop system response is shown below in FIG S17.9.

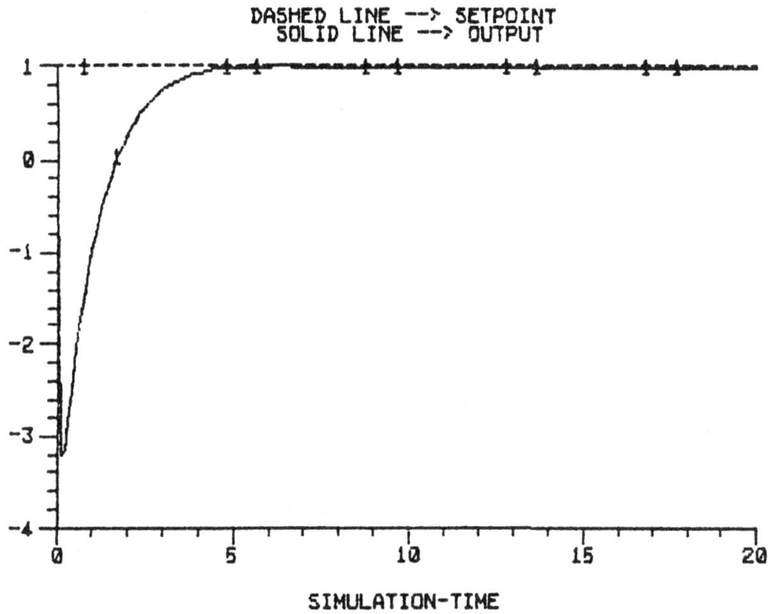

FIG S 17.9 PID Control of IR System.

17.4 The consolidated process transfer function is:

$$g(s) = \frac{3(s^2-3s+2)}{7s^3+29s^2+25s+3}$$

(a) The required root locus diagram is shown in FIG S 17.10. Note the two zeros located in the Right half of the complex plane.

From the program used to generate the root locus diagram, obtain that $K_c = 2.4036$ when the root locus crosses the imaginary axis; also obtain the roots at this point as $0 \pm 0.6936j$ (the other root is stable).

Thus the range of K_c for stability is $0 < K_c < 2.4036$; the frequency of oscillation at the verge of instability is $\omega = 0.6936$.

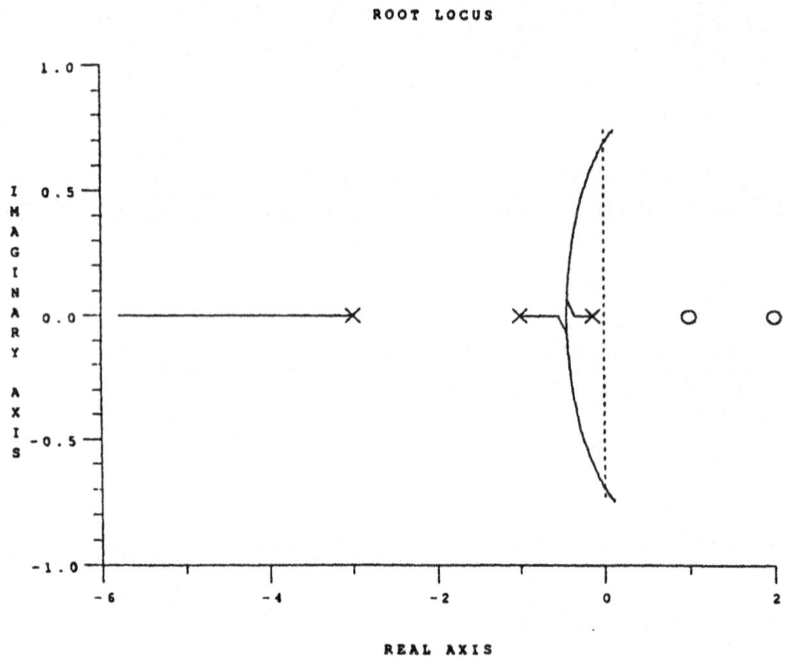

FIG S 17.10 Root Locus Diagram.

(b) Use Ziegler-Nichols tuning rules, introducing the stability information obtained in part (a).

$$K_u = 2.4036 \Rightarrow K_c = K_u/2.22 = 1.08$$

$$P_u = 2\pi/\omega_{co} = 2\pi/0.6936 = 9.058$$

$$\Rightarrow \tau_I = P_u/1.2 = 7.549.$$

The closed loop response using this PI controller, is shown in FIG S17.11. Given that the process has two right half plane zeros, the closed loop performance is quite reasonable, even if somewhat oscillatory. (Controllers designed on the basis of Z-N tuning rules tend to be naturally oscillatory.)

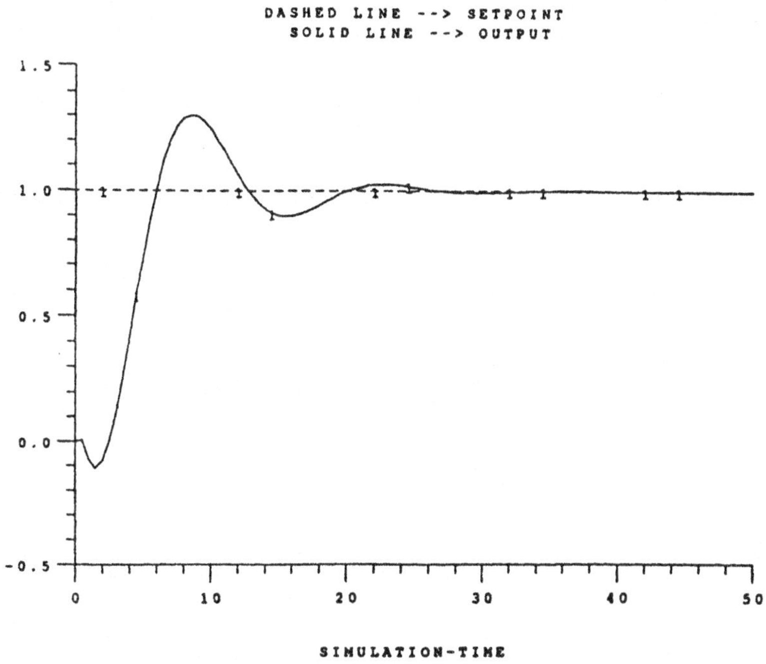

FIG S 17.11

(c) As in the discussion in Section 17.3.3 in the text, let the IR compensator be represented by $g'(s)$. The objective is to choose g' such that

$$g(s) + g'(s) = g^*(s)$$

has no RHP zero. In this case

$$g(s) = g^0(s)\cdot(1-\eta_1 s)(1-\eta_2 s) \qquad (17.8)$$

where

$$g^0(s) = \frac{1}{(\tau_1 s+1)(\tau_2 s+1)(\tau_3 s+1)} \qquad (17.9)$$

Again, as in Eq (17.31) in the text, choose $g'(s)$ as $g^0(s)\lambda s$, where $g^0(s)$ is as given in Eq (17.9) above. Under these conditions, we have

$$g^*(s) = g^o(s)\left[(1-\eta_1 s)(1-\eta_2 s) + \lambda s\right]$$

or

$$g^*(s) = g^o(s)\left\{1 + [\lambda - (\eta_1+\eta_2)]s + \eta_1\eta_2 s^2\right] \qquad (17.10)$$

For $g^*(s)$ not to have a RHP zero, we now require that

$$\lambda - (\eta_1+\eta_2) \geq 0$$

or

$$\boxed{\lambda \geq \eta_1 + \eta_2} \qquad (17.11)$$

(cf. Eq (17.39) in the text). Thus, the required IR compensator is given by

$$g'(s) = \frac{\lambda s}{(\tau_1 s+1)(\tau_2 s+1)(\tau_3 s+1)} \qquad (17.12)$$

with λ satisfying the condition in Eq (17.11) above.

17.5 (a) Under proportional feedback control, the characteristic equation is:

$$1 + \frac{5K_c(1-0.5s)}{(2s+1)(0.5s+1)} = 0$$

which reduces to

$$s^2 + (2.5 - 2.5K_c)s + 1 + 5K_c = 0$$

Stability requires

$$2.5 - 2.5K_c > 0 \quad \Rightarrow \quad K_c < 1$$

and

$$1 + 5K_c > 0 \quad \Rightarrow \quad K_c > -1/5$$

The required K_c range is therefore

$$-1/5 < K_c < 1 .$$

(b) With the new controller, the characteristic equation becomes

$$1 + \frac{5K_c(0.5s+1)(1-0.5s)}{(\tau_L s+1)(2s+1)(0.5s+1)} = 0$$

or

$$2\tau_L s^2 + (2+\tau_L - 2.5K_c)s + 5K_c = 0$$

and now stability requires (apart from $K_c > 0$),

$$2 + \tau_L - 2.5K_c > 0$$

or $$K_c < \frac{2+\tau_L}{2.5} \qquad (17.13)$$

In order to obtain a stability range $0 < K_c < 4$ choose τ_L in (17.13) such that

$$\frac{2+\tau_L}{2.5} = 4$$

$$\Rightarrow \boxed{\tau_L = 8} \qquad (17.14)$$

(c) The closed-loop transfer function in this case is given by

$$\psi(s) = \frac{5K_c(1-0.5s)}{(8s+1)(2s+1) + 5K_c(1-0.5s)}.$$

$$\lim_{s\to 0} \psi(s) = 5K_c/(1+5K_c) \qquad (17.15)$$

and regardless of specific values chosen for τ_L or τ_z, there will be a non-zero offset.

17.6 The characteristic equation is:

$$1 + \frac{2g_c}{(5s+1)(2s-1)} = 0 \qquad (17.16)$$

1. For $g_c = K_c$, proportional-only control, (17.16) becomes

$$10s^2 - 3s + (2K_c - 1) = 0 \qquad (17.17)$$

and for all values of K_c the system is unstable because of the presence of -3 as the coefficient of s.

For a PI controller, $g_c = K_c\left(1 + \frac{1}{\tau_I s}\right)$, (17.16) becomes

$$10\tau_I s^3 - 3\tau_I s^2 + (2K_c - 1)\tau_I s + 1 = 0 \qquad (17.18)$$

and, irrespective of τ_I, no value of K_c will stabilize this system, ESTABLISHING POINT 1.

2. For $g_c = K_c(1 + \tau_D s)$, a PD controller, (17.16) becomes

$$10s^2 + (2K_c\tau_D - 3)s + 2K_c = 0 \qquad (17.19)$$

and now for $2K_c\tau_3 > 3$, the system is stabilized.

Similarly, for the PID controller, $g_c = K_c\left(1 + \frac{1}{\tau_I s} + \tau_D s\right)$ (17.16) becomes

$$10\tau_I s^3 + \tau_I(2K_c\tau_D - 3)s^2 + \tau_I(2K_c - 1)s + 2K_c = 0 \qquad (17.20)$$

Now, given a cubic equation

$$a_0 s^3 + a_1 s^2 + a_2 s + a_3 = 0 \qquad (17.21)$$

one may use the Routh array to establish that it will have no roots in the RHP if, and only if,

$$a_0, a_1, a_2, a_3 > 0 \qquad (17.22a)$$

and

$$a_1 a_2 > a_0 a_3 \qquad (17.22b)$$

Employing these results on (17.20), we obtain the following as the required conditions for stability:

(i) $a_0 > 0 \Rightarrow \tau_I > 0$ (trivially satisfied)

(ii) $a_1 > 0 \Rightarrow \tau_I (2K_c\tau_D - 3) > 0$

or

$$\boxed{\tau_D > \frac{3}{2K_c}.} \qquad (17.23)$$

(iii) $a_2 > 0 \Rightarrow \tau_I (2K_c - 1) > 0$

or

$$\boxed{K_c > 1/2} \qquad (17.24)$$

(iv) $a_3 > 0 \Rightarrow K_c > 0$, redundant in the face of Eq (17.24) above

(v) $a_1 a_2 > a_0 a_3 \Rightarrow \tau_I^2 (2K_c\tau_D - 3)(2K_c - 1) > 20\tau_I K_c$

Because $\tau_I > 0$, this condition rearranges to give

$$\boxed{\tau_I > \frac{20K_c}{(2K_c\tau_D - 3)(2K_c - 1)}.} \qquad (17.25)$$

thus confirming <u>all</u> the statements.

17.7 Solution not provided.

17.8 For a time delay process with transfer function

$$g(s) = g^* e^{-\alpha s}$$

the transfer function to be implemented in the minor loop of the Smith Predictor, as shown in Fig 17.4 in the text is

$$g_K = g^*(1 - e^{-\alpha s}) \tag{17.26}$$

With a first-order Padé approximation for $e^{-\alpha s}$, (17.26) above becomes

$$g_K^{Pade} = g^*\left(1 - \frac{1-\frac{\alpha}{2}s}{1+\frac{\alpha}{2}s}\right)$$

$$\boxed{g_K^{Pade} = g^*\left(\frac{\alpha s}{1+\frac{\alpha}{2}s}\right)} \tag{17.27}$$

Now, if we start with the time delay process itself and approximate the delay with a 1st order Padé approximation, we obtain

$$g(s) = g^*\left(\frac{1-\frac{\alpha}{2}s}{1+\frac{\alpha}{2}s}\right) \tag{17.28}$$

which is the transfer function of an inverse response system, because of the RHP zero. Eq (17.28) may now be represented as

$$g(s) = g^0(s)(1-\eta s)$$

with
$$\eta = \frac{\alpha}{2} \tag{17.29}$$
and
$$g^0(s) = \left(\frac{g^*}{1+\eta s}\right) \tag{17.30}$$

An IR compensator for this system is obtained as

$$g'(s) = g^0(s)\,\lambda s$$

or
$$g'(s) = \frac{g^* \lambda s}{1+\eta s} \tag{17.31}$$

If we now choose $\lambda = 2\eta$, then (17.31) becomes

$$g'(s) = \frac{2\eta s\, g^*(s)}{(1+\eta s)} \tag{17.32}$$

and from (17.29), we obtain a simplified (17.32) as

$$\boxed{g'(s) = g^*(s)\left(\frac{\alpha s}{1+\frac{\alpha}{2}s}\right)} \tag{17.33}$$

which is identical to the expression in (17.27), establishing the required result.

CHAPTER 18

18.1 (a) From the given reactor model, the equations to be solved for the steady state conditions are:

$$0 = 3(6-\xi_1^*) - 1.4\xi_1^*\sqrt{\xi_2^*} \quad (18.1)$$

$$0 = 1 - 3\xi_2^* \quad (18.2)$$

$$0 = 0.001\xi_1^*\sqrt{\xi_2^*} + 0.035\xi_2^* - 10\xi_3^* \quad (18.3)$$

$$0 = 130\,\xi_1^*\sqrt{\xi_2^*} - 10\xi_4^* \quad (18.4)$$

$$\eta^* = \xi_4^*/\xi_3^* \quad (18.5)$$

And from (18.2) obtain

$$\xi_2^* = 1/3\,; \quad (18.6a)$$

introduce this into (18.1) and solve for ξ_1^* to obtain

$$\xi_1^* = 4.7265 \quad (18.6b)$$

So that from (18.3) ξ_3^* may be obtained as

$$\xi_3^* = 0.00144 \quad (18.6c)$$

and from (18.4),

$$\xi_4^* = 35.475 \quad (18.6d)$$

and finally, $\eta^* = 24{,}644.1$

(b) The "process reaction curve" is shown in FIG S18.1 for a step increase in μ from 0.02 to 0.025. The approximate characterizing parameters (obtained directly from the response curve) are

$$K = -520.0\,; \quad \tau = 0.04\text{ h}; \quad \alpha = 0.1\text{ hr.}$$

To confirm this approximate characterization, the response of the first-order-plus-delay model to the same input is shown in the dashed line in FIG S18.2.

The controller settings derived from this approximate model are given below:

ZIEGLER - NICHOLS	ITAE
$K_c = -9.23 \times 10^{-6}$	$K_c = -6.0713 \times 10^{-6}$
$\tau_I = 0.2$	$\tau_I = 0.5267$
$\tau_D = 0.01$	$\tau_D = 0.034$

and the ITAE-designed controller is more conservative: the gain value is almost 1½ times smaller.

(c) The response is shown in FIG S18.3.

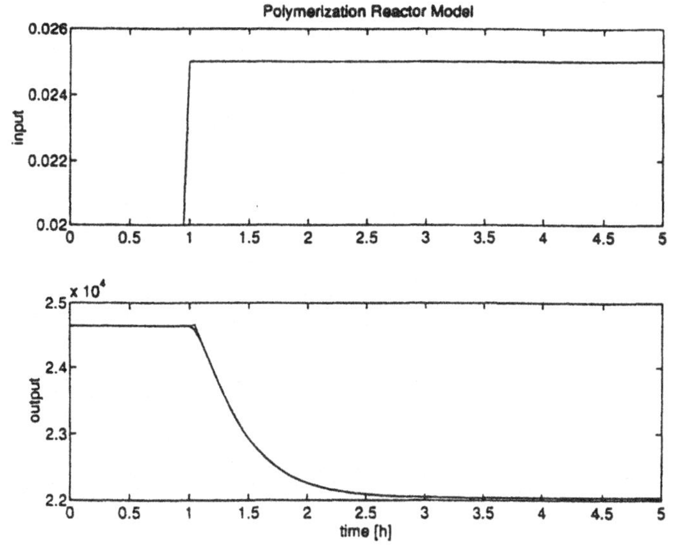

FIG S18.1 The Process Reaction Curve

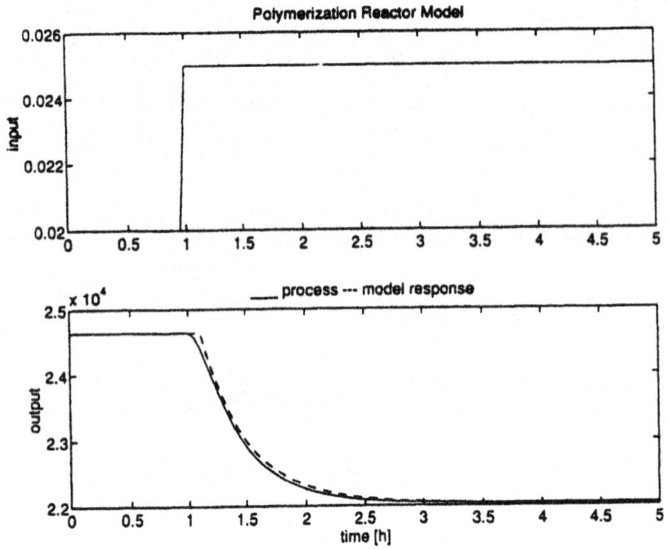

FIG S18.2. Approximate model/Process response.

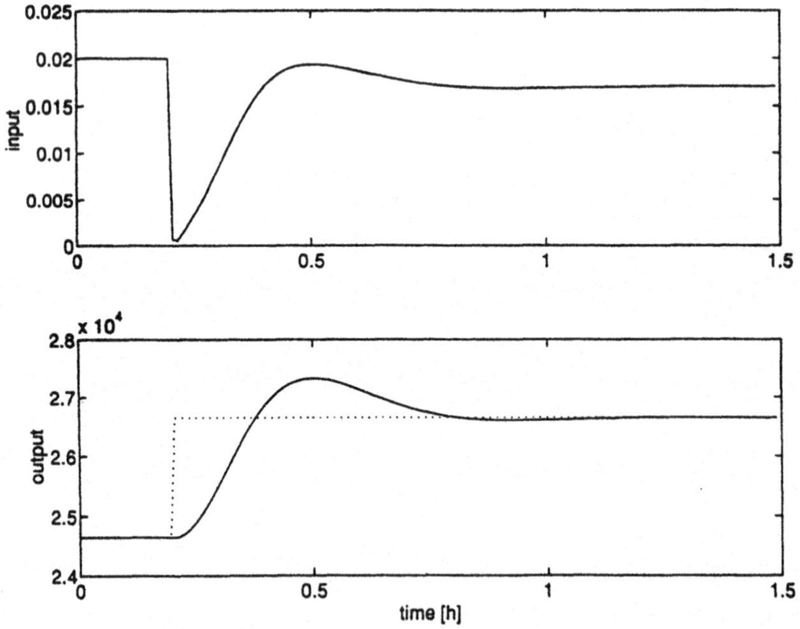

FIG S18.3 Closed loop response of Polymerization Reactor.

18.2 (a) Use the fact that given

$$\frac{d\xi_i}{dt} = f_i(\underline{\xi}, \mu) \quad ; \quad \begin{aligned} x_i &= \xi_i - \xi_i^* \\ u &= \mu - \mu^* \end{aligned}$$

upon linearization yields

$$\frac{dx_i}{dt} = \left.\frac{\partial f_i}{\partial \xi_1}\right|_* x_1 + \left.\frac{\partial f_i}{\partial \xi_2}\right|_* + \cdots + \left.\frac{\partial f_i}{\partial \xi_i}\right|_* x_i + \left.\frac{\partial f_i}{\partial \mu}\right|_* u$$

to obtain the following linearization of the polymer reactor model:

$$\frac{dx_1}{dt} = \left(-3 - 1.4\sqrt{\xi_2^*}\right)x_1 - \left(\frac{1.4}{2}\xi_1^*\xi_2^{*\,-1/2}\right)x_2 \qquad (18.7)$$

$$\frac{dx_2}{dt} = 50u - 3x_2 \qquad (18.8)$$

$$\frac{dx_3}{dt} = \left(0.001\sqrt{\xi_2^*}\right)x_1 + \left(\frac{0.001\xi_1^*\xi_2^{*\,-1/2}}{2} + 0.035\right)x_2 - 10x_3 \qquad (18.9)$$

$$\frac{dx_4}{dt} = \left(130\sqrt{\xi_2^*}\right)x_1 + \left(\frac{130}{2}\xi_1^*\xi_2^{*\,-1/2}\right)x_2 - 10x_4 \qquad (18.10)$$

$$y = \left(-\xi_4^*\xi_3^{*\,-2}\right)x_3 + \left(\xi_3^{*\,-1}\right)x_4 \qquad (18.11)$$

And upon introducing numerical values, obtain

$$\dot{x}_1 = -3.80829x_1 - 5.73058x_2 \qquad (18.12a)$$

$$\dot{x}_2 = 50u - 3x_2 \qquad (18.12b)$$

$$\dot{x}_3 = 5.7735\times10^{-4}x_1 + 0.03909x_2 - 10x_3 \qquad (18.12c)$$

$$\dot{x}_4 = 75.05553\,x_1 + 532.12498x_2 - 10x_4 \qquad (18.12d)$$

$$y = -17{,}107{,}928.24x_3 + 694.44x_4 \qquad (18.12e)$$

upon taking Laplace transforms and rearranging, obtain

$$(s+3.80829)x_1 = -5.73058\, x_2 \quad (18.13)$$

$$(s+3)\, x_2 = 50u \quad (18.14)$$

$$(s+10)\, x_3 = (5.7735\times 10^{-4})\, x_1 + 0.03909 x_2 \quad (18.15)$$

$$(s+10)\, x_4 = 75.05553\, x_1 + 532.12498\, x_2 \quad (18.16)$$

along with equation (18.12e) (in transform form).

From (18.15) and (18.16) obtain, respectively,

$$x_3 = \left(\frac{a_{31}}{s+10}\right)x_1 + \left(\frac{a_{32}}{s+10}\right)x_2 \;;\; a_{31} = 5.7735\times 10^{-4} \;;\; a_{32} = 0.03909 \quad (18.17)$$

$$x_4 = \left(\frac{a_{41}}{s+10}\right)x_1 + \left(\frac{a_{42}}{s+10}\right)x_2 \;;\; a_{41} = 75.05553 \;;\; a_{42} = 532.12498 \quad (18.18)$$

From (18.13) and (18.14) obtain

$$x_2 = \left(\frac{50}{s+3}\right)u \;;\quad x_1 = \frac{(-5.73058)}{(s+3.80829)}\cdot\left(\frac{50}{s+3}\right)u$$

Substitute these for x_1 and x_2 in (18.17) and (18.18) to eliminate x_1 and x_2; consolidate, introduce into (18.12e) and simplify the expression, and obtain

$$y(s) = \frac{(-3.27s - 14.9)\,50\times 10^5}{(s+3.81)(s+10)(s+3)}$$

or $$y(s) = \frac{-6.51\times10^{5}(0.219s+1)}{(0.262s+1)(0.1s+1)(0.333s+1)}\,u(s) \qquad (18.19)$$

as the approximate transfer function model.

Owing to the presence of the zero, the phase angle for this transfer function never crosses −180°. Thus, the Bode stability criterion cannot, in principle be applied for this transfer function; there is no finite K_u. (See FIG S18.4)

The controller may thus be designed by any suitable method. (e.g. Direct Synthesis: see chapter 19). One possible set of controller parameters ($K_c = -1\times10^{-5}$; $\tau_I = 0.93$; $\tau_D = 0$) gives rise to the response shown in FIG S18.5.

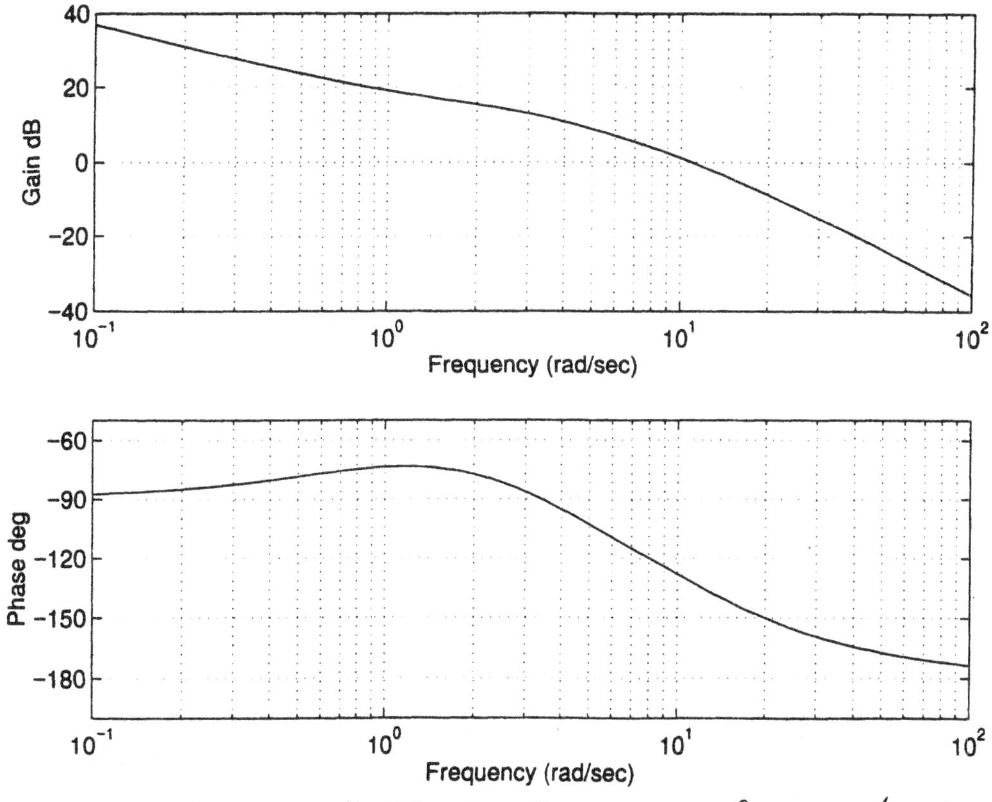

FIG S18.4 Bode Diagram for Transfer function in (18.19).

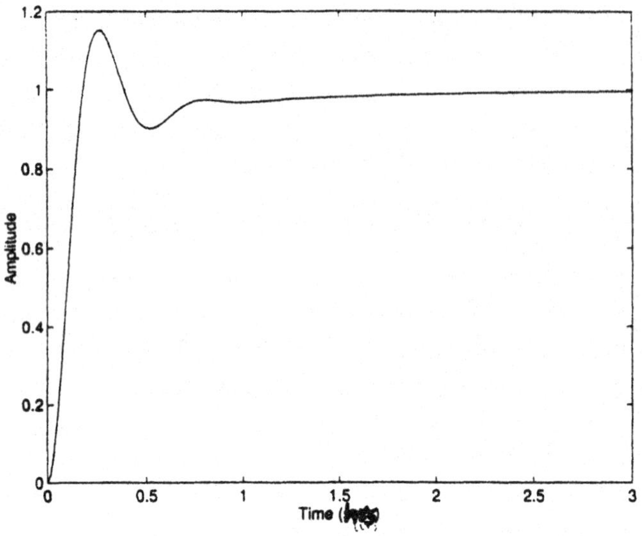

FIG S 18.5 Closed loop response of Polymer Reactor.

(b) In practice, if a process model is unavailable, the procedure illustrated in Problem 18.1 will be preferable provided process operations permit step tests. If process upsets are unacceptable, then a process model must be developed and then the linearization approach may be used.

18.3 Solution not provided

18.4 Solution not provided

18.5 Solution not provided

18.6 Rearrange the model as:

$$\frac{dx}{dt} = -\frac{c}{\pi}\frac{x^{-1/2}}{(2R-x)} + \frac{1}{\pi}\frac{u}{(2Rx-x^2)} \qquad (18.20)$$

with $x = h$ and $u = F_i$

Now let $c_1 = -\frac{c}{\pi}$; $c_2 = \frac{1}{\pi}$

$$f_1(x) = \frac{x^{-1/2}}{(2R-x)} \; ; \; f_2(x,u) = \frac{u}{x(2R-x)}$$

Then from Eq (18.18) in the text, we have that the required transformation is given by:

$$z = \exp\left\{\int \frac{dx}{f_1(x)}\right\} = \exp\left\{\int (2Rx^{1/2} - x^{3/2})\,dx\right\}$$

or

$$\boxed{z = \exp\left\{x^{3/2}\left(\frac{4R}{3} - \frac{2x}{5}\right)\right\}}. \qquad (18.21)$$

And from Eq (18.19) in the text,

$$v = \frac{z f_2(x,u)}{f_1(x)}$$

or

$$\boxed{v = \frac{zu}{x^{1/2}}} \qquad (18.22)$$

These transformations will convert the model in (18.20) above into $\frac{dz}{dt} = az + bv$.

(See Problem 10.4!)

For the alternate transformation, from Eq (18.20) in the text, obtain the required transformation as

$$\boxed{z = x^{3/2}\left(\frac{4R}{3} - \frac{2x}{5}\right)}$$

and, from Eq (18.21) of the text, obtain

$$\boxed{v = \frac{f_2(x,u)}{f_1(x)} = \frac{u}{x^{1/2}}}$$ in this case.

18.7 (a) The ordinary differential equation to be solved is:

$$\frac{d\xi}{dt} = \alpha\xi(\mu^* - \xi) \qquad (18.23)$$

Separating the variables and partial fraction expansion gives

$$\frac{1}{\mu^*}\left[\frac{1}{\xi} + \frac{1}{\mu^* - \xi}\right]d\xi = \alpha dt$$

Integrating both sides now yields

$$\ln\xi - \ln(\mu^* - \xi) = \alpha\mu^* t + c$$

or $$\ln\left(\frac{\mu^*}{\xi} - 1\right) = -\alpha\mu^* t - c \qquad (18.24)$$

The arbitrary integration constant, c, is obtained from the given initial conditions: $t = 0$ $\xi = \xi_0$ so that

$$\ln\left\{\frac{\left(\frac{\mu^*}{\xi} - 1\right)}{\left(\frac{\mu^*}{\xi_0} - 1\right)}\right\} = -\alpha\mu^* t$$

Rearranging easily to give

$$\xi = \frac{\mu^*}{1 + \left(\frac{\mu^*}{\xi_0} - 1\right) e^{-\alpha\mu^* t}} \qquad (18.25)$$

as required.

As $t \to \infty$ $e^{-\alpha\mu^* t} \to 0$ and $\xi \to \mu^*$ in (18.25) above.
Also, observe directly from the original ODE that at steady state, $d\xi/dt = 0$ and

$$0 = \alpha\xi^*(\mu^* - \xi^*)$$

with the only non-trivial solution: $\mu^* = \xi^*$.
All this indicates that the process "gain" is 1.

(b) The linearized model is obtained as follows:

$$\frac{dx}{dt} = \left.\frac{\partial f}{\partial \xi}\right|_* x + \left.\frac{\partial f}{\partial \mu}\right|_* u$$

with $f = \alpha\xi(\mu - \xi)$; i.e.

$$\frac{dx}{dt} = [\alpha(\mu^* - 2\xi^*)]\, x + \alpha\xi^* u$$

Laplace transformation and simple algebraic manipulations give

$$x(s) = \frac{\alpha\xi^*}{s + \alpha(2\xi^* - \mu^*)} u(s)$$

or, since $\xi^* = \mu^*$, this simplifies further to give

$$x(s) = \left[\frac{1}{\left(\frac{1}{\alpha\xi^*}\right) s + 1} \right] u(s)$$

From where it is obvious that, regardless of ξ^* and μ^* the steady state gain is always 1. The indicated apparent time constant is $(1/\alpha\xi^*)$.

(c) The particular process characteristics give rise to an approximate transfer function with $\tau = 1/\alpha\xi^* = 1/5 = 0.2$. The actual process response is shown in the solid line in FIG S 18.6; the approximate model response is shown in the dashed line.

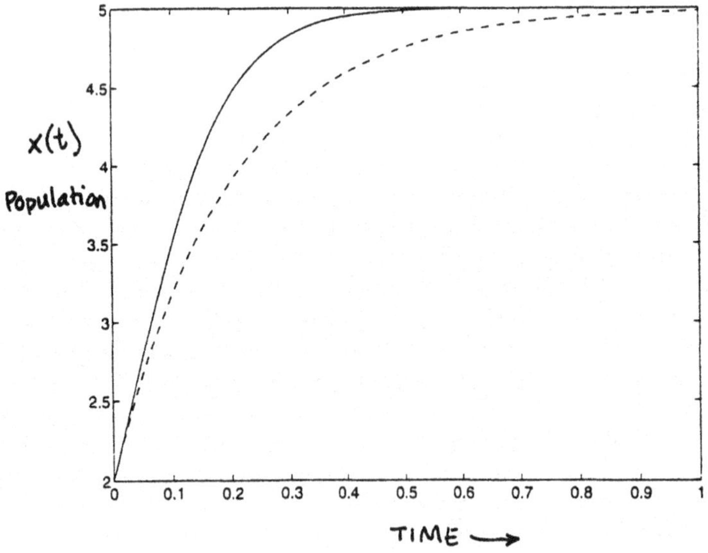

FIG S 18.6 Biological System Step Responses.

(d) In this case, the approximate linear transfer function becomes

$$\frac{e^{-0.1s}}{(0.2s+1)}$$

Using the integral error tuning method the controller parameters are obtained as: $K_c = 1.745$; $\tau_I = 0.277$ and $\tau_D = 0.0324$. The closed-loop behavior of the biological system and this controller is shown in FIG S18.7

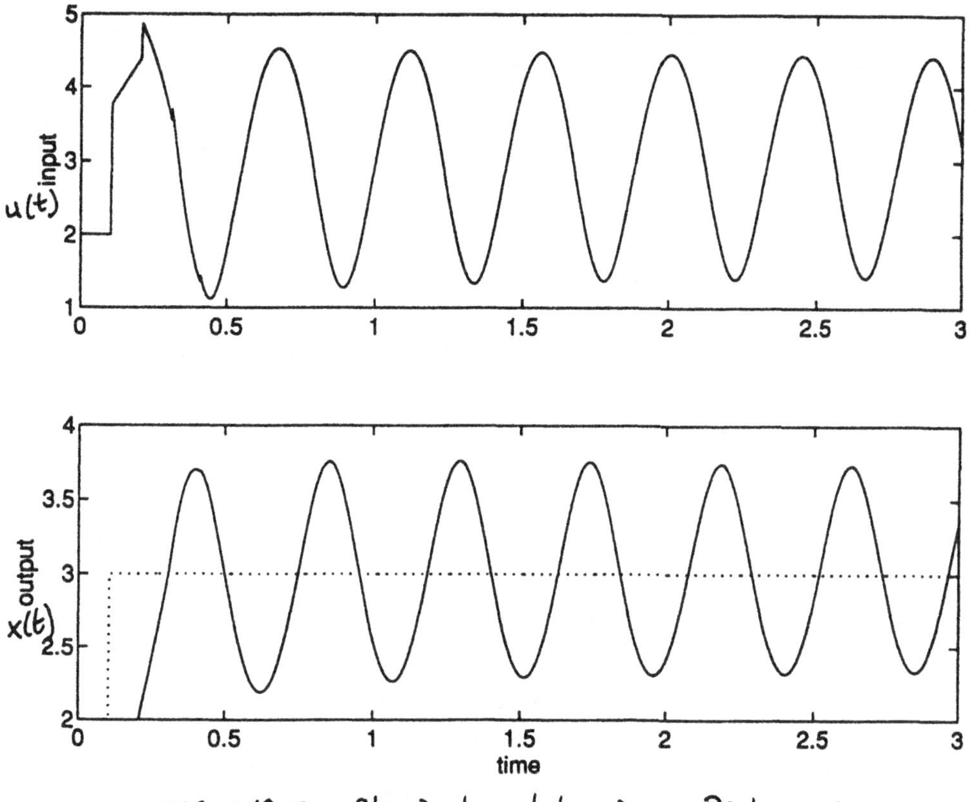

FIG S18.7. Closed-loop behavior of Biological System.

Observe that the closed-loop system is just barely stable. By reducing the controller parameters, it is possible to obtain better performance. For example, using a PI controller with $K_c = 1.356$, $\tau_I = 0.26$, the response shown in FIG S18.8 is obtained.

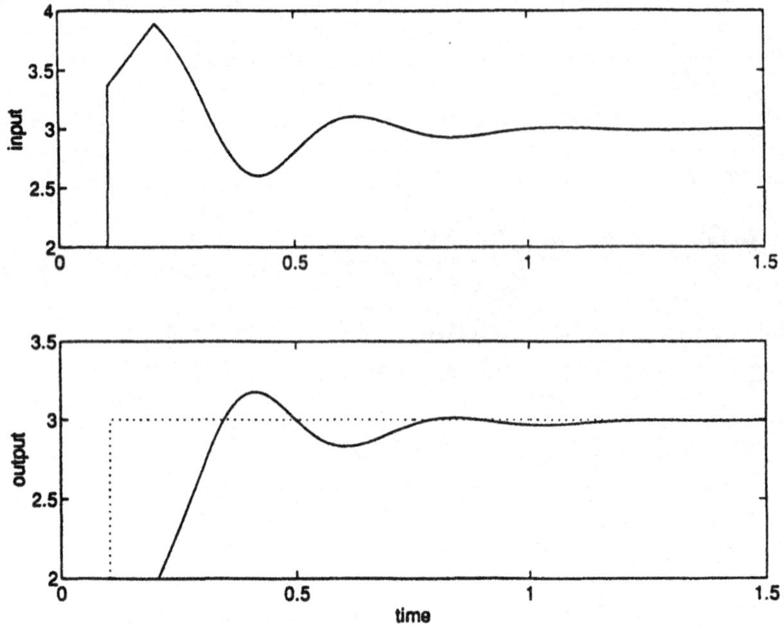

FIG S 18.8 Closed-loop behavior of Biological System:
(More conservative controller tuning).

CHAPTER 19

19.1 When the controller

$$g_c = \frac{1}{\tau_r s} \cdot \frac{1}{g}$$

is implemented as in Fig 19.2 (in the text), but on a process with "true" transfer function g_p, the resulting closed-loop transfer function relation is

$$y = \frac{\frac{g_p}{g} \frac{1}{\tau_r s}}{1 + \frac{g_p}{g} \cdot \frac{1}{\tau_r s}} \, y_d \qquad (19.1a)$$

or

$$y = \psi \, y_d \qquad (19.1b)$$

where ψ simplifies to

$$\psi = \frac{g_p}{\tau_r s g + g_p} \qquad (19.2)$$

Now, so long as ψ has no pole in the RHP, (i.e. $\tau_r s g + g_p = 0$ has no roots in the RHP) then at steady state,

$$y = \left\{ \lim_{s \to 0} \psi(s) \right\} y_d \qquad (19.3)$$

Observe from (19.2) above that $\lim_{s \to 0} \psi = g_p / g_p = 1$

regardless of g_p; thus, (19.3) reduces to

$$y = y_d$$

at steady state and there will be no offset, <u>regardless</u> of g_p, so long as $\tau_r s g + g_p$ has no roots in the RHP.

19.2 The required direct synthesis controller is as given in Eq (19.6) in the text; with g as given in the problem, we obtain immediately

$$g_c = \frac{1}{\tau_r s} \frac{(\tau_1 s+1)(\tau_2 s+1)}{K(\xi s+1)} \qquad (19.4)$$

which may be rearranged first as:

$$g_c = \frac{1}{K\tau_r}\left[\frac{1}{s}(\tau_1\tau_2)s^2 + (\tau_1+\tau_2)s + 1\right]\left[\frac{1}{\xi s+1}\right]$$

or

$$g_c = \left(\frac{\tau_1+\tau_2}{K\tau_r}\right)\left[1 + \frac{1}{(\tau_1+\tau_2)s} + \left(\frac{\tau_1\tau_2}{\tau_1+\tau_2}\right)s\right]\left[\frac{1}{\xi s+1}\right] \qquad (19.5)$$

as required.

Alternatively, Eq (19.4) above could be rearranged as

$$g_c = \frac{1}{K\tau_r}\left[\frac{\tau_1 s+1}{s}\right]\left[\frac{\tau_2 s+1}{\xi s+1}\right]$$

finally yielding

$$g_c = \frac{\tau_1}{K\tau_r}\left(1 + \frac{1}{\tau_1 s}\right)\left(\frac{\tau_2 s+1}{\xi s+1}\right)$$

the commercial PID form.

19.3 For the open loop unstable process with transfer function

$$g(s) = \frac{K}{(\sigma s+1)(\tau s-1)}$$

a reference trajectory as in Eq (19.40) of the text gives rise to $(q/1-q)$ as in Eq (19.41) of the text, and a direct synthesis controller

$$g_c = \frac{(\sigma s+1)(\tau s-1)}{K} \cdot \left(\frac{q}{1-q}\right)$$

By choosing $-\tau$ as in Eq (19.42) of the text, obtain

$$g_c = \frac{(\sigma s+1)\cancel{(\tau s-1)}}{K} \frac{(\eta_r s+1)}{\left(\frac{\tau_{r_1}\tau_{r_2}}{\tau}\right) s \cancel{(\tau s-1)}}$$

$$= \frac{\tau}{K\tau_{r_1}\tau_{r_2}}\left[\frac{1}{s}(\sigma s+1)(\eta_r s+1)\right]$$

which rearranges to give

$$g_c = \frac{\tau(\sigma+\eta_r)}{K(\tau_{r_1}\tau_{r_2})}\left[1 + \frac{1}{(\sigma+\eta_r)s} + \left(\frac{\sigma\eta_r}{\sigma+\eta_r}\right)s\right]$$

as required.

19.4 (a) The resulting direct synthesis controller is as in Eq (19.6) of the text:

$$g_c = \frac{1}{\tau_r s} \cdot \frac{1}{g}$$

and the closed-loop transfer function in this case

is given by:

$$\psi = \frac{(g_p/g)(1/\tau_r s)}{1 + (g_p/g)(1/\tau_r s)}$$

$$= \frac{g_p}{\tau_r s g + g_p}$$

so that the closed-loop characteristic equation is

$$\tau_r s g + g_p = 0. \qquad (19.6)$$

Introduce the transfer function expressions to obtain

$$\frac{K\tau_r s}{\tau s + 1} + \frac{K_p}{\tau_p s + 1} = 0$$

rearrange to obtain

$$K\tau_r \tau_p s^2 + (K\tau_r + K_p \tau) s + K_p = 0$$

or

$$\tau_r \tau_p s^2 + \left(\tau_r + \frac{K_p}{K}\tau\right) s + \frac{K_p}{K} = 0 \qquad (19.7)$$

For this quadratic equation to have <u>all</u> roots in the LHP, all the coefficients must be positive. Given that τ_r, τ_p, and τ are all positive, the stability condition will be satisfied when

$$\frac{K_p}{K} > 0 \quad \Rightarrow K_p \text{ and } K \text{ have the same sign.}$$

(b) For the new g_p as in Eq (P19.4) in the problem the closed-loop characteristic equation (Eq (19.6) above) becomes:

$$\frac{K\tau_r s}{\tau s + 1} + \frac{K_p}{(\tau_p s + 1)(\tau_n s + 1)} = 0$$

Rearrange to obtain

$$\tau_r s(\tau_p s + 1)(\tau_n s + 1) + \rho_p(\tau s + 1) = 0 \quad ; \quad \rho_p = \frac{K_p}{K}$$

or:

$$\boxed{\tau_r \tau_p \tau_n s^3 + \tau_r(\tau_p + \tau_n)s^2 + (\tau_r + \rho_p \tau)s + \rho_p = 0} \quad (19.8)$$

Now, conditions for stability when the system characteristic equation is the cubic:

$$a_0 s^3 + a_1 s^2 + a_2 s + a_3 = 0$$

are (using the Routh array, or otherwise)

(i) $a_0, a_1, a_2, a_3 > 0$

and (ii) $a_1 a_2 > a_0 a_3$

Applying these conditions to Eq (19.8) above yields

(i) $\rho_p > 0$ (given that $\tau_r, \tau_p, \tau_n, \tau$ are all > 0)

(ii) $(\tau_r + \rho_p \tau)[\tau_r(\tau_p + \tau_n)] > \rho_p \tau_r \tau_p \tau_n$

or

$$(\tau_r + \rho_p \tau) > \rho_p \frac{\tau_p \tau_n}{(\tau_p + \tau_n)}$$

which rearranges to give

$$\tau_r > \rho_p \left[\frac{\tau_p \tau_n}{\tau_p + \tau_n} - \tau\right]$$

as required.

(c) In this case $\tau = 3$; $K = 2.0$; $\tau_p = 10$; $\tau_n = 6$; $K_p = 4$ so that $\rho_p = 2.0$. Thus the condition for stability is

$$\tau_r > 2\left[\frac{60}{16} - 3\right] = 1.5$$

as required.

From Eq (19.11) in the text, for a $g(s)$ as given and $\tau_r = 2.5$, the resulting PI controller is:

$$g_c = \frac{3}{5}\left[1 + \frac{1}{3s}\right] \quad \text{with } K_c = 0.6;\ \tau_I = 3.$$

Implementing this controller on the "process" with transfer function g_p as given in Eq (P19.7) gives the closed-loop response shown in the solid line in FIG S19.1. The dashed line is the desired trajectory. Observe that even though the system is stable, the response is too oscillatory. Choosing $\tau_r = 20$ gives the closed loop response shown in the dotted line, far less oscillatory.

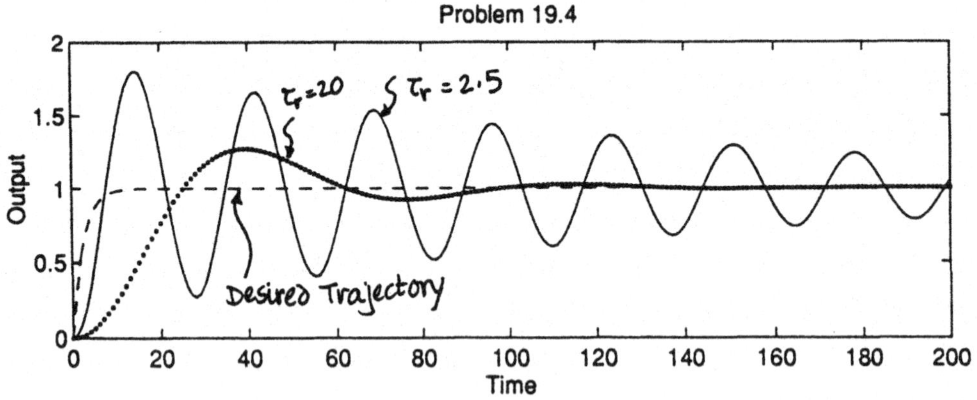

FIG S 19.1 Closed-loop system responses;
$\tau_r = 2.5$ (solid line)
$\tau_r = 20$ (dotted line)

19.5 (a) In general, the resulting controller will be:

$$g_c = \left(\frac{\tau_1+\tau_2}{K\tau_r}\right)\left[1 + \frac{1}{(\tau_1+\tau_2)s} + \left(\frac{\tau_1\tau_2}{\tau_1+\tau_2}\right)s\right]$$

In this specific case, the result is

$$g_c = \frac{2}{K}\left[1 + \frac{1}{8s} + \frac{15}{8}s\right] \qquad (19.9)$$

(b) closed-loop stability is determined from

$$1 + g_p g_c = 0$$

in this case:

$$1 + \frac{K_p(-\xi s+1)}{(3s+1)(5s+1)} \cdot \frac{2}{K}\left[1 + \frac{1}{8s} + \frac{15s}{8}\right] = 0$$

with the "worst case" condition $K_p = 2K$, the characteristic equation becomes

$$1 + \frac{4(-\xi s+1)}{(3s+1)(5s+1)}\left[\frac{8s + 1 + 15s^2}{8s}\right] = 0$$

or

$$s(8-4\xi) + 4 = 0$$

and stability requires $8 > 4\xi$, or

$$\boxed{\xi < 2} \qquad (19.10)$$

(c) The controller in this case is

$$g_c = \frac{8}{K\tau_r}\left[1 + \frac{1}{8s} + \frac{15s}{8}\right]$$

with τ_r yet undetermined, and the closed-loop characteristic equation becomes:

$$1 + \frac{K_p[-\xi s + 1]}{(3s+1)(5s+1)} \cdot \frac{8}{K\tau_r}\left[1 + \frac{1}{8s} + \frac{15s}{8}\right] = 0$$

In the worst case ($K_p = 2K$; $\xi = 3.5$), this reduces to

$$1 + \frac{16(-3.5s+1)}{8s\tau_r} = 0$$

or

$$(8\tau_r - 56)s + 16 = 0$$

and closed-loop stability requires $8\tau_r > 56$

$$\text{or} \quad \boxed{\tau_r > 7} \qquad (19.11)$$

19.6 (a) Under the given circumstances,

$$g = \frac{2}{(3s+1)(5s+1)}$$

whereas g_p is as given in the problem statement. With $\tau_r = 4.0$, the direct synthesis controller is

$$g_c = \left[1 + \frac{1}{8s} + \frac{15s}{8}\right]$$

The response of the process, g_p, in conjunction with this controller in the closed-loop is shown in the solid line in FIG S19.2; the desired trajectory is shown in the dashed line.

(b) With $\tau_r = 6$, the resulting controller is

$$g_c = \frac{2}{3}\left[1 + \frac{1}{8s} + \frac{15s}{8}\right]$$

and the closed-loop response using this new controller

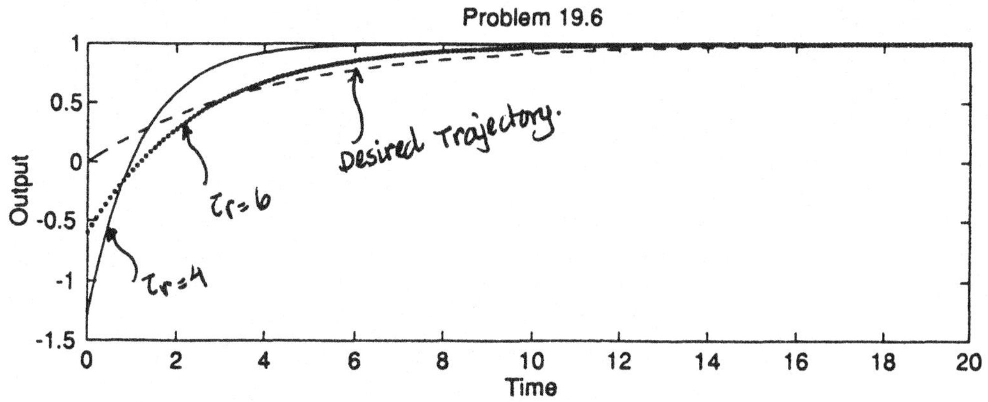

FIG S19.2 Closed-loop system responses:
$\tau_r = 4$ (solid line)
$\tau_r = 6$ (dotted line)
Desired Trajectory (dashed line).

is shown in the dotted line. Observe that the initial inverse response is less pronounced with the larger τ_r value. The model used for controller design does not exhibit inverse response. One of the effects of the larger τ_r value is to "reduce" the effect of plant/model mismatch on closed-loop behavior. The price for this is increased sluggishness.

19.7 Solution not provided; problem involves choices.

19.8 (a) The controller in this case is

$$g_c = \left(\frac{15s+1}{10.3}\right) \cdot \frac{1}{\cancel{e^{-7.5s}}} \; \frac{\cancel{e^{-7.5s}}}{5s+1-e^{-7.5s}}$$

and upon introducing the Padé approximation,

$$e^{-7.5s} \approx \frac{1-\frac{7.5}{2}s}{1+\frac{7.5}{2}s}$$

the controller becomes

$$g_c = \frac{15s+1}{(10.3)(12.5)} \frac{\left[\left(7.5/2\right)s+1\right]}{s(1.5s+1)}$$

or
$$g_c = 0.1456\left[1 + \frac{1}{18.75s} + 3s\right] \cdot \frac{1}{(1.5s+1)} \qquad (19.12)$$

which may be approximated by the PID controller

$$g_c = 0.1456\left[1 + \frac{1}{18.75s} + 3\right] \qquad (19.13)$$

by neglecting the $\frac{1}{1.5s+1}$ term. This approximation is reasonable since $\tau^* = 1.5$ in the neglected term is small compared to $\alpha = 7.5$ (the delay), and $\tau = 15$, the process time constant.

(b) The closed-loop response is shown in FIGS19.3.

(c) The result obtained depends on what is chosen for the desired trajectory $q(s)$.

Starting with

$$g_2(s) = \frac{10.3\left(1-\frac{7.5}{2}s\right)}{(15s+1)\left(1+\frac{7.5}{2}s\right)} \qquad (19.14)$$

by choosing
$$q(s) = \frac{1-\eta_r s}{(\tau_r s+1)(1+\eta_r s)} \qquad (19.15)$$

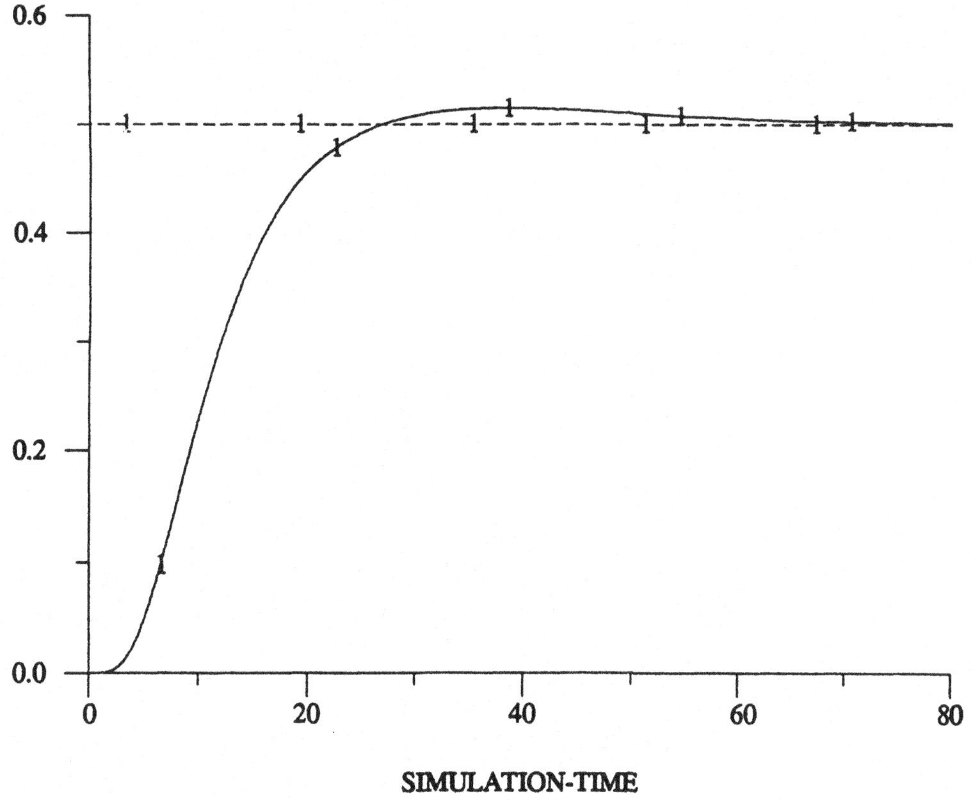

FIG S19.3 Closed-loop system response

along with $\eta_r = 7.5/2$ and $\tau_r = 5$ as given, obtain the resulting direct synthesis controller as (see Table 19.2 or see Eq (19.33) in the text)

$$g_c = \frac{\tau_1+\tau_2}{K(2\eta+\tau_r)}\left[1+\frac{1}{(\tau_1+\tau_2)s}+\left(\frac{\tau_1\tau_2}{\tau_1+\tau_2}\right)s\right]\cdot\frac{1}{\tau^* s+1}$$

with $\tau_1 = 15$, $\tau_2 = 7.5/2$; $\eta = 7.5/2$; $K = 10.3$, simplifying to

$$g_c = 0.1456\left[1+\frac{1}{18.75s}+3.0s\right]\cdot\frac{1}{1.5s+1}$$

precisely the same as in Eq (19.12) above. It may then be approximated by the same PID controller in Eq (19.13) for precisely the same reason.

However, it is also possible to choose

$$q_r(s) = \frac{-\eta_r s + 1}{\tau_r s + 1}$$

in which case, the controller will be

$$g_c = \frac{1}{g} \cdot \frac{\dfrac{-\eta_r s + 1}{\tau_r s + 1}}{1 - \left(\dfrac{-\eta_r s + 1}{\tau_r s + 1}\right)}$$

$$= \frac{1}{g} \frac{-\eta_r s + 1}{s(\tau_r + \eta_r)}$$

or
$$g_c = \frac{1}{10.3} \frac{(15s+1)\left(1 + \frac{7.5}{2}s\right)}{\left(1 - \frac{7.5}{2}s\right)} \cdot \frac{-\eta_r s + 1}{s(\tau_r + \eta_r)}. \qquad (19.16)$$

Now, by choosing $\eta_r = \frac{7.5}{2}$ and τ_r as given (5.0) this controller reduces to

$$g_c = 0.208\left(1 + \frac{1}{18.75s} + 3s\right) \qquad (19.17)$$

a PID controller, directly, with no need for a simplifying approximation.

19.9 From $g(s) = \dfrac{Ke^{-\alpha s}}{\tau s + 1}$

1. Introducing the Padé approximation yields

$$\tilde{g}(s) = \frac{K\left(1 - \frac{\alpha}{2}s\right)}{(\tau s + 1)\left(1 + \frac{\alpha}{2}s\right)} \qquad (19.18)$$

2. This may be factored as:

$$\tilde{g}_- = \frac{K}{(\tau s+1)(1+\frac{\alpha}{2}s)} \quad ; \quad \tilde{g}_+ = \left(1 - \frac{\alpha}{2}s\right)$$

3. With $f(s) = \frac{1}{\lambda s + 1}$, obtain the IMC controller

$$c(s) = \frac{1}{\tilde{g}_-(s)} \cdot f(s)$$

as

$$c(s) = \frac{(\tau s+1)(\frac{\alpha}{2}s+1)}{K(\lambda s+1)} \qquad (19.19)$$

4. The required equivalent feedback controller is obtained from

$$g_c(s) = \frac{c(s)}{1 - c(s)\tilde{g}(s)}$$

which, upon introducing Eqs (19.18) and (19.19) above simplifies to

$$g_c(s) = \frac{1}{K}\left[\frac{(\tau s+1)(\frac{\alpha}{2}s+1)}{(\lambda + \frac{\alpha}{2})s}\right]$$

This is easily rearranged into

$$g_c(s) = \frac{2\tau+\alpha}{K(2\lambda+\alpha)}\left[1 + \frac{1}{(\tau+\frac{\alpha}{2})s} + \left(\frac{\alpha\tau}{2\tau+\alpha}\right)s\right] \qquad (19.20)$$

a PID controller, as required. The procedure thus recommends the following controller parameters:

$$K_c = \frac{2\tau+\alpha}{K(2\lambda+\alpha)} \quad ; \quad \tau_I = \left(\tau + \frac{\alpha}{2}\right) \quad ; \quad \tau_D = \frac{\alpha\tau}{2\tau+\alpha}$$

19.10 Based on the transfer function in Eq (P19.15), the model parameters are:

$$\alpha = 7.5 \; ; \; \tau = 15 \; ; \; K = 10.3$$

The IMC procedure of Problem 19.9 yielded the PID controller in Eq (19.20) above. Upon introducing these numerical values, we obtain

$$g_c = 0.208\left[1 + \frac{1}{18.75s} + 3.0s\right] \qquad (19.21)$$

Observe that this is identical to the direct synthesis controller with $q = \left(1 - \frac{7.5}{2}s\right)/(5s+1)$ obtained in Eq (19.17) above. When the alternative q is used, the resulting direct synthesis controller has the same τ_I and τ_D as this IMC-designed controller, but with a different gain (and the "extra" $\left(1/1.5s+1\right)$ term has been omitted).

The similarities between the IMC design results and those of the direct synthesis approach are clear.

19.11 Solution not provided.

19.12 Solution not provided.

CHAPTER 20

20.1 (a) The eigenvalues are determined from the solution to the characteristic equation

$$\begin{vmatrix} (-3-\lambda) & 2 \\ 1 & (-4-\lambda) \end{vmatrix} = 0$$

or

$$\lambda^2 + 7\lambda + 10 = 0$$

$$(\lambda+2)(\lambda+5) = 0$$

$$\Rightarrow \quad \lambda_1 = -2$$

$$\lambda_2 = -5$$

The eigenvector $\underset{\sim}{x}_1$ corresponding to $\lambda_1 = -2$ is obtained from any non-trivial column of $\text{adj}\,(\underset{\sim}{A} - \lambda_1 \underset{\sim}{I})$ (or any scalar multiple thereof).

i.e. since

$$(\underset{\sim}{A} - \lambda_1 \underset{\sim}{I}) = \begin{pmatrix} -3 & 2 \\ 1 & -4 \end{pmatrix} - \begin{pmatrix} -2 & 0 \\ 0 & -2 \end{pmatrix}$$

$$= \begin{pmatrix} -1 & 2 \\ 1 & -2 \end{pmatrix}$$

$$\text{adj}\,(\underset{\sim}{A} - \lambda_1 \underset{\sim}{I}) = \begin{pmatrix} -2 & -2 \\ -1 & -1 \end{pmatrix}$$

and

$$\underset{\sim}{x}_1 = \begin{pmatrix} 2 \\ 1 \end{pmatrix}$$

Similarly obtain $\underset{\sim}{x}_2$ from adj $(\underset{\sim}{A} - \lambda_2 \underset{\sim}{I})$ as

$$\underset{\sim}{x}_2 = \begin{pmatrix} 1 \\ -1 \end{pmatrix}$$

Thus

$$\underset{\sim}{M} = \begin{bmatrix} 2 & 1 \\ 1 & -1 \end{bmatrix}$$

and therefore $$\underset{\sim}{M}^{-1} = \begin{bmatrix} 1/3 & 1/3 \\ 1/3 & -2/3 \end{bmatrix}$$

And now,

$$\underset{\sim}{M}\,\underset{\sim}{\Lambda}\,\underset{\sim}{M}^{-1} = \begin{bmatrix} 2 & 1 \\ 1 & -1 \end{bmatrix}\begin{bmatrix} -2 & 0 \\ 0 & -5 \end{bmatrix}\begin{bmatrix} 1/3 & 1/3 \\ 1/3 & -2/3 \end{bmatrix}$$

$$= \begin{bmatrix} -4 & -5 \\ -2 & 5 \end{bmatrix}\begin{bmatrix} 1/3 & 1/3 \\ 1/3 & -2/3 \end{bmatrix} = \begin{bmatrix} -3 & 2 \\ -1 & -4 \end{bmatrix} = \underset{\sim}{A}$$

as required.

(b) The corresponding transfer function model is

$$\underset{\sim}{y}(s) = \underline{G}(s)\,\underset{\sim}{u}(s)$$

where

$$\underline{G}(s) = \underset{\sim}{C}(s\underset{\sim}{I} - \underset{\sim}{A})^{-1}\underset{\sim}{B}$$

$$= \begin{pmatrix} s+3 & -2 \\ -1 & s+4 \end{pmatrix}^{-1}\begin{pmatrix} 2 & 0 \\ 0 & 1 \end{pmatrix}$$

which, upon performing the indicated matrix operations, reduces to

$$\underset{\sim}{G}(s) = \begin{bmatrix} \frac{2(s+4)}{(s+5)(s+2)} & \frac{2}{(s+5)(s+2)} \\ \frac{2}{(s+5)(s+2)} & \frac{s+3}{(s+5)(s+2)} \end{bmatrix} \qquad (20.1)$$

20.2 (a) With $\underset{\sim}{G}(t)$ as given, use Eq (20.35) of the text

$$e^{\underset{\sim}{A}t} = \underset{\sim}{M} e^{\underset{\sim}{\Lambda}t} \underset{\sim}{M}^{-1}$$

to obtain.

$$\underset{\sim}{G}(t) = \left(\underset{\sim}{M} e^{\underset{\sim}{\Lambda}t} \underset{\sim}{M}^{-1}\right) \underset{\sim}{B} \qquad (20.2)$$

Introduce results from problem 20.1 (expressions for $\underset{\sim}{M}$, $e^{\underset{\sim}{\Lambda}t}$, $\underset{\sim}{M}^{-1}$ and the given $\underset{\sim}{B}$) to obtain

$$G(t) = \begin{pmatrix} 2 & 1 \\ 1 & -1 \end{pmatrix} \begin{bmatrix} e^{-2t} & 0 \\ 0 & e^{-5t} \end{bmatrix} \begin{bmatrix} 1/3 & 1/3 \\ 1/3 & -2/3 \end{bmatrix} \begin{bmatrix} 2 & 0 \\ 0 & 1 \end{bmatrix}$$

Perform indicated matrix operations carefully to obtain

$$G(t) = \begin{bmatrix} \frac{2}{3}\left(2e^{-2t} + e^{-5t}\right) & \frac{2}{3}\left(e^{-2t} - e^{-5t}\right) \\ \frac{2}{3}\left(e^{-2t} - e^{-5t}\right) & \frac{1}{3}\left(e^{-2t} + 2e^{-5t}\right) \end{bmatrix} \qquad (20.3)$$

(b) From the impulse response model, the required response is obtained from

$$y(t) = \int_0^t \underset{\sim}{G}(t-\sigma)\begin{bmatrix}1\\0\end{bmatrix} d\sigma$$

or:

$$y_1(t) = \int_0^t \frac{2}{3}\left(2e^{-2(t-\sigma)} + e^{-5(t-\sigma)}\right] d\sigma \qquad (20.4)$$

$$y_2(t) = \int_0^t \frac{2}{3}\left[e^{-2(t-\sigma)} - e^{-5(t-\sigma)}\right] d\sigma \qquad (20.5)$$

Perform the required integration in Eq (20.4) above to obtain

$$y_1(t) = \frac{2}{3}\left(1-e^{-2t}\right) + \frac{2}{15}\left(1-e^{-5t}\right) \qquad (20.6)$$

and similarly, from Eq (20.5) above, obtain

$$y_2(t) = \frac{1}{3}\left(1-e^{-2t}\right) - \frac{2}{15}\left(1-e^{-5t}\right) \qquad (20.7)$$

These same equations are obtainable directly from the transfer function matrix given in Eq (20.1) in part (b) of Problem 20.1.

20.3 (a). The eigenvalues of the system matrix $\underset{\sim}{A}$ are obtained from

$$\begin{vmatrix}(4-\lambda) & -1\\ 5 & (-2-\lambda)\end{vmatrix} = 0 \;;\text{ or } \lambda^2 - 2\lambda - 3 = 0$$

as $\lambda_1 = +3$; $\lambda_2 = -1$

and the positive eigenvalue confirms that the system is unstable at this operating condition.

(b) The required transfer function is obtained from

$$\underset{\sim}{G}(s) = \begin{bmatrix} (s-4) & 1 \\ -5 & (s+2) \end{bmatrix}^{-1} \begin{bmatrix} 1 & 0 \\ 0 & 2 \end{bmatrix}$$

or

$$\underset{\sim}{G}(s) = \frac{1}{(s-3)(s+1)} \begin{bmatrix} (s+2) & -2 \\ 5 & 2(s-4) \end{bmatrix}.$$

The poles are located at $s = 3$; $s = -1$ (same as the eigenvalues of $\underset{\sim}{A}$, of course); there are no zeros.

(c) With the given controller expression $\underset{\sim}{u} = -\underset{\sim}{K}_c \underset{\sim}{x}$, the model equation becomes

$$\dot{\underset{\sim}{x}} = (\underset{\sim}{A} - \underset{\sim}{B}\underset{\sim}{K}_c)\underset{\sim}{x}$$

in the closed loop; or,

$$\dot{\underset{\sim}{x}} = \underset{\sim}{A}_{CL}\underset{\sim}{x} \qquad (20.8a)$$

where

$$\underset{\sim}{A}_{CL} = (\underset{\sim}{A} - \underset{\sim}{B}\underset{\sim}{K}_c) \qquad (20.8b)$$

Given $K_{c_2} = 2$,

$$\underset{\sim}{A}_{CL} = \begin{bmatrix} (4 - K_{c_1}) & -1 \\ 5 & -6 \end{bmatrix} \qquad (20.9)$$

The eigenvalues of $\underset{\sim}{A}_{CL}$ are obtained from

$$\begin{vmatrix} (4-K_{c_1})-\lambda & -1 \\ 5 & -6-\lambda \end{vmatrix} = 0$$

or

$$\lambda^2 + (K_{c_1}+2)\lambda + (6K_{c_1}-19) = 0 \qquad (20.10)$$

For the closed-loop system to be stable, <u>all</u> eigenvalues of $\underset{\sim}{A}_{CL}$ must lies in the LHP, requiring that <u>all</u> the coefficients of the quadratic equation in (20.10) above be positive. Thus for closed-loop stability

$$\boxed{K_{c_1} > 19/6.}$$

20.4 (a) The poles of the transfer function matrix are located at

$$s = -3;\ -4;\ -1;\ -7$$

the collection of the poles of the individual elements of the matrix.

To obtain the zeros, find the roots of

$$|G(s)| = 0$$

i.e

$$\frac{1}{(s+3)(s+4)} - \frac{1}{(s+1)(s+7)} = 0$$

or

$$\frac{(s-5)}{(s+3)(s+4)(s+1)(s+7)} = 0 \qquad (20.11)$$

And the zero is located at $s = +5$, i.e. it is

a RHP zero!

(b) In response to step inputs $m_1 = 1$ and $m_2 = 1$, obtain

$$y_1(s) = \left(\frac{2/3}{1/3 s + 1}\right) \cdot \frac{1}{s} - \left(\frac{1}{s+1}\right)\frac{1}{s} \qquad (20.12)$$

and

$$y_2(s) = \left(\frac{-1/7}{1/7 s + 1}\right)\frac{1}{s} + \left(\frac{1/8}{\frac{1}{4}s + 1}\right)\frac{1}{s} \qquad (20.13)$$

From where Laplace inversion yields

$$y_1(t) = \frac{2}{3}\left(1 - e^{-3t}\right) - \left(1 - e^{-t}\right) \qquad (20.14)$$

$$y_2(t) = -\frac{1}{7}\left(1 - e^{-7t}\right) + \frac{1}{8}\left(1 - e^{-4t}\right). \qquad (20.15)$$

And the response plots are shown in FIG S20.1. Note the inverse response indicated in the y_1 response and the "strong zero" response indicated in y_2.

It is instructive to consolidate Eqs (20.12) and (20.13) above as follows:

$$y_1(s) = \left[\frac{2/3}{1/3 s + 1} - \frac{1}{s+1}\right]\frac{1}{s}$$

$$y_2(s) = \left[\frac{-1/7}{1/7 s + 1} + \frac{1/8}{1/4 s + 1}\right]\frac{1}{s}$$

giving

$$y_1(s) = \frac{\frac{1}{3}(s-1)}{\left(\frac{1}{3}s+1\right)(s+1)} \cdot \frac{1}{s}$$

$$y_2(s) = \frac{\frac{1}{56}(s+1)}{\left(\frac{1}{7}s+1\right)\left(1/4 s+1\right)} \cdot \frac{1}{s}$$

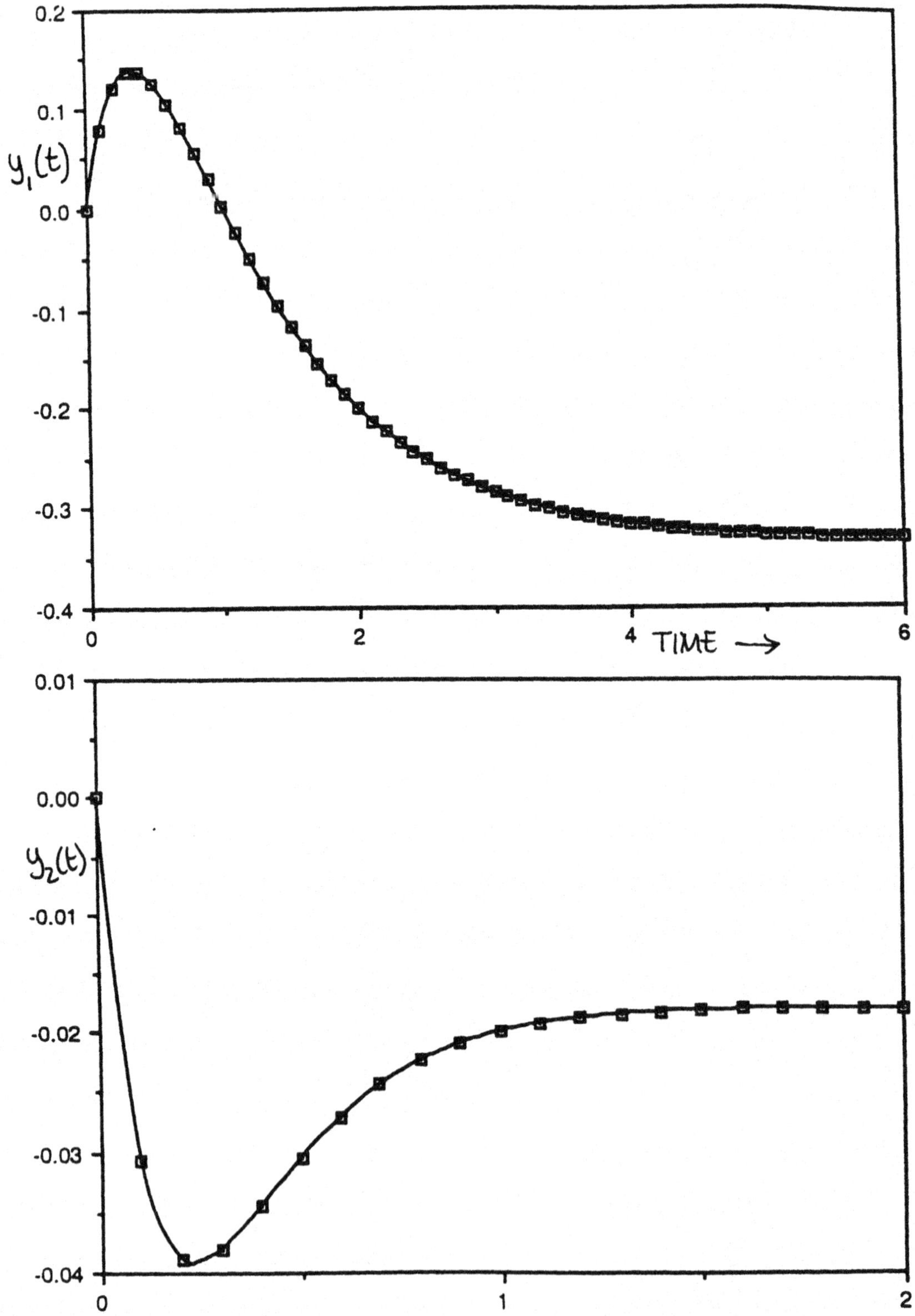

FIG S 20.1 step responses of multivariable system

20.5 From each of the "state" variables defined in Eqs (P20.8) – (P20.11), obtain the following differential equations immediately:

$$8.5\frac{dx_1}{dt} + x_1(t) = -2.16\, m_1(t-1)$$

$$7.05\frac{dx_2}{dt} + x_2(t) = 1.26\, m_2(t-0.3)$$

$$8.2\frac{dx_3}{dt} + x_3(t) = -2.75\, m_1(t-1.8)$$

$$9.0\frac{dx_4}{dt} + x_4(t) = 4.28\, m_2(t-0.35)$$

along with

$$y_1 = x_1 + x_2$$

$$y_2 = x_3 + x_4$$

20.6 If

$$g_{ij}(s) = \frac{K_{ij}e^{-\alpha_{ij}s}}{\tau_{ij}s+1}$$

then upon Laplace inversion, we obtain

$$g_{ij}(t) = \frac{K_{ij}}{\tau_{ij}}e^{-\frac{1}{\tau_{ij}}(t-\alpha_{ij})} \qquad t > \alpha_{ij}.$$

$$= 0 \qquad t \leq \alpha_{ij}.$$

Apply this result to the given transfer function matrix $G(s)$, along with the hint to obtain the following required result:

The required impulse response matrix $\underline{G}(t)$ has the following 4 elements:

$$g_{11}(t) = \begin{cases} -0.254\, e^{-0.118(t-1)} & ; \ t > 1 \\ 0 & ; \ t \le 1 \end{cases}$$

$$g_{12}(t) = \begin{cases} 0.179\, e^{-0.142(t-0.3)} & ; \ t > 0.3 \\ 0 & ; \ t \le 0.3 \end{cases}$$

$$g_{21}(t) = \begin{cases} 0.335\, e^{-0.122(t-1.8)} & ; \ t > 1.8 \\ 0 & ; \ t \le 1.8 \end{cases}$$

$$g_{22}(t) = \begin{cases} 0.476\, e^{-0.111(t-0.35)} & ; \ t > 0.35 \\ 0 & ; \ t \le 0.35 \end{cases}$$

20.7 To facilitate the analysis, introduce labels for the block diagram signals as shown below:

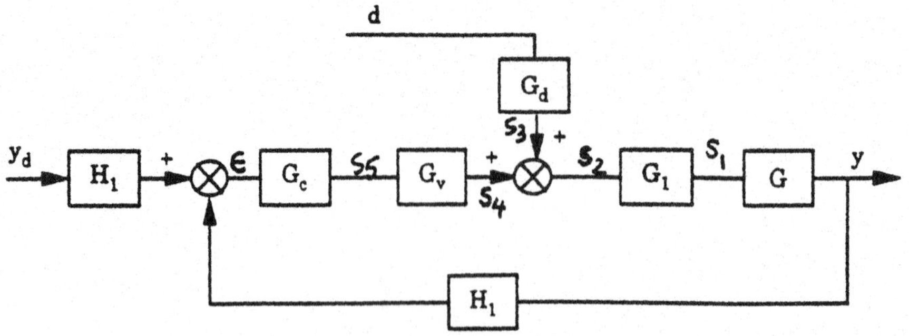

Figure P20.1

Proceed systematically around the block diagram as follows:

$$\begin{aligned}
\underset{\sim}{y} &= \underset{\sim}{G}\underset{\sim}{S}_1 \\
&= \underset{\sim}{G}\underset{\sim}{G}_1\underset{\sim}{S}_2 = \underset{\sim}{G}\underset{\sim}{G}_1(S_3 + S_4) \\
&= \underset{\sim}{G}\underset{\sim}{G}_1(\underset{\sim}{G}_d\underset{\sim}{d} + \underset{\sim}{G}_v\underset{\sim}{S}_5) \\
&= \underset{\sim}{G}\underset{\sim}{G}_1\underset{\sim}{G}_v\underset{\sim}{G}_c\underset{\sim}{\epsilon} + \underset{\sim}{G}\underset{\sim}{G}_1\underset{\sim}{G}_d\underset{\sim}{d} \\
&= \underset{\sim}{G}\underset{\sim}{G}_1\underset{\sim}{G}_v\underset{\sim}{G}_c(\underset{\sim}{H}_1\underset{\sim}{y}_d - \underset{\sim}{H}_1\underset{\sim}{y}) + \underset{\sim}{G}\underset{\sim}{G}_1\underset{\sim}{G}_d\underset{\sim}{d} \\
&= \underset{\sim}{G}\underset{\sim}{G}_1\underset{\sim}{G}_v\underset{\sim}{G}_c\underset{\sim}{H}_1(\underset{\sim}{y}_d - \underset{\sim}{y}) + \underset{\sim}{G}\underset{\sim}{G}_1\underset{\sim}{G}_d\underset{\sim}{d}
\end{aligned}$$

$$(\underset{\sim}{I} + \underset{\sim}{G}\underset{\sim}{G}_1\underset{\sim}{G}_v\underset{\sim}{G}_c\underset{\sim}{H}_1)\underset{\sim}{y} = \underset{\sim}{G}\underset{\sim}{G}_1\underset{\sim}{G}_v\underset{\sim}{G}_c\underset{\sim}{H}_1\underset{\sim}{y}_d + \underset{\sim}{G}\underset{\sim}{G}_1\underset{\sim}{G}_d\underset{\sim}{d}$$

Finally obtain the required transfer function matrix relationship:

$$\underset{\sim}{y} = (\underset{\sim}{I} + \underset{\sim}{G}\underset{\sim}{G}_1\underset{\sim}{G}_v\underset{\sim}{G}_c\underset{\sim}{H}_1)^{-1}\left[(G G_1 G_v G_c H_1) y_d + (G G_1 G_d) d\right]$$

The characteristic equation is

$$\left|\underset{\sim}{I} + \underset{\sim}{G}\underset{\sim}{G}_1\underset{\sim}{G}_v\underset{\sim}{G}_c\underset{\sim}{H}_1\right| = 0 \qquad (20.13)$$

20.8 (a)

$$\underset{\sim}{K} = \begin{bmatrix} 1 & 0.0001 \\ 100 & 1 \end{bmatrix}$$

(b) Eigenvalues of $\underset{\sim}{K}$ are $\lambda_1 = 1.1$; $\lambda_2 = 0.9$

Singular values of K are $\sigma_1 = 100.01$; $\sigma_2 = 0.0099$ with a condition number

$$K = 1.0102 \times 10^4$$

implying a very poorly conditioned matrix, and by extension, an equally poorly conditioned process.

(c) By comparing first the elements of column 1 and column 2, observe that the second input variable is 4 orders of magnitude less effective in influencing output 1 as is the first input variable; in influencing y_2, m_2 is two orders of magnitude less effective. Thus m_1 the first input variable is overwhelmingly more influential than m_2. Next observe that regardless of the input variable in question, y_2 is affected more than y_1 overwhelmingly.

Thus we have a system in which y_1 is significantly less sensitive to process inputs and in which m_2 is overwhelmingly the most influential input.

20.9 (a) The closed-loop transfer function between $\underset{\sim}{y}$ and $\underset{\sim}{y}_d$ is obtained from

$$\underset{\sim}{y} = \underset{\sim}{G}\underset{\sim}{u} = \underset{\sim}{G}\underset{\sim}{V}\underset{\sim}{\mu} = \underset{\sim}{G}\underset{\sim}{V}\underset{\sim}{G}_c^{\Sigma}\epsilon$$

$$= \underset{\sim}{G}\underset{\sim}{V}\underset{\sim}{G}_c^{\Sigma}(\eta_d - \eta)$$

$$= \underset{\sim}{G}\underset{\sim}{V}\underset{\sim}{G}_c^{\Sigma}(\underset{\sim}{W}^T\underset{\sim}{y}_d - \underset{\sim}{W}^T\underset{\sim}{y})$$

finally rearranging to give:

$$\underset{\sim}{y} = \left(\underset{\sim}{I} + \underset{\sim}{G}\underset{\sim}{V}\underset{\sim}{G}_c^{\Sigma}\underset{\sim}{W}^T\right)^{-1}\left(\underset{\sim}{G}\underset{\sim}{V}\underset{\sim}{G}_c^{\Sigma}\underset{\sim}{W}^T\right)\underset{\sim}{y}_d \qquad (20.16)$$

(b) The characteristic equation is:

$$\left|\underset{\sim}{I} + \underset{\sim}{G}\underset{\sim}{V}\underset{\sim}{G}_c^{\Sigma}\underset{\sim}{W}^T\right| = 0 \qquad (20.17)$$

Introduce the given expression into Eq (20.17) above for $\underset{\sim}{G}$ to obtain

$$\left| \underset{\sim}{I} + \underset{\sim}{W}\underset{\sim}{\Sigma}\underset{\sim}{V}^T\underset{\sim}{V}\underset{\sim}{G}_c^{\Sigma}\underset{\sim}{W}^T \right| = 0$$

but since $\underset{\sim}{V}$ is a unitary matrix, $\underset{\sim}{V}^T\underset{\sim}{V} = \underset{\sim}{I}$ and the characteristic equation becomes

$$\left| \underset{\sim}{I} + \underset{\sim}{W}\underset{\sim}{\Sigma}\underset{\sim}{G}_c^{\Sigma}\underset{\sim}{W}^T \right| = 0 \qquad (20.18)$$

Now, because $\underset{\sim}{W}$ is a unitary matrix, $\underset{\sim}{W}\underset{\sim}{W}^T = \underset{\sim}{I} = \underset{\sim}{W}^T\underset{\sim}{W}$ it is now convenient to introduce $\underset{\sim}{W}\underset{\sim}{W}^T$ for I in Eq (20.18) above to obtain

$$\left| \underset{\sim}{W}\underset{\sim}{W}^T + \underset{\sim}{W}\underset{\sim}{\Sigma}\underset{\sim}{G}_c^{\Sigma}\underset{\sim}{W}^T \right| = 0$$

Now factor out $\underset{\sim}{W}$ and $\underset{\sim}{W}^T$ to obtain

$$\left| \underset{\sim}{W}\left(\underset{\sim}{I} + \underset{\sim}{\Sigma}\underset{\sim}{G}_c^{\Sigma}\right)\underset{\sim}{W}^T \right| = 0 \qquad (20.19)$$

and since $|\underset{\sim}{A}\underset{\sim}{B}\underset{\sim}{C}| = |\underset{\sim}{A}||\underset{\sim}{B}||\underset{\sim}{C}|$, Eq (20.19) above becomes

$$|\underset{\sim}{W}|\,\left|\underset{\sim}{I} + \underset{\sim}{\Sigma}\underset{\sim}{G}_c^{\Sigma}\right|\,|\underset{\sim}{W}^T| = 0 \qquad (20.20)$$

And now either by

(i) Invoking the fact that $\underset{\sim}{W}$ is <u>not</u> singular so that $|\underset{\sim}{W}| \neq 0$ and by the same token $|\underset{\sim}{W}^T| \neq 0$ obtain the required result that the characteristic equation becomes

$$\left|\underset{\sim}{I} + \underset{\sim}{\Sigma}\underset{\sim}{G}_c^{\Sigma}\right| = 0$$

or (ii) by noting that determinants are scalars and can therefore be multiplied in any order, so that

Eq (20.20) above may be rewritten as:

$$|\underline{W}||\underline{W}^T||\underline{I} + \Sigma \underline{G}_c^{\Sigma}| = 0 \qquad (20\text{-}21)$$

and since $|\underline{W}| = \dfrac{1}{|\underline{W}^T|}$, (20.21) above reduces immediately to

$$|\underline{I} + \Sigma \underline{G}_c^{\Sigma}| = 0$$ as required.

20.10 (a) The characteristic equation to be used in investigating stability is

$$|\underline{I} + \underline{G}\underline{G}_c| = 0$$

In this case, with $\underline{G}$ as given in eqn (P20.14),

$$(\underline{I} + \underline{G}\underline{G}_c) = \begin{bmatrix} \dfrac{2K_{c_1}}{(s+1)} + 1 & \dfrac{K_{c2}}{(3s+1)} \\ \dfrac{4K_{c_1}}{(s+1)} & 1 + \dfrac{K_{c_2}}{(2s+1)} \end{bmatrix}$$

and for $K_{c_1} = 1$, $K_{c_2} = 5$, obtain

$$(\underline{I} + \underline{G}\underline{G}_c) = \begin{bmatrix} \dfrac{s+3}{s+1} & \dfrac{5}{3s+1} \\ \dfrac{4}{s+1} & \dfrac{2s+6}{2s+1} \end{bmatrix}$$

Upon taking the determinant and rearranging, obtain the characteristic equation:

$$3s^3 + 19s^2 + 13s - 1 = 0 \qquad (20.22)$$

and the presence of the -1 term is enough to indicate that the system with Eq (20.22) above as its characteristic equation is UNSTABLE.

(b) When the pairing is switched, the process model becomes

$$\bar{G}(s) = \begin{bmatrix} \dfrac{1}{3s+1} & \dfrac{2}{s+1} \\ \dfrac{1}{2s+1} & \dfrac{4}{s+1} \end{bmatrix}$$

and in this case, for the same $K_{c_1} = 1$, $K_{c_2} = 5$,

$$\left(\underset{\sim}{I} + \bar{G}G_c\right) = \begin{bmatrix} \dfrac{3s+2}{3s+1} & \dfrac{10}{s+1} \\ \dfrac{1}{2s+1} & \dfrac{s+21}{s+1} \end{bmatrix}$$

with the characteristic equation:

$$6s^3 + 133s^2 + 119s + 32 = 0 \qquad (20.23)$$

For this system to be stable, Routh's criterion indicates that the following condition must hold:

$$133 \times 119 > 6 \times 32$$

Since this inequality is true, we conclude that the system is indeed stable.

It is interesting to note that the RG parameter for the original y_1-m_1/y_2-m_2 pairing is -1.

CHAPTER 21

21.1 (a) A study of the transfer function matrix shows that y_1 is affected only by m_1, but y_2 is affected by <u>both</u> m_1 and m_2. Thus there is evidence of <u>one-way</u> interaction. Clearly, y_2 is expected to be more susceptible; it is affected by <u>both</u> input variables.

(b) From the given $\underline{G}(s)$, obtain

$$\underset{\sim}{K} = \begin{bmatrix} 0.81 & 0 \\ 0.55 & -0.013 \end{bmatrix} \qquad (21.1)$$

For this 2x2 system,

$$\lambda = \frac{1}{1-\zeta} \ ; \quad \zeta = \frac{K_{12}K_{21}}{K_{11}K_{22}}$$

$$\Rightarrow \qquad \lambda = \frac{K_{11}K_{22}}{K_{11}K_{22} - K_{12}K_{21}} \qquad (21.2)$$

$$= 1$$

so that the required RGA is

$$\Lambda = \begin{bmatrix} 1 & 0 \\ 0 & 1 \end{bmatrix}$$

suggesting that the $y_1 - m_1 / y_2 - m_2$ pairing will be "ideal", but not in the sense of a total absence of interaction. This is a case in which the interaction goes "one-way", hence the perfect RGA (See Eqns (21.43) and (21.44) and the text that follows in the main text)

21.2 (a) For loop 1, (Production rate controlled by monomer flow rate) the "process transfer function" used for the controller design is:

$$g_{11}(s) = \frac{0.81}{2.6s + 1}$$

Observe that there is no time delay. Thus use Cohen and Coon settings with the minimum possible α/τ ratio, i.e, $\alpha/\tau = 0.1$. From Table 15.3 on p 537 in the text obtain

$$K_{c_1} = \frac{1}{0.81}\left(\frac{1}{0.1}\right)\left[0.9 + \frac{1}{12}(0.1)\right] = 11.21$$

Combining the "assumed" α/τ ratio of 0.1 with the actual transfer function time constant, obtain the "design value" of (0.1×2.6) for α; use in Table 15.3 to obtain

$$\tau_{I_1} = 0.26\left[\frac{30 + 3(0.1)}{9 + 20(0.1)}\right] = 0.716$$

Follow the same procedure, with

$$g_{22}(s) = \frac{-0.013}{3.5s + 1}$$

for loop 2 and obtain $K_{c_2} = -699$; $\tau_{I_2} = 0.964$

(b) The overall closed-loop system response to the production rate set-point change is shown below in FIGS S21.1(a),(b). The effect of loop 1 interaction on loop 2 (molecular weight loop) is seen to be relatively small. However, from the plot of the control action (FIG S21.1b) observe that this good control is achieved with the "expenditure" of (significant use of) chain transfer agent.

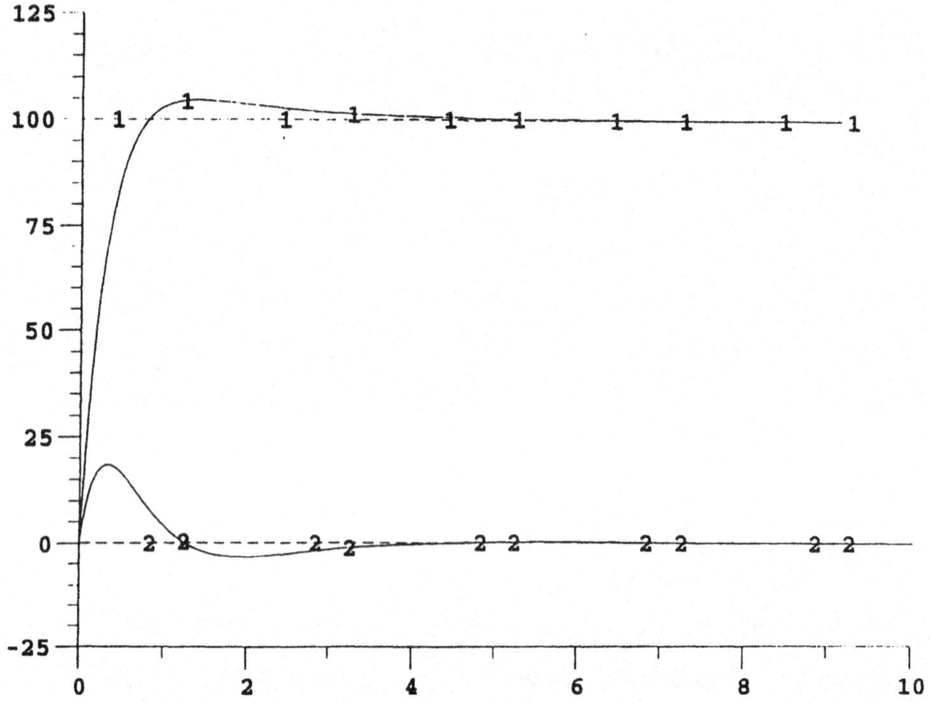

FIG S21.1(a) Process Outputs & Set points

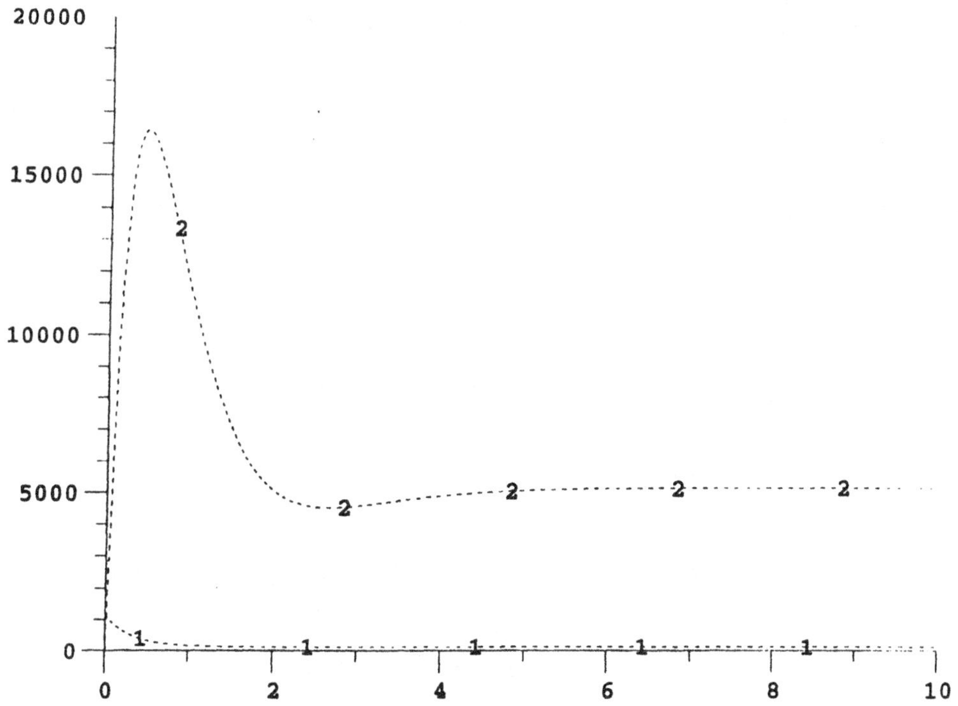

FIG S21.1(b) Manipulated Variables.

21.3 The steady state gain matrix for the process is

$$\underset{\sim}{K} = \begin{bmatrix} 0.7 & -0.005 \\ -34.7 & 0.9 \end{bmatrix} \qquad (21.3)$$

from which the RGA is immediately obtained as:

$$\Lambda = \begin{bmatrix} \textcircled{1.38} & -0.38 \\ -0.38 & \textcircled{1.38} \end{bmatrix}$$

indicating a $y_1 - m_1$, $y_2 - m_2$ pairing, and a resulting control system that will experience some interactions.

21.4 (a) Assuming a pre-set $y_2 - m_2$ pairing, obtain from the given $\underset{\sim}{G}(s)$ that for the relevant subsystem

$$G(0) = \begin{bmatrix} -0.012 & -0.00175 \\ -0.0043\,I & 8.6\times10^{-5}\,I \end{bmatrix} \qquad (21.4)$$

where $I = \lim_{s\to0} \frac{1}{s}$

The RGA computed from this $\underline{G}(0)$ is:

$$\begin{bmatrix} 0.12 & \textcircled{0.88} \\ \textcircled{0.88} & 0.12 \end{bmatrix}$$

indicating a $y_1 - m_3$ and $y_3 - m_1$ pairing.

(b) The required responses are shown in FIGS S21.2 (process outputs) and S21.3 (manipulated variables) The response under the "new configuration" is shown in the <u>solid lines</u>. Observe that in either case, the same steady state conditions are attained; the difference lies in the transient path taken to this steady state.

(c) The "new configuration" clearly affords much tighter control of the critical T_{36} variable; however, this is achieved at the expense of good level control. This is obvious from FIG S21.2. But good level control is not as critical as tight control of T_{36}; the new configuration is therefore preferable under these circumstances.

The RGA recommended the "old configuration"

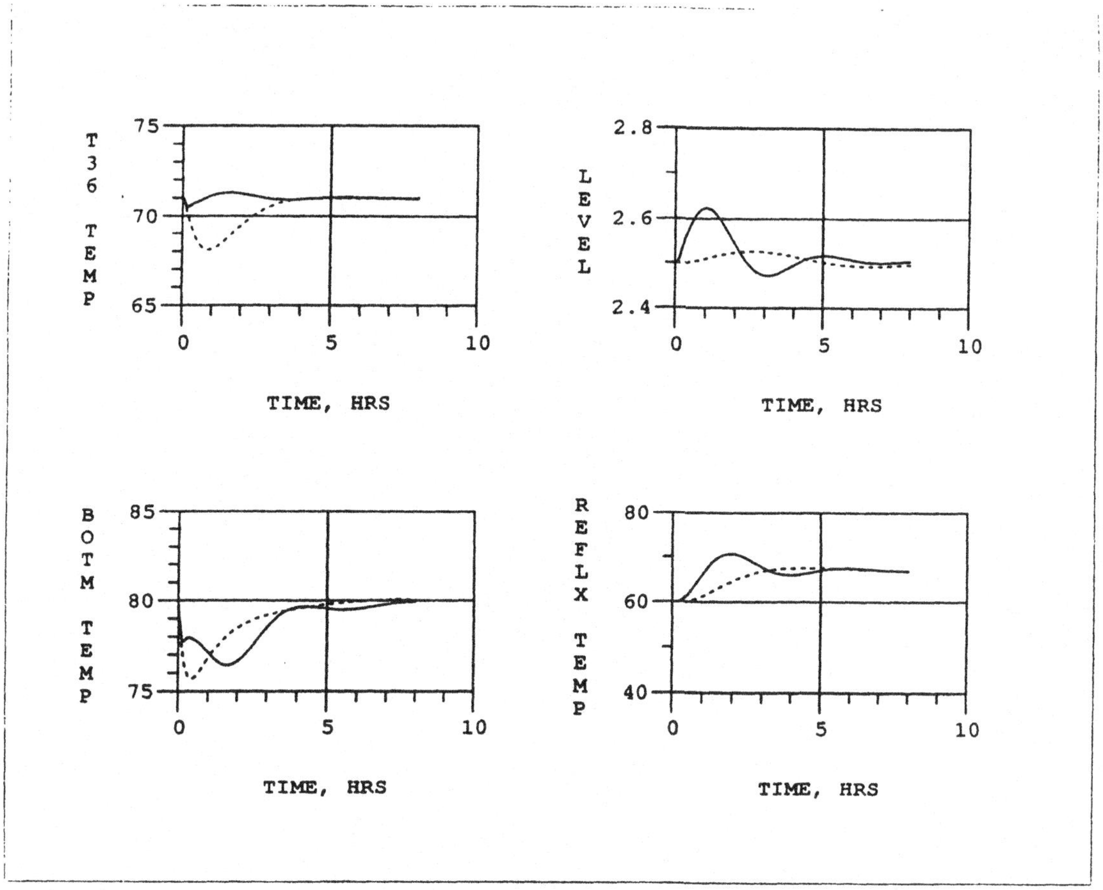

FIG S21.2 Distillation process output variables
—— New Configuration
- - - - Original Configuration

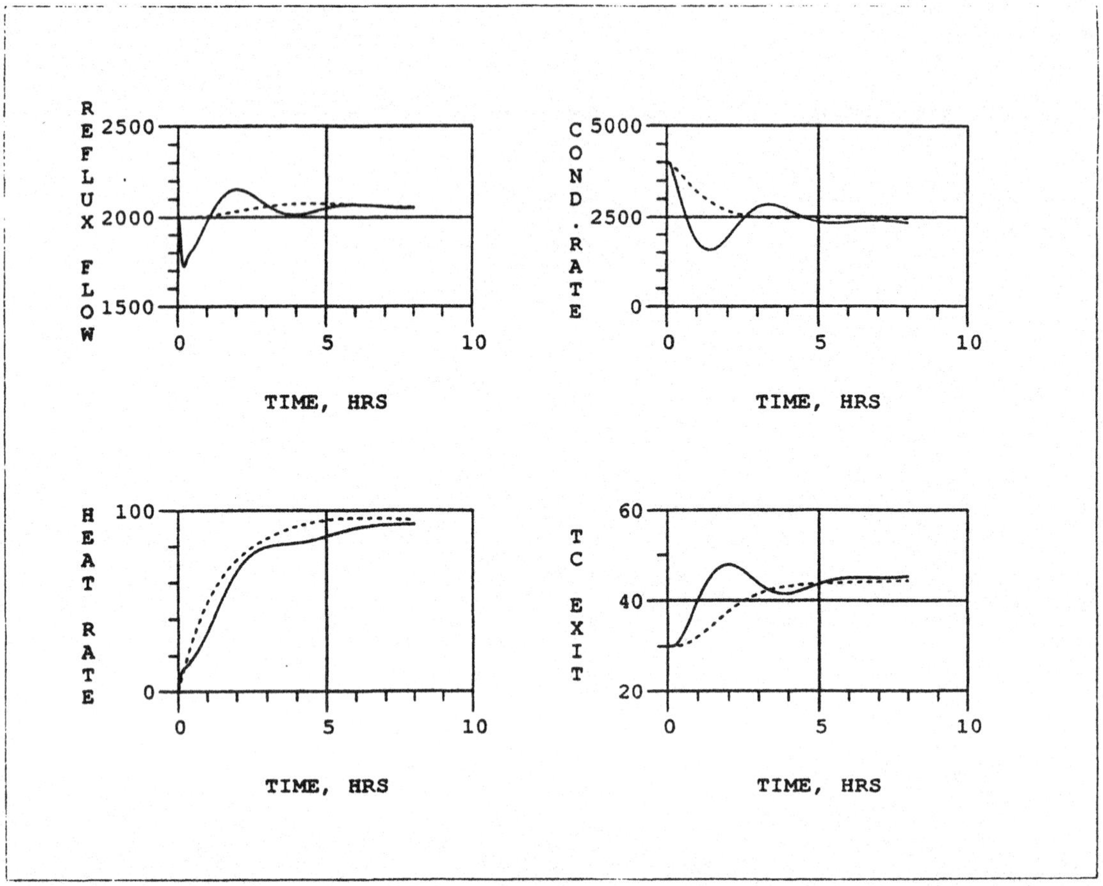

FIG S21.3 Distillation process manipulated variables
—— New Configuration
----- Original Configuration

because all variables are considered equally important in computing the RGA. In this particular case, it is perfectly acceptable to achieve tighter control of T_{36} at the expense of poor level control.

The reflux flow rate will clearly be a more effective manipulated variable to use in controlling T_{36}: its effect on T_{36} will be immediate and significant. The effect on the condenser cooling rate on T_{36} is indirect and sluggish as a result of the physics of the process. Observe that a change in the cooling rate will (i) cause more "overhead" to be condensed, raising the level in the reflux receiver, however (ii) because the reflux receiver is a 3-ft diameter drum, it will take a significant amount of cooling before there is a noticeable change in the reflux drum level; but (iii) since the reflux flow rate will only increase to reduce the level in the reflux drum, very litte change will be observed in the reflux flow rate; (iv) any serious change in T_{36} will only arise from a change in the reflux flow rate.

Observe therefore how relatively ineffective the condenser cooling rate will be if tight control of T_{36} is desired; on the other hand, such a configuration will make tight level control easy to achieve.

(d). The required responses are shown in FIGS S21.4 (process outputs) and S21.5 (manipulated variables) with the "new configuration" response shown in the solid lines. Clearly, the "new configuration" provides tighter T_{36} control but at the expense of an offset in Level, and sluggish control of Bottoms Temp.

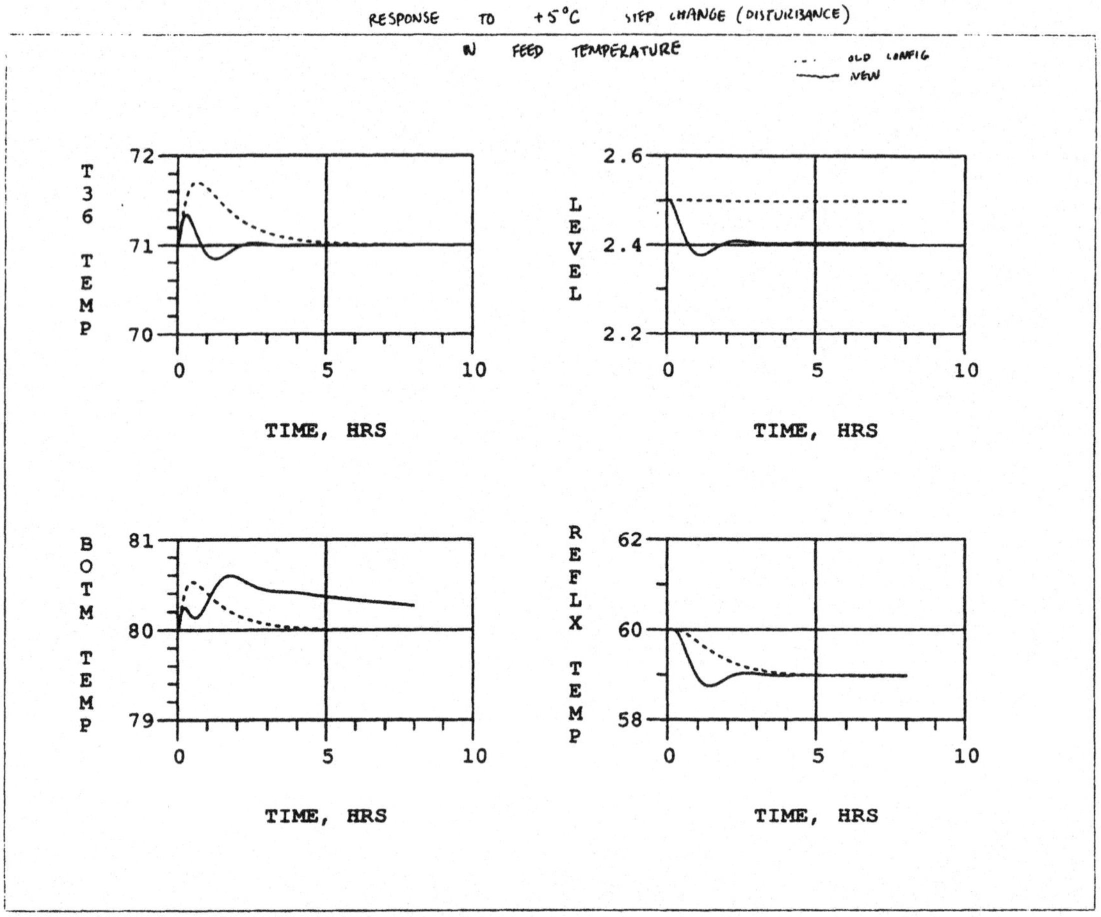

FIG S21.4 Distillation Process Outputs: Response to +5°C Step change (Disturbance) in Feed Temperature.

—— New Configuration

- - - - Original Configuration.

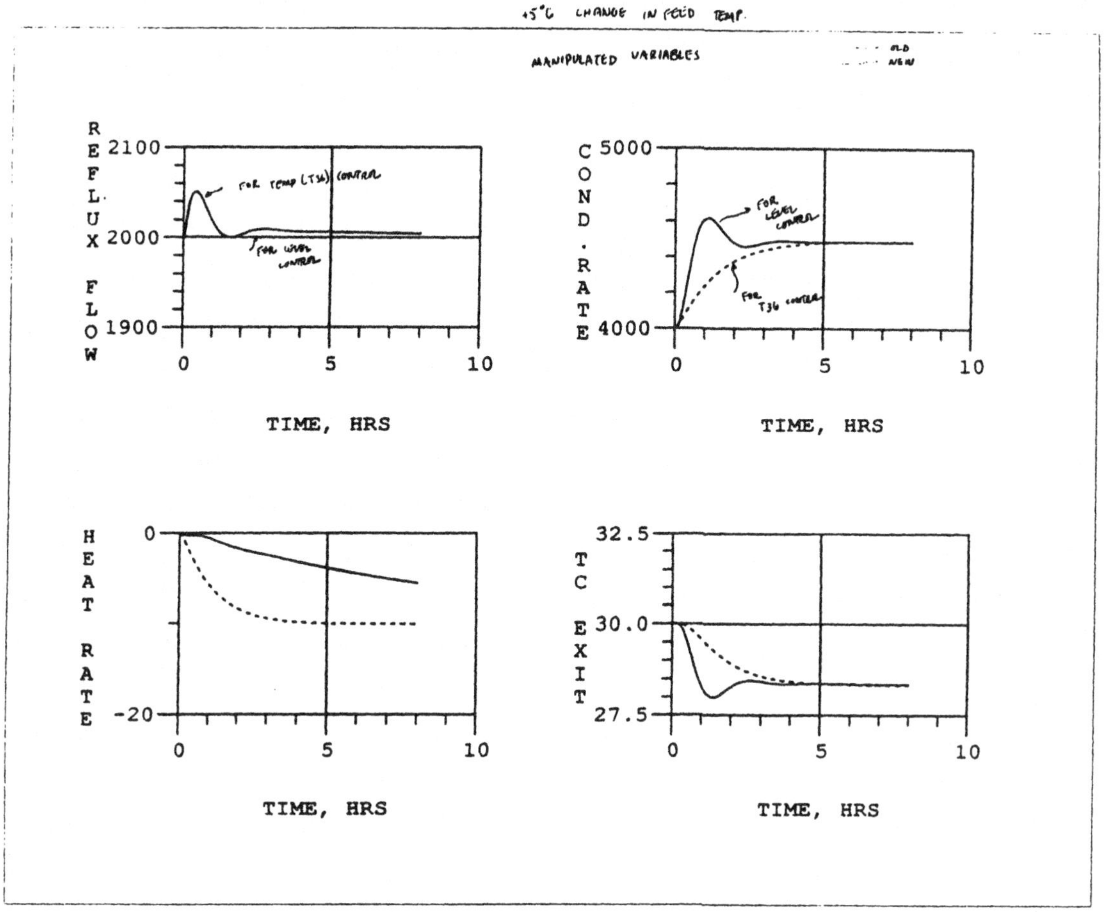

FIG S21.5 Distillation Process Manipulated Variables Corresponding to FIG S21.4

—— New Configuration

- - - - Original Configuration

21.5 By inspecting the transfer function, observe that y_2 is affected only by m_4, clearly indicating that the $y_2 - m_4$ pairing is not only desirable, it is the only means by which y_2 is controllable.

The resulting process gain matrix, after eliminating the obvious $y_2 - m_4$ coupling is:

$$\begin{bmatrix} y_1 \\ y_3 \end{bmatrix} = \begin{bmatrix} 0.5 & 0.07 & 0.04 \\ 0.004 & -0.003 & -0.006 \end{bmatrix} \begin{bmatrix} m_1 \\ m_2 \\ m_3 \end{bmatrix}$$

From this 2x3 system, one is able to obtain three subsystems:

1. $$\begin{bmatrix} y_1 \\ y_3 \end{bmatrix} = \underset{\sim}{K}_I \begin{bmatrix} m_1 \\ m_2 \end{bmatrix} \quad \text{with} \quad \underset{\sim}{K}_I = \begin{bmatrix} 0.5 & 0.07 \\ 0.004 & -0.003 \end{bmatrix}$$

and a corresponding RGA

$$\underset{\sim}{\Lambda}_1 = \begin{bmatrix} 0.843 & 0.157 \\ 0.157 & 0.843 \end{bmatrix}$$

2. $$\begin{bmatrix} y_1 \\ y_3 \end{bmatrix} = \underset{\sim}{K}_2 \begin{bmatrix} m_1 \\ m_3 \end{bmatrix} \quad \text{with} \quad \underset{\sim}{K}_2 = \begin{bmatrix} 0.5 & 0.04 \\ 0.004 & -0.006 \end{bmatrix}$$

and a corresponding RGA

$$\underset{\sim}{\Lambda}_2 = \begin{bmatrix} 0.949 & 0.051 \\ 0.051 & 0.949 \end{bmatrix}$$

and finally,

3. $$\begin{bmatrix} y_1 \\ y_3 \end{bmatrix} = \underset{\sim}{K}_3 \begin{bmatrix} m_2 \\ m_3 \end{bmatrix} \quad \text{with} \quad \underset{\sim}{K}_3 = \begin{bmatrix} 0.07 & 0.04 \\ -0.003 & -0.006 \end{bmatrix}$$

and the corresponding RGA

$$\underset{\sim}{\Lambda}_3 = \begin{bmatrix} 1.4 & -0.4 \\ -0.4 & 1.4 \end{bmatrix}$$

Since the "best" of these 3 RGA's is $\underset{\sim}{\Lambda}_2$ for subsystem 2, and since this RGA recommends a $y_1 - m_1$, $y_3 - m_3$ pairing, then the overall recommendation is a $y_1 - m_1 / y_2 - m_4 / y_3 - m_3$ pairing. It is necessary to check this configuration for structural stability. In this case, the resulting full system gain matrix is given by

$$\begin{bmatrix} y_1 \\ y_2 \\ y_3 \end{bmatrix} = \begin{bmatrix} 0.5 & 0.01 & 0.04 \\ 0 & 0.5 & 0 \\ 0.004 & 0 & -0.006 \end{bmatrix} \begin{bmatrix} m_1 \\ m_4 \\ m_3 \end{bmatrix}$$

for which $$\frac{|K|}{\prod_{i=1}^{3} k_{ii}} = \frac{-0.00158}{-0.0015} = 1.05 \text{ which is } > 0;$$

hence the system is stable.

21.6 From the original 2x4 system, form the 6 possible 2x2 subsystems and obtain the corresponding RGA's as follows:

1. For m_1/m_2 ;
$$\underset{\sim}{K}_1 = \begin{bmatrix} 4.05 & 1.77 \\ 4.38 & 4.42 \end{bmatrix}$$
so that
$$\underset{\sim}{\Lambda}_1 = \begin{bmatrix} 1.76 & -0.76 \\ -0.76 & 1.76 \end{bmatrix}$$

2. For m_1/m_3 ;
$$\underset{\sim}{K}_2 = \begin{bmatrix} 4.05 & 5.88 \\ 4.38 & 7.2 \end{bmatrix}$$
so that
$$\underset{\sim}{\Lambda}_2 = \begin{bmatrix} 8.56 & -7.56 \\ -7.56 & 8.56 \end{bmatrix}$$

3. For m_1/m_4 ;
$$\underset{\sim}{K}_3 = \begin{bmatrix} 4.05 & 1.44 \\ 4.38 & 1.26 \end{bmatrix}$$
so that
$$\underset{\sim}{\Lambda}_3 = \begin{bmatrix} -4.24 & 5.24 \\ 5.24 & -4.24 \end{bmatrix}$$

4. For m_2/m_3 ;
$$\underset{\sim}{K}_4 = \begin{bmatrix} 1.77 & 5.88 \\ 4.42 & 7.2 \end{bmatrix}$$
so that
$$\underset{\sim}{\Lambda}_4 = \begin{bmatrix} -0.96 & 1.96 \\ 1.96 & -0.96 \end{bmatrix}$$

5. For m_2/m_4; $\underset{\sim}{K}_5 = \begin{bmatrix} 1.77 & 1.44 \\ 4.42 & 1.26 \end{bmatrix}$

so that $\underset{\sim}{\Lambda}_5 = \begin{bmatrix} -0.54 & 1.54 \\ 1.54 & -0.54 \end{bmatrix}$

6. For m_3/m_4; $\underset{\sim}{K}_6 = \begin{bmatrix} 5.88 & 1.44 \\ 7.2 & 1.26 \end{bmatrix}$

and $\underset{\sim}{\Lambda}_6 = \begin{bmatrix} -2.5 & 3.5 \\ 3.5 & -2.5 \end{bmatrix}$

The "best" RGA is $\underset{\sim}{\Lambda}_5$ for subsystem 5 involving (y_1, y_2) and (m_2, m_4); with the recommended pairing $y_1 - m_4 / y_2 - m_2$.

21.7 (a) The main property of the RGA required here is:

$$\sum_j \lambda_{ij} = \sum_i \lambda_{ij} = 1.$$

In this particular case, for column 2,

$$\lambda_{12} + 0.8 - 0.2 = 1$$

$$\Rightarrow \quad \lambda_{12} = 0.4 \qquad (21.5)$$

Now for row 1, $\lambda_{11} + \lambda_{12} - 0.6 = 1$

From (21.5), obtain $\lambda_{11} = 1.2$ (21.6)

Similarly use (21.6) and column 1 to obtain

$$\lambda_{21} = 0 \qquad (21.7)$$

Next use (21.7) and row 2 to obtain

$$\lambda_{23} = 0.2 \qquad (21.8)$$

Finally, either from column 3 and (21.8), or row 3 independently, obtain

$$\lambda_{33} = 1.4$$

so that the complete RGA is given by.

$$\begin{bmatrix} \textcircled{1.2} & 0.4 & -0.6 \\ 0 & \textcircled{0.8} & 0.2 \\ -0.2 & -0.2 & \textcircled{1.4} \end{bmatrix}$$

indicating a $y_1 - m_1 / y_2 - m_2 / y_3 - m_3$ recommended pairing. However, it may still be necessary to check the Nierderlinski index for structural stability. For this the steady state gain matrix $\underset{\sim}{K}$ is needed, but this is unavailable and <u>cannot</u> be back calculated from the RGA.

(b) use the same approach as in (a) above to obtain

$\lambda_{22} = -0.43$ (from row 2)

$\lambda_{42} = -2.03$ (column 2)

$\lambda_{43} = 0.97$ (row 4)

$\lambda_{13} = -0.53$ (column 3)

$\lambda_{11} = 1.47$ (row 1)

$\lambda_{31} = -0.68$ (column 1)

$\lambda_{34} = -1.9$ (row 3)

so that the desired, complete RGA is obtained as:

$$\Lambda = \begin{bmatrix} \circled{1.47} & 0.15 & -0.53 & -0.09 \\ -0.01 & -0.43 & 0.29 & \circled{1.15} \\ -0.68 & \circled{3.31} & 0.27 & -1.9 \\ 0.22 & -2.03 & \circled{0.97} & 1.84 \end{bmatrix}$$

suggesting a $y_1 - m_1 / y_2 - m_4 / y_3 - m_2 / y_4 - m_3$ pairing. Once more, a check on the Nierderlinski index is needed but cannot be done because of the unavailability of K, the process steady state gain matrix.

21.8 By mere inspection of the process transfer function matrix observe:

(i) m_4 affects <u>only</u> y_3, and

(ii) y_4 is affected <u>only</u> by m_5

indicating a $y_3 - m_4$, and $y_4 - m_5$ pairing as obvious. The remaining 2x3 system, after eliminating the obvious pairings has the following steady state relationship:

$$\begin{bmatrix} y_1 \\ y_2 \end{bmatrix} = \begin{bmatrix} 0.34 & 0.21 & 0.5 \\ -0.41 & 0.66 & -0.3 \end{bmatrix} \begin{bmatrix} m_1 \\ m_2 \\ m_3 \end{bmatrix}$$

From here obtain 3 2x2 subsystems, & corresponding RGA's.

1. (using m_1/m_2) $\Lambda_1 = \begin{bmatrix} 0.723 & 0.277 \\ 0.277 & 0.723 \end{bmatrix}$

2. (using m_1/m_3); $\Lambda_2 = \begin{bmatrix} -0.99 & 1.99 \\ 1.99 & -0.99 \end{bmatrix}$

3. (using m_2/m_3); $\Lambda_3 = \begin{bmatrix} 0.16 & 0.84 \\ 0.84 & 0.16 \end{bmatrix}$

The "best" RGA is Λ_3, and the recommended pairing is $y_1 - m_3$; $y_2 - m_2$. The resulting overall 4×4 pairing is therefore: $y_1 - m_3 / y_2 - m_2 / y_3 - m_4 / y_4 - m_5$. The gain matrix for this configuration is:

$$K = \begin{bmatrix} 0.5 & 0.21 & 0 & 6.46 \\ -0.3 & 0.66 & 0 & -3.72 \\ -0.71 & 0.49 & -0.20 & -4.71 \\ 0 & 0 & 0 & 1.03 \end{bmatrix}$$

For which the Nierderlinski index $\dfrac{|K|}{\prod_{i=1}^{4} k_{ii}} = \dfrac{-0.0810}{-0.06798} = 1.19$

which is >0, hence the pairing is acceptable.

21.9 (a) The three RGA's are easily obtained as:

$\Lambda_A = \begin{bmatrix} 0.5 & 0.5 \\ 0.5 & 0.5 \end{bmatrix}$; $\Lambda_B = \begin{bmatrix} 0.5 & 0.5 \\ 0.5 & 0.5 \end{bmatrix}$; $\Lambda_C = \begin{bmatrix} 115 & -114 \\ -114 & 115 \end{bmatrix}$

suggesting significant interactions for systems A and B, and extremely poor performance for system C.

(b) $\underset{\sim}{K}_C$ is very nearly singular, with row 2 almost exactly ½ of row 1. The condition number for this matrix is 736.55. The near linear dependency indicates poor conditioning which is reflected in the unusually high value for λ_{11}.

(c) System B should be easier to control than system A; the singular values for $\underset{\sim}{K}_A$ and $\underset{\sim}{K}_B$ are

$$\sigma(\underset{\sim}{K}_A) = 14.142, \quad 0.0566 \Rightarrow \kappa(\underset{\sim}{K}_A) = 250$$

$$\sigma(\underset{\sim}{K}_B) = 1.4142, \quad 0.5657 \Rightarrow \kappa(\underset{\sim}{K}_B) = 2.5$$

$\underset{\sim}{K}_B$ is "better conditioned" than $\underset{\sim}{K}_A$. This fact is not reflected in the identical RGA's, but a close inspection of the original gain matrices reveals this fundamental structural difference.

21.10 (a) The RGA for the transfer function is:

$$\Lambda = \begin{bmatrix} 1.159 & -0.159 \\ -0.159 & 1.159 \end{bmatrix}$$

Observe that the time constants are not too far apart; and the delays in y_1 are not too large. Thus, the RGA ought to provide a reasonable assessment of the overall interactions.

The assessment is that "OGZ" interactions with Hydrochlorothiazide will not be severe, primarily because of the closeness of 1.159 to 1. Recommended pairing: HCT to regulate Blood Pressure, and

OGZ to regulate urine production.

(b). From the given transfer function, one can compute these final steady state values as follows:

Given $u_1 = 25$; $u_2 = 0.5(y_{d_2} - y_2)$, obtain

$$y_1^* = (-0.04)(25) + 0.0005(-y_2^*)$$

$$y_2^* = (0.22)(25) - 0.02(-y_2^*)$$

since $y_{d_2} = 0$. Solve equations for y_2^* and y_1^* to obtain

$$y_1^* = -1.002$$

$$y_2^* = 5.556,$$ the max for y_2.

The actual simulation results are shown in FIG S21.6

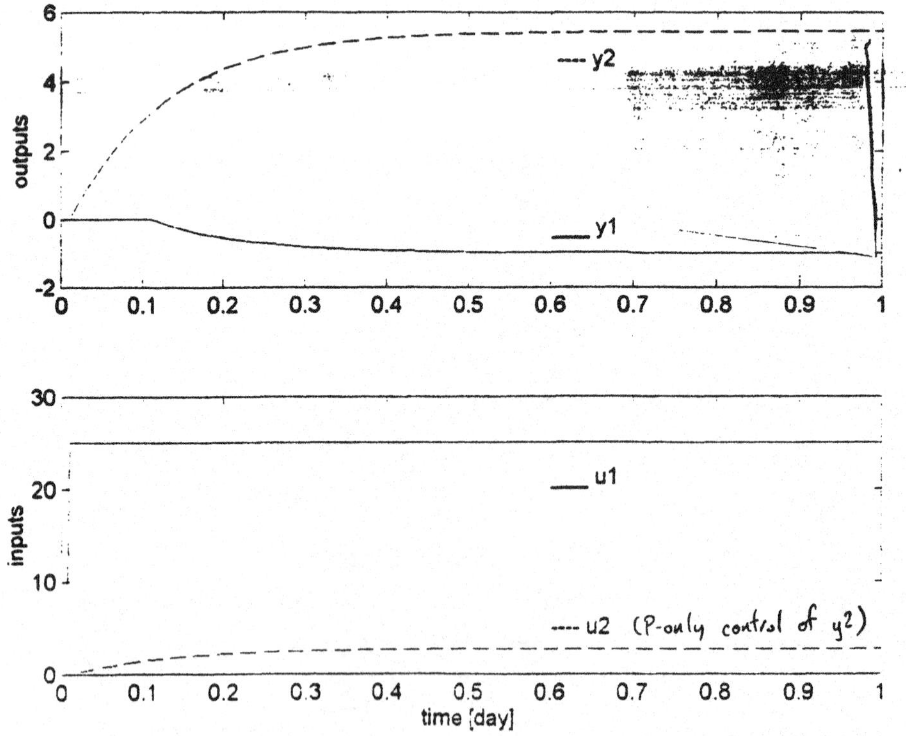

FIG S21.6 System response to HCT Step.

(c) Representing the HCT infusion as a delta function produces the response shown in FIG S21.7. Observe that while in FIG S21.6 the blood pressure drop is sustained, in this "new" case, the drop is temporary, lasting about 3/10th of a day. The urine production as a result hits a maximum value of about 44 and gradually returns to the initial value over a period of about 1/2 day.

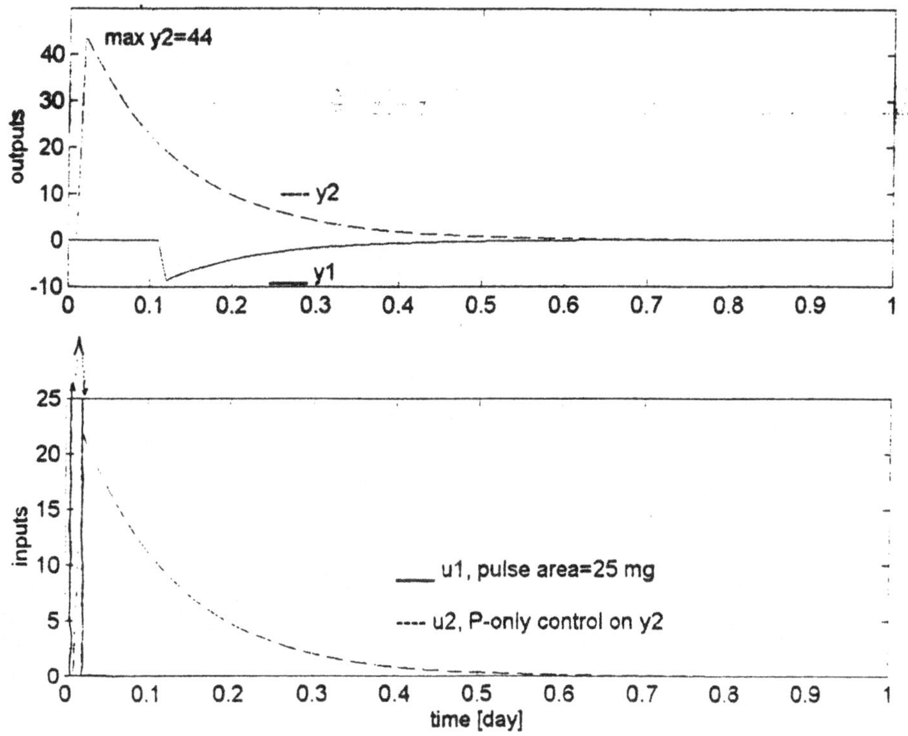

FIG S21.7 System response to HCT impulse.

CHAPTER 22

22.1 (a) For the given transfer function matrix model, the required simplified dynamic decoupler (as obtained) in Example 22.1 is:

$$\underset{\sim}{G_I} = \begin{bmatrix} 1 & g_{I_1} \\ g_{I_2} & 1 \end{bmatrix} \quad \text{with} \quad g_{I_1} = 1.48\frac{(16.7s+1)}{(21.0s+1)}e^{-2s}$$

$$\text{and} \quad g_{I_2} = 0.34\frac{(14.4s+1)}{(10.9s+1)}e^{-4s}$$

Implementing this along with the two supplied single loop PI controllers (see Fig 22.1, p775 of the main text), yields the overall closed-loop response shown in FIG S22.1 (for a step change of 0.1 in the y_1 set point). The y_1 response is "swifter" than that shown in FIG 22.7 (main text); but the most significant difference lies in the y_2 response: with dynamic decoupling, y_2 remains entirely unaffected; with only steady state decoupling, the y_2 response shows the effect of residual interactions.

(b) Implementing the decoupler of part (a) and the controller this time on the "process" modelled by Eqn (P22.1) gives rise to the response shown in FIG S22.2, which is UNSTABLE. Thus the indicated plant/model mismatch is sufficient to destabilize the process. [SHOWN IN FIG S22.3 IS THE PROCESS RESPONSE <u>WITHOUT</u> THE DECOUPLER, using only the PI controllers!]

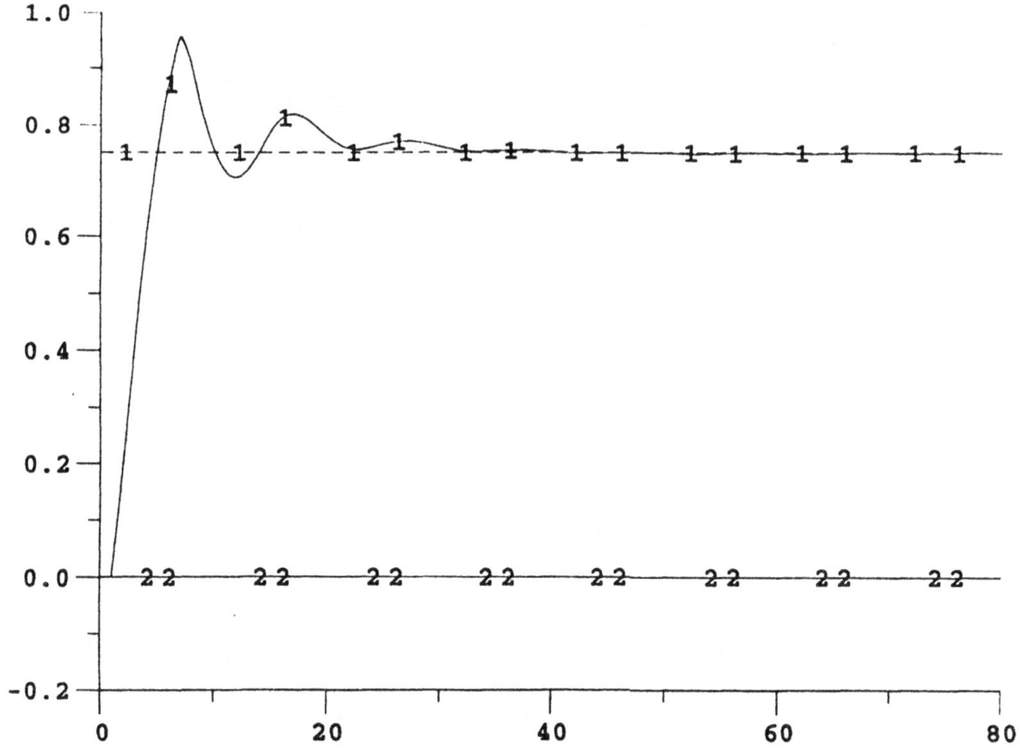

FIG S22.1. Closed-loop response: Perfect dynamic decoupling

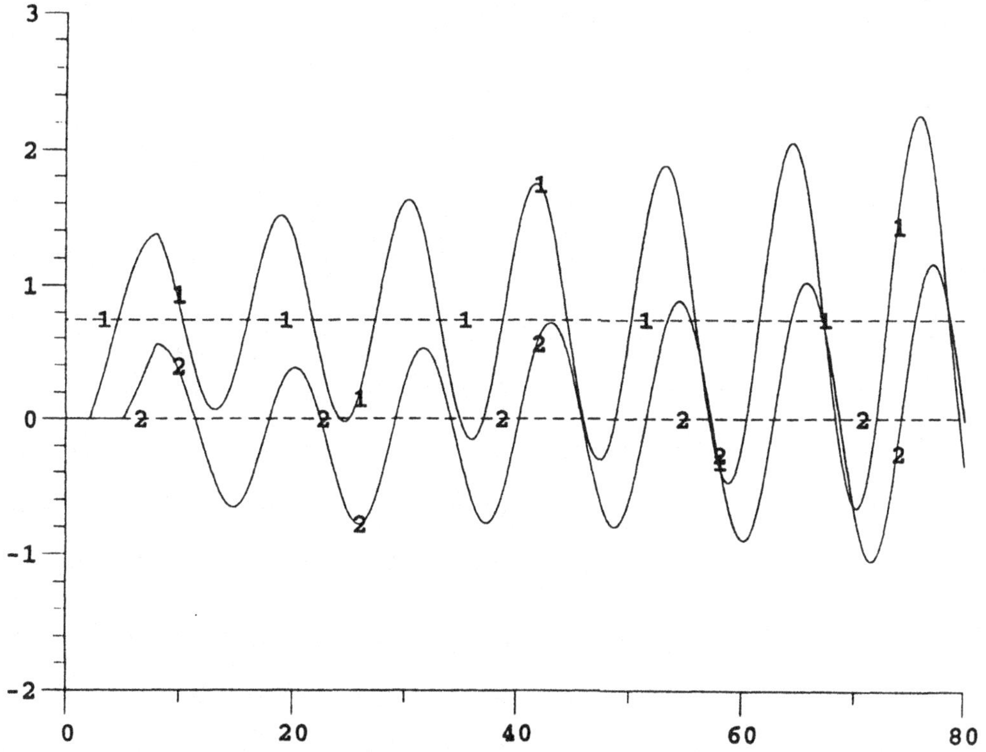

FIG S22.2. Closed-loop response: Effect of plant/model mismatch

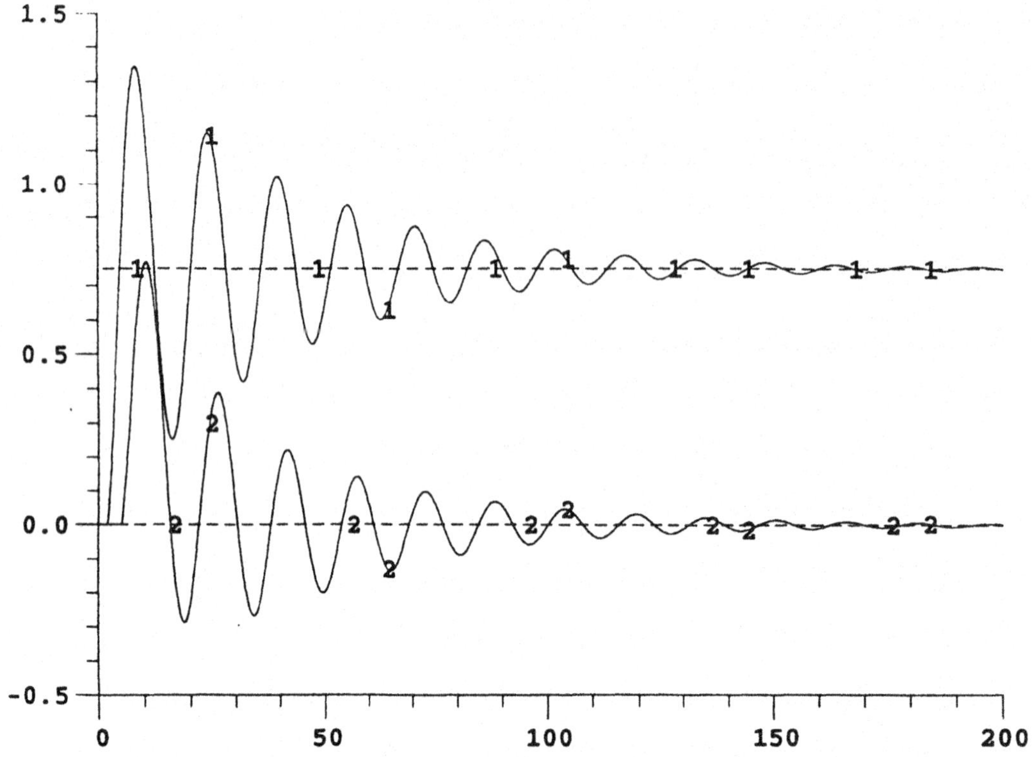

FIG S22.3 Closed-loop response: No decoupling (but with significant plant/model mismatch).

22.2 (a) The required RGA is

$$\Lambda = \begin{bmatrix} \boxed{3.676} & -1.922 & -0.754 \\ -2.382 & \boxed{2.967} & 0.415 \\ -0.294 & -0.045 & \boxed{1.340} \end{bmatrix} \qquad (22\text{-}1)$$

confirming the 1-1/2-2/3-3 pairing implied in the main text's Fig 22.6. The indication here is that there will be significant steady state interaction among the loops with loop 3 experiencing the least amount.

(b) and (c)   SOLUTION NOT PROVIDED.

22.3 (a) The three singular values are (by loop number)
$$\sigma_1 = 194$$
$$\sigma_2 = 8.99 \times 10^{-5}$$
$$\sigma_3 = 1542$$

with the condition number $\kappa = \dfrac{1542}{8.99 \times 10^{-5}} = 1.714 \times 10^7$

implying a very ill-conditioned process.

(b) The new scaling gives rise to a steady state gain matrix

$$\underset{\sim}{K} = \begin{bmatrix} 119 & 153 & -21 \\ 370 & 767 & -50 \\ 930 & -667 & -1033 \end{bmatrix} \quad (22.2)$$

with a new condition number $\kappa = 77.5$, indicating substantial improvement in the conditioning.

(c). The complete SVD:
$$\underset{\sim}{K} = \underset{\sim}{U}\underset{\sim}{\Sigma}\underset{\sim}{V}^T \quad ; \quad \underset{\sim}{\Sigma} = \begin{bmatrix} 1544 & 0 & 0 \\ 0 & 870 & 0 \\ 0 & 0 & 19.9 \end{bmatrix}$$

$$\underset{\sim}{U} = \begin{bmatrix} 8.1134 \times 10^{-3} & -0.225 & 0.9749 \\ -0.0689 & -0.97 & -0.22 \\ 0.998 & -0.0654 & -0.023 \end{bmatrix}$$

and
$$\underset{\sim}{V} = \begin{bmatrix} 0.58 & -0.51 & 0.63 \\ -0.46 & -0.85 & -0.26 \\ -0.67 & 0.14 & 0.73 \end{bmatrix}$$

[REST OF THE SOLUTION NOT PROVIDED].

22.4 (a) Using the simplified approach, obtain the desired decoupler as:

$$\underset{\sim}{G}_I = \begin{bmatrix} 1 & g_{I_1}(s) \\ g_{I_2}(s) & 1 \end{bmatrix} \quad \text{with } g_{I_1}(s) = \frac{-g_{12}}{g_{11}} = -e^{+2s}$$

$$\text{and } g_{I_2}(s) = \frac{-g_{21}}{g_{22}} = 0$$

so that
$$\underset{\sim}{G}_I = \begin{bmatrix} 1 & -e^{+2s} \\ 0 & 1 \end{bmatrix} \qquad (22.3)$$

With the generalized approach, from Eq.(22.29) of the main text,

$$\underset{\sim}{G}_I = \underset{\sim}{G}^{-1}\underset{\sim}{G}_R$$

& choosing $\underset{\sim}{G}_R$ as $\text{Diag}[\underset{\sim}{G}(s)]$, upon inverting $\underset{\sim}{G}$, obtain

$$\underset{\sim}{G}_I = \begin{bmatrix} -(10s+1)e^{2s} & -(60s+1)e^{12s} \\ 0 & +(60s+1)e^{10s} \end{bmatrix} \begin{bmatrix} \frac{-e^{-2s}}{10s+1} & 0 \\ 0 & \frac{e^{-10s}}{60s+1} \end{bmatrix}$$

finally obtain

$$\underline{G}_I = \begin{bmatrix} 1 & -e^{+2s} \\ 0 & 1 \end{bmatrix}$$

precisely as in (22.3) above.

The decoupler is NOT realizable because of the e^{+2s} term requiring a prediction. Its implementation will require the value of $u_2(t+2)$ at time t.

(b) Ignoring the delay term in the (1,1) element results in

$$\tilde{G} = \begin{bmatrix} -\frac{1}{10s+1} & \frac{-1}{10s+1} \\ 0 & \frac{e^{-10s}}{60s+1} \end{bmatrix}$$

so that
$$\tilde{G}^{-1} = \begin{bmatrix} -(10s+1) & -e^{10s}(60s+1) \\ 0 & e^{10s}(60s+1). \end{bmatrix} \qquad (22.4)$$

The dynamic decoupler, $\tilde{G}_I$ in this case becomes

$$\tilde{G}_I(s) = \tilde{G}^{-1}\tilde{G}_R = \tilde{G}^{-1}\begin{bmatrix} \frac{-1}{(10s+1)} & 0 \\ 0 & \frac{e^{-10s}}{60s+1} \end{bmatrix}$$

$$\tilde{G}_I(s) = \begin{bmatrix} 1 & -1 \\ 0 & 1 \end{bmatrix} \qquad (22.5)$$

and now, either by using the generalized approach

$$\underset{\sim}{G}_I^{ss} = \underset{\sim}{G}(0)^{-1} G_R(0) = \begin{bmatrix} -1 & -1 \\ 0 & 1 \end{bmatrix}^{-1} \begin{bmatrix} -1 & 0 \\ 0 & 1 \end{bmatrix}$$

$$\underline{G}_I^{ss} = \begin{bmatrix} 1 & -1 \\ 0 & 1 \end{bmatrix} \qquad (22.6)$$

or the simplified approach:

$$\underline{G}_I^{ss} = \begin{bmatrix} 1 & g_{I_1}^{ss} \\ g_{I_2}^{ss} & 1 \end{bmatrix} \quad \text{where } g_{I_1}^{ss} = -1, \quad g_{I_2}^{ss} = 0$$

so that

$$\underline{G}_I^{ss} = \begin{bmatrix} 1 & -1 \\ 0 & 1 \end{bmatrix} \qquad (22.7)$$

observe that $\underline{G}_I^{ss}$ obtained is precisely as obtained in (22.5).

Implementing this decoupler along with the supplied controllers gives rise to the closed-loop response shown in FIG S22.4. Note that the decoupler is simple, and the "residual" interaction effect on y_1 (the normalized boiler underpressure) is not serious.

(c) In this case,

$$\underline{G}_m = \underline{G}\underline{D} = \begin{bmatrix} \dfrac{e^{-2s}}{10s+1} & \dfrac{e^{-2s}}{10s+1} \\ 0 & \dfrac{e^{-12s}}{60s+1} \end{bmatrix} \qquad (22.8)$$

and now designing the decoupler with $\underset{\sim}{G}_m$ gives

$$\underset{\sim}{G}_{I_M} = \begin{bmatrix} 1 & -1 \\ 0 & 1 \end{bmatrix} \qquad (22.9)$$

which is, of course realizable. (Also note that it is identical to the decouplers obtained earlier in (22·5), (22·6), (22·7)).

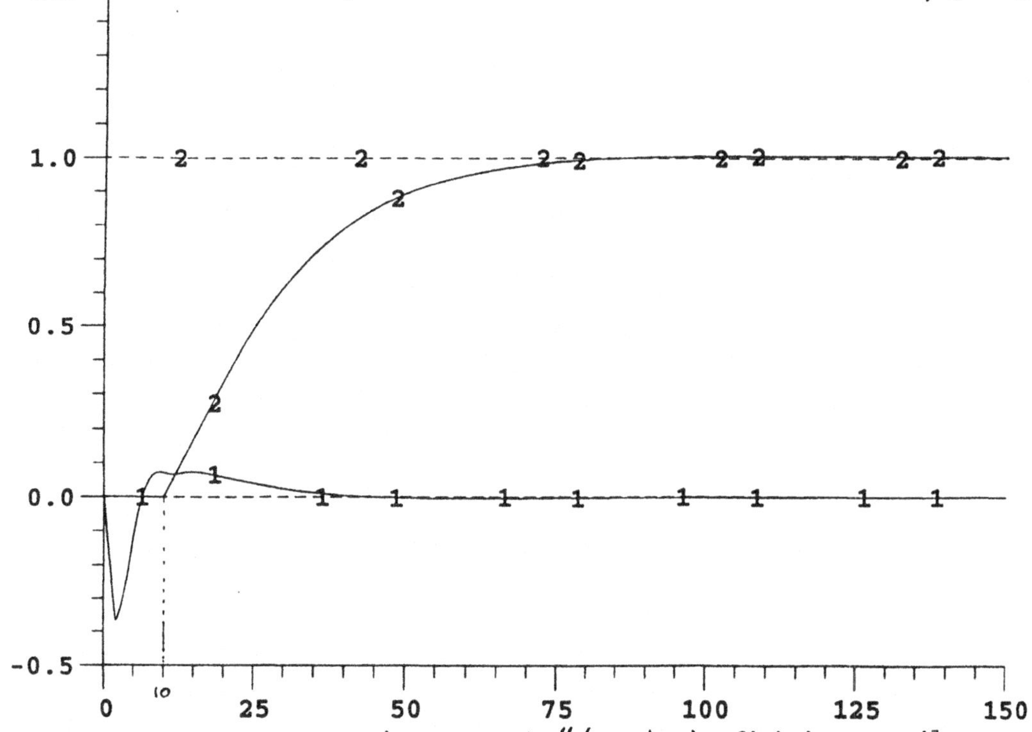

FIG S22.4. "Approximate" (or steady state) Decoupling: Closed-loop process response.

(d) Implementing this decoupler on the augmented process (see Fig 22·5, p 786 main text) gives rise to the response shown in FIG S22.5. The advantage here is clear: the dynamic decoupler is realizable, and eliminates <u>all</u> interactions; the disadvantage is that an additional 2 seconds delay is introduced into the y_2 response.

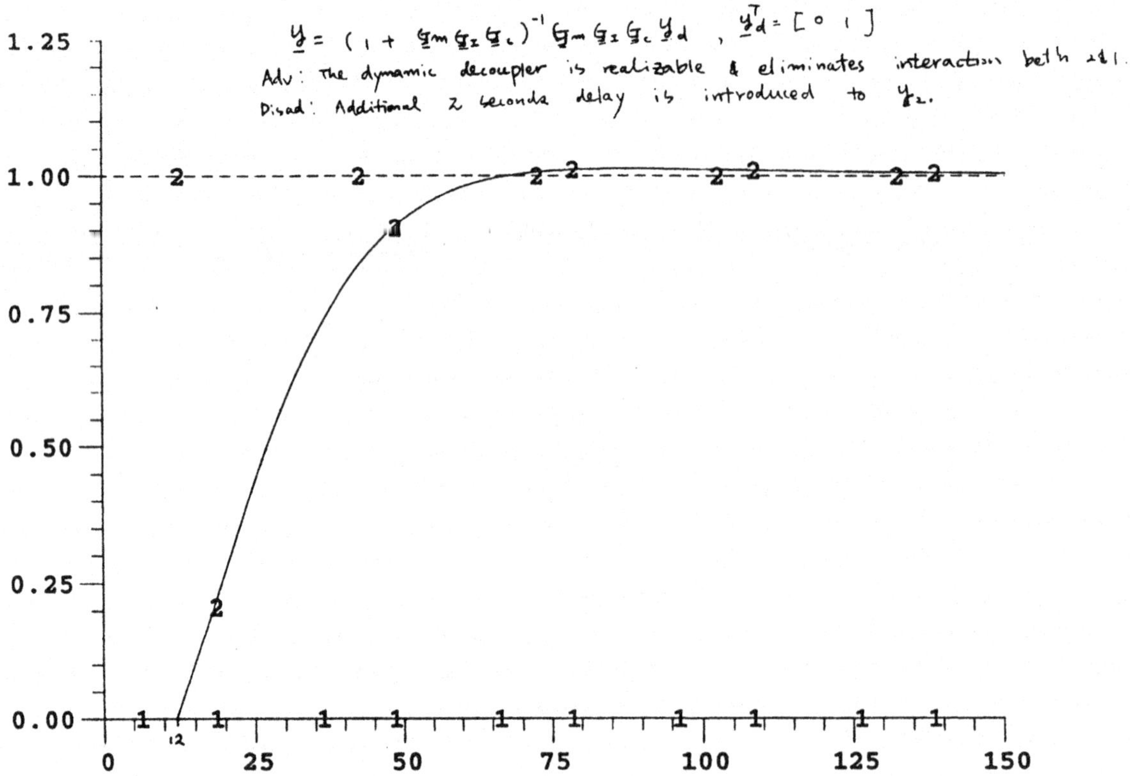

FIG S22.5. Dynamic Decoupling of "Augmented" Process: Closed-loop process response.

$$\underline{y} = (I + \underline{G}_m \underline{G}_z \underline{G}_c)^{-1} \underline{G}_m \underline{G}_z \underline{G}_c \underline{y}_d, \quad \underline{y}_d^T = [0 \; 1]$$

Adv: The dynamic decoupler is realizable & eliminates interaction both ±1.

Disad: Additional 2 seconds delay is introduced to y_2.

22.5 SOLUTION NOT PROVIDED.

22.6 SOLUTION NOT PROVIDED.

22.7 (a) For steady state decoupling, examine the characteristics of the process gain matrix

$$\underline{K} = \begin{bmatrix} -33.89 & -18.85 \\ 32.63 & 34.84 \end{bmatrix} \quad (22.10)$$

The determinant, $|\tilde{K}| = -565.7$ is not close to 0; its singular values are:

$$\sigma_1 = 60.8 \; ; \quad \sigma_2 = 9.3$$

so that its condition number, $\kappa = 6.54$, is quite reasonable. The indication therefore is that steady-state decoupling ought to be feasible.

(b) From Table 19.1, use transfer function element g_{11} to obtain the following controller parameters for loop 1:

$$K_{c_1} = -0.058$$
$$\tau_{I_1} = 98.44$$
$$\tau_{D_1} = 0.42$$

Similarly, use the g_{22} element to obtain, for loop 2,

$$K_{c_2} = 0.063$$
$$\tau_{I_2} = 110.53$$
$$\tau_{D_2} = 0.03$$

The steady state decoupler, via the generalized approach is obtained as:

$$\underset{\sim}{G}_I(0) = \underset{\sim}{K}^{-1}\underset{\sim}{G}_R(0) \; ; \quad \text{with } G_R(0) = \text{Diag}[\underset{\sim}{K}]$$

$$= \begin{bmatrix} 2.1 & 2.01 \\ 1.11 & 2.09 \end{bmatrix}$$

The closed-loop implementation of this decoupler and the two PID controllers yields the response in FIGS S22.6(a) and (b), indicating quite acceptable performance.

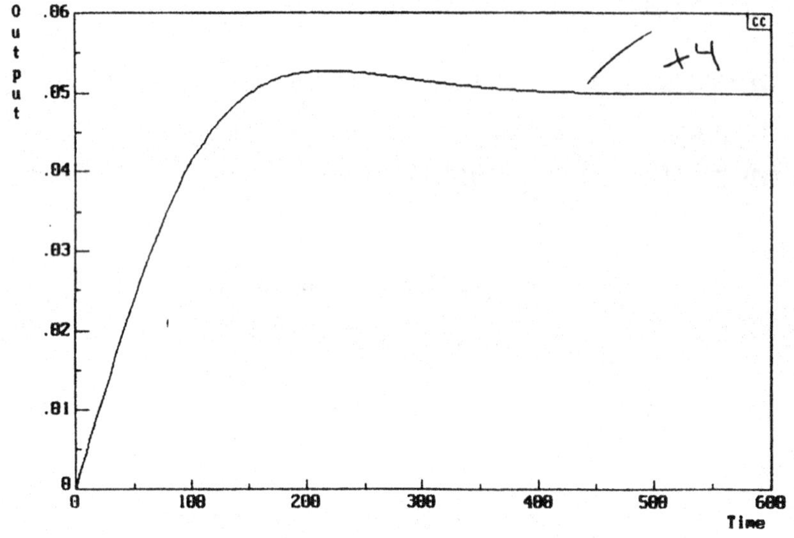

FIG S22.6(a) Response of y_2 under PID control with SS Decoupling.

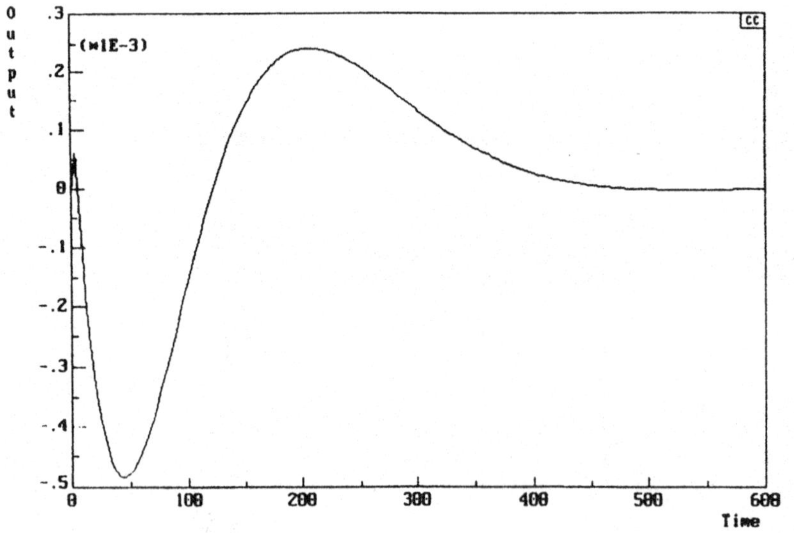

FIG S22.6(b) Response of y_1 under PID control with SS Decoupling

22.8 (a) The required % paralysis "process reaction curve" is shown in FIG S22.7, from where the approximate characterization is obtained as:

$$\frac{Ke^{-\alpha s}}{\tau s+1} \quad \text{with} \quad K = 1;\ \alpha = 2.3;\ \tau = 29.05$$

The Cohen-Coon tuning rules now provide the following controller parameters:

$$K_{c_1} = 17.09\ ; \quad \tau_{I_1} = 5.48\ ; \quad \tau_{D_1} = 0.82.$$

Similarly, for loop 2, using the g_{22} element directly, obtain the following tuning parameters:

$$K_{c_2} = -0.44\ ; \quad \tau_{I_2} = 0.95\ ; \quad \tau_{D_2} = 0.15$$

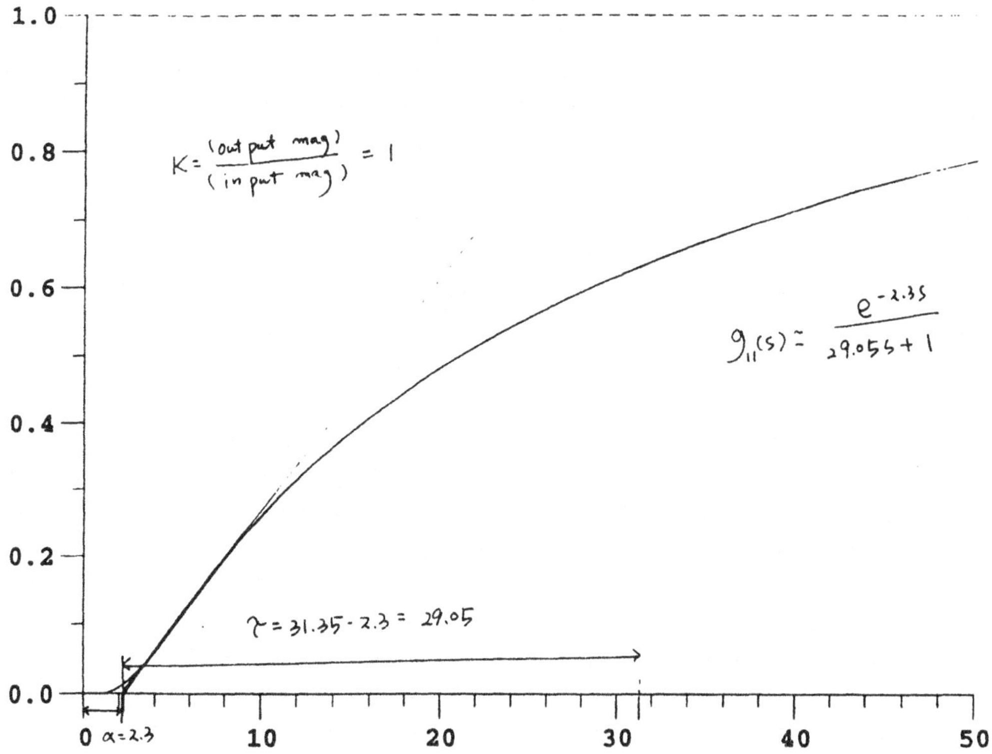

FIG S22.7. The "% PARALYSIS" process reaction curve and approximate characterization.

The closed loop system response to a −10 mm Hg step in MAP set-point, using these two single loop PID controllers, is shown in FIG S22.8.

The interaction effect suffered by loop 1 is not particularly severe. Note, however, the oscillatory response of y_2 (MAP). This is typical of controllers designed with the Cohen-Coon rules. (cf. FIG 15.11 and FIG 15.12, pp 540 and 541 main text!)

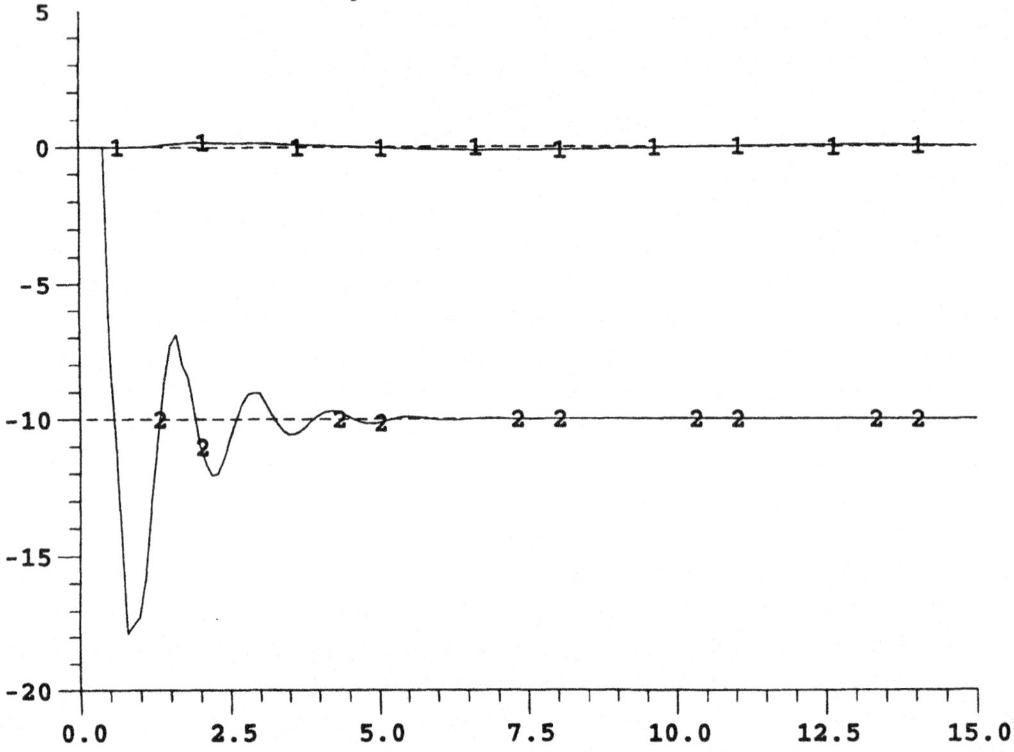

FIG S22.8 Closed-loop Anaesthesia control response: No Decoupling; Cohen-Coon designed PID controllers.

(b) The dynamic decoupler is given by

$$\underset{\sim}{G}_I = \begin{bmatrix} 1 & g_{I_1}(s) \\ g_{I_2}(s) & 1 \end{bmatrix} \quad \text{with} \quad g_{I_1}(s) = \frac{0.27(3.08s+1)(4.81s+1)(34.4s+1)}{(1.283s+1)(1.25s+1)(10.6s+1)} \tag{22.11}$$

with $g_{I_2}(s) = 0$.

And now, there are various alternatives for simplifying Eq (22.11) down to a lead/lag element. Here are two of them.

(i) Clearly, the dominance of the $(34.4s+1)$ term in the numerator and that of the $(10.6s+1)$ term in the denominator indicates that

$$g_{I_1}(s) \approx \frac{0.27\,(34.4s+1)}{(10.6s+1)} \qquad (22.12)$$

is a reasonable lead/lag approximation.

(ii) Alternatively, expand out the cubic polynomials and "ignore" the higher frequency terms indicated by the s^3 and s^2 terms; the result:

$$g_{I_2}(s) \approx \frac{0.27\,(42.29s+1)}{(13.133s+1)} \qquad (22.13)$$

is not much different from that in (22.12) above, and either will be acceptable.

Using now the approximation in Eq (22.13), implementing this decoupler along with the controllers in part (a) gives rise to the closed-loop response shown in FIG S22.9.

Observe that the decoupler offers no discernible improvement in the closed loop performance.

The reasons for this are:

(i) Loop 2 (which sees the most action) actually does not suffer from any interaction from loop 1; (note the one-way interaction). Loop 1 suffers only a negligible amount of interaction;

(ii) the Cohen-Coon controller 2 is too active; the entire system behavior is driven by this: interaction is NOT the problem.

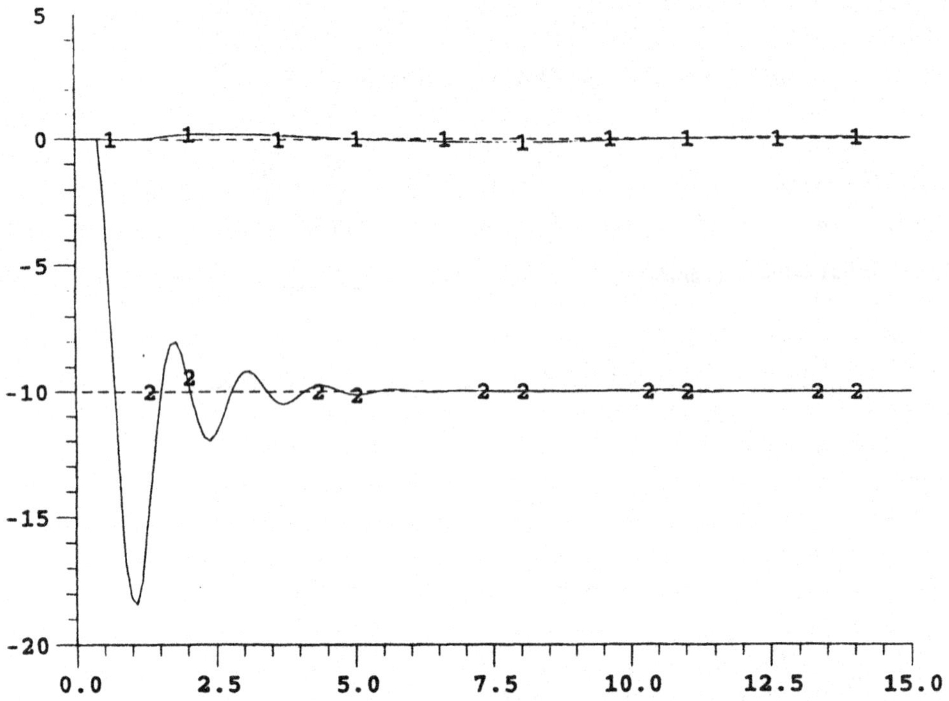

FIG S22.9 Closed-loop Anesthesia control response: Approximate Dynamic Decoupling and Cohen-Coon designed PID controllers.

22.9 (a) The steady state gain matrix for this process has a condition number

$$\kappa = 3.4 \times 10^3 \tag{22.14}$$

indicating that care must be taken if steady-state decoupling is to be applied.

(b) Using the indicated generalized approach, obtain

$$\underset{\sim}{G}_I = \underset{\sim}{K}^{-1} \text{Diag}\, \underset{\sim}{G}(0) = \begin{bmatrix} 1.945 & 1.412 & 0.00726 \\ 0.7289 & 1.9 & -0.00425 \\ 36.61 & -46.17 & 1.506 \end{bmatrix} \tag{22.15}$$

From Example 21.3 in the main text, the 1-1/2-2/3-3 pairing is recommended. For this configuration, the Nierderlinski index is

$$N = \frac{-0.508}{-1.355} = 0.375 > 0$$

so that this pairing will <u>not</u> result in structural instability.

Any tuning technique will do for obtaining controller parameters; in the specific case of using the CONSYD program MTUNE, obtain the following parameters:

LOOP #	K_c	$1/\tau_I$
1	1.035	1.968×10^{-4}
2	−0.189	0.263
3	3.027	0.1304

Implementing the decoupler of Eq (22.15) along with these 3 single loop controllers, the overall closed-loop system response under the stipulated temperature set-point change is shown in FIG S22.10.

The observed performance appears acceptable.

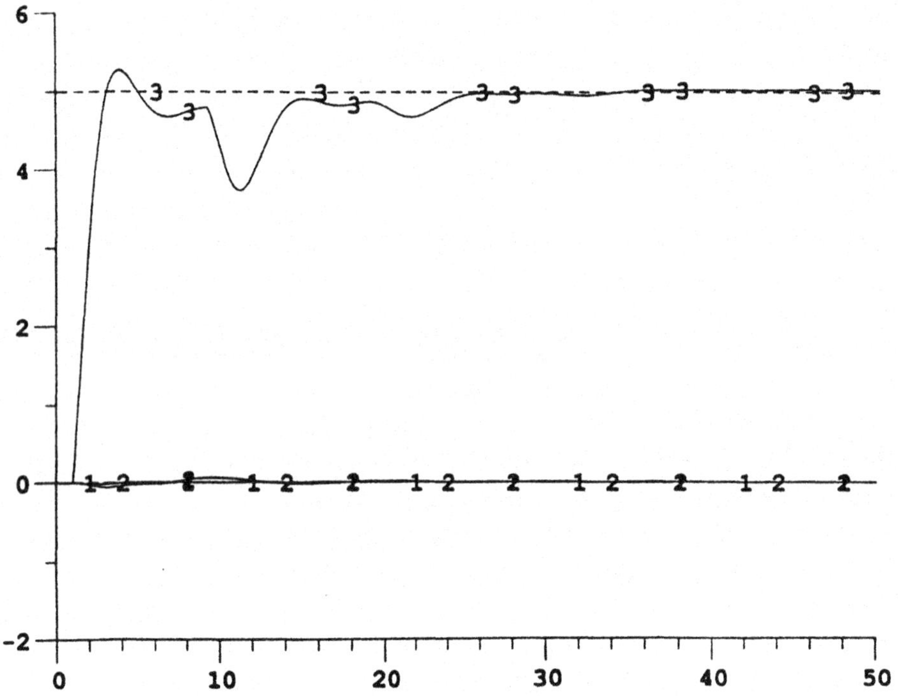

FIG S22.10 Closed-loop response to Temp. set-point change: steady-state decoupling and 3 PID controllers.

22.10 SOLUTION NOT PROVIDED.

22.11 (a) The required response is shown in the solid lines in FIG S22.11.

(b) In this case,

$$\underset{\sim}{\Gamma} = \begin{bmatrix} -0.1263 & 0.0220 \\ -0.0635 & 0.0371 \end{bmatrix}$$

and for the given $\underset{\sim}{\Theta}$ matrix, obtain the controller matrices

$$\underset{\sim}{K}_c = \begin{bmatrix} -2.2534 & 0.6662 \\ -3.8543 & 3.8317 \end{bmatrix} \quad ; \quad \underset{\sim}{K}_I = \begin{bmatrix} -0.4399 & 0.1184 \\ -0.2284 & 0.2538 \end{bmatrix}$$

The corresponding closed-loop performance of this MV controller is shown in the dashed lines in FIG S22.11. The y_1 response is a little more sluggish, but y_2 settles down somewhat quicker.

By choosing $\underset{\sim}{\Theta} = \begin{bmatrix} 3 & 0 \\ 0 & 3 \end{bmatrix}$ or $3\underset{\sim}{I}$, the resulting MV controller gives rise to the closed-loop response shown in the dashed-dotted lines in FIG S22.11; finally when $\underset{\sim}{\Theta} = \begin{bmatrix} 3.7 & 0 \\ 0 & 5.0 \end{bmatrix}$, the resulting closed-loop response is shown in the dotted lines.

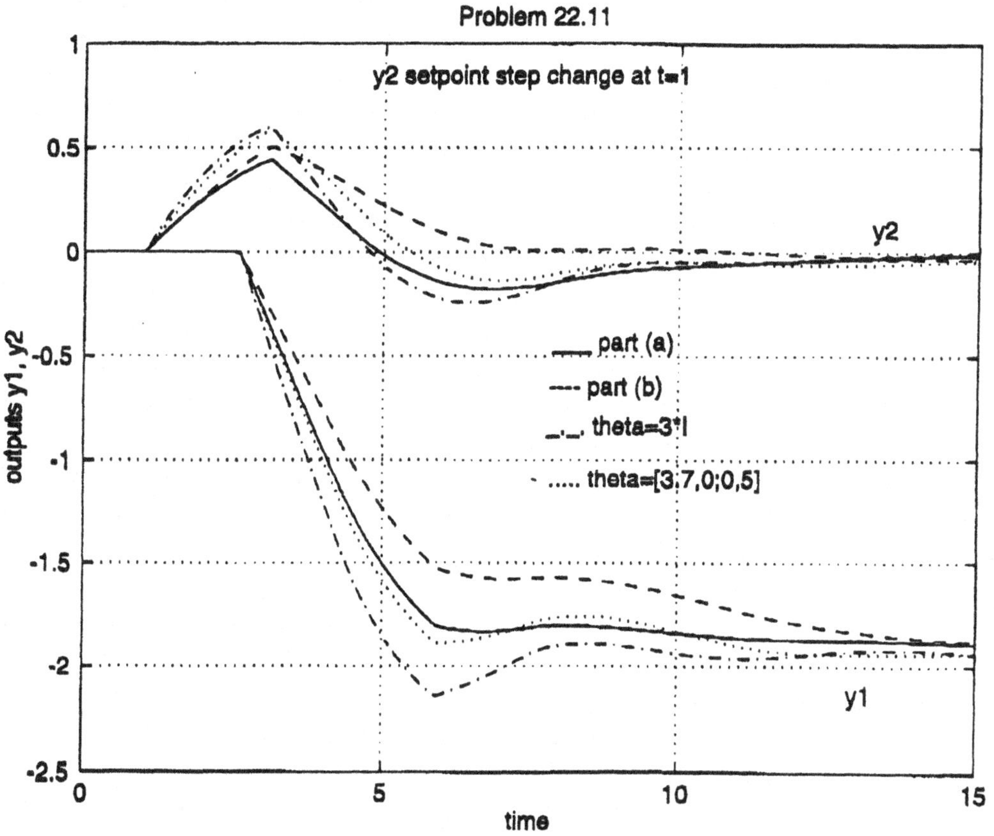

FIG S22.11 Closed-loop system response under MV PI control using various designs.

22.12 SOLUTION NOT PROVIDED

22.13 SOLUTION NOT PROVIDED

22.14 (a) From the provided information, construct the following process model:

$$\begin{bmatrix} y_{fet} \\ y_{bet} \end{bmatrix} = \begin{bmatrix} \dfrac{1.66}{39s+1} & \dfrac{-1.74e^{-2s}}{4.4s+1} \\ \dfrac{0.34e^{-s}}{8.9s+1} & \dfrac{1.40e^{-s}}{3.8s+1} \end{bmatrix} \begin{bmatrix} u_{fuel} \\ u_{damper} \end{bmatrix} \qquad (22.16)$$

so that the steady state gain matrix is:

$$\underset{\sim}{K} = \begin{bmatrix} 1.66 & -1.74 \\ 0.34 & 1.40 \end{bmatrix} \qquad (22.17)$$

from which the RGA is easily obtained as:

$$\underset{\sim}{\Lambda} = \begin{bmatrix} \boxed{0.797} & 0.203 \\ 0.203 & \boxed{0.797} \end{bmatrix} \qquad (22.18)$$

The recommended pairing is thus:

u_{fuel} to y_{fet}

u_{damper} to y_{bet}

The indication from this RGA is that two single-loop

controllers will interfere with each other to a certain degree, but the interactions are not expected to be severe. Nevertheless, keep in mind that (22.18) is based on steady-state considerations only; The time constant associated with the $y_{fet} - u_{fuel}$ loop is significantly longer than all the others; by the same token, this is the only element with no delays. Dynamics could well play a significant role in determining the true extent of interactions.

(b) For this problem,

$$\underset{\sim}{\Gamma} = \begin{bmatrix} \frac{1.66}{39} & \frac{-1.74}{4.4} \\ \frac{0.34}{8.9} & \frac{1.40}{3.8} \end{bmatrix}$$

and with the given $\underset{\sim}{\Theta}$, apply the equations (P22.10) and (P22.12) to obtain

$$\underset{\sim}{K_c} = \begin{bmatrix} 0.398 & 3.21 \\ -0.041 & 0.346 \end{bmatrix}; \quad \underset{\sim}{K_I} = \begin{bmatrix} 0.016 & 0.149 \\ -0.004 & 0.142 \end{bmatrix}$$

The required closed-loop response using this MV PI controller is shown in FIG S22.12.

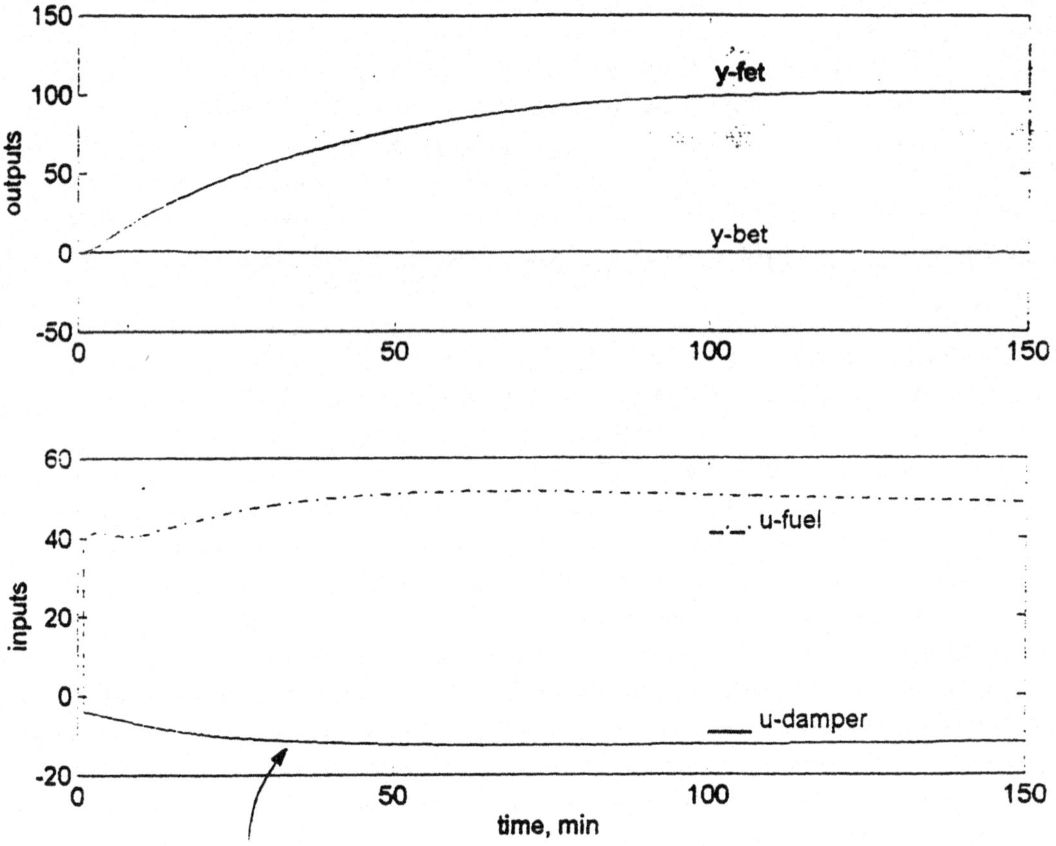

FIG S22.12. Multivariable PI control of Industrial lime kiln.

Chapter 23

23.1 (a) The plot of the raw data, and the three sets of filtered data [TABLE S23.1] is shown in FIG S23.1. The true value of the relative viscosity is about 50. (see last column of TABLE S23.1).

TABLE S23.1

	Raw data	beta = 0.3	beta = 0.5	beta = 0.9
1	49.763	49.763	49.763	49.763
2	50.050	49.964	49.863	49.773
3	51.079	50.744	50.304	49.826
4	49.310	49.740	50.022	49.846
5	49.855	49.821	49.921	49.853
6	51.056	50.685	50.303	49.898
7	49.887	50.127	50.215	49.930
8	49.806	49.902	50.059	49.943
9	49.735	49.785	49.922	49.941
10	50.799	50.495	50.208	49.967
11	50.699	50.638	50.423	50.013
12	50.469	50.520	50.471	50.059
13	50.007	50.161	50.316	50.085
14	50.341	50.287	50.301	50.106
15	49.798	49.945	50.123	50.108
16	50.767	50.520	50.322	50.129
17	48.678	49.231	49.776	50.094
18	50.372	50.030	49.903	50.075
19	50.782	50.556	50.230	50.090
20	49.648	49.920	50.075	50.089
21	49.889	49.898	49.987	50.079
22	49.669	49.738	49.862	50.057
23	49.554	49.609	49.736	50.025
24	50.472	50.213	49.974	50.020
25	50.837	50.650	50.312	50.049

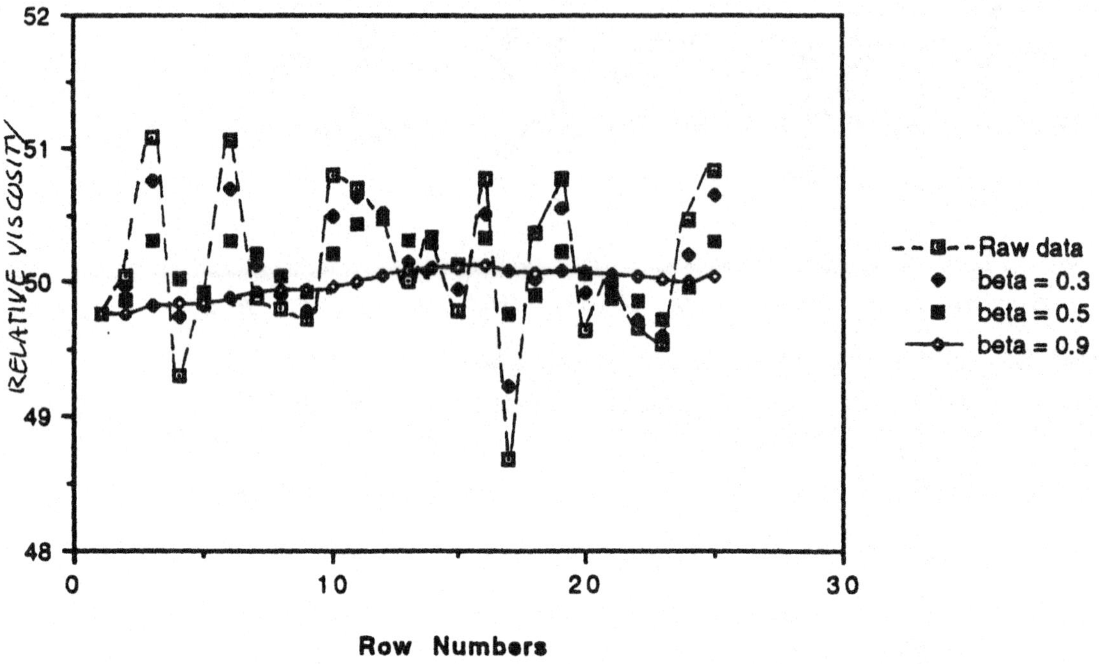

FIG S23.1 Hourly Viscosity: Raw and filtered data.

(b) The raw data incorporating the indicated shift, and the corresponding two sets of filtered data, are all shown in TABLE S23.2. The accompanying plot of these data sets is shown in FIG S23.2.

The raw data is clearly noisy, and the filtered data set ($\beta = 0.3$) is only barely improved; however this filtered data set is able to indicate the abrupt change quite distinctly.

The filtered data set with $\beta = 0.9$, on the other hand successfully smooths out the noise evident in the raw data. However, this smoothness is achieved at the expense of the ability to "track" abrupt changes in the raw data. Heavy filtering with a first order filter always results in the "smearing out" of all abrupt changes, spurious as well as real.

TABLE S23.2

	Raw data	beta = 0.3	beta = 0.9
1	49.763	49.763	49.763
2	50.050	49.964	49.783
3	51.079	50.744	49.879
4	49.310	49.740	49.865
5	49.855	49.821	49.861
6	51.056	50.685	49.943
7	49.887	50.127	49.962
8	49.806	49.902	49.956
9	49.735	49.785	49.939
10	50.799	50.495	49.994
11	50.699	50.638	50.059
12	50.469	50.520	50.105
13	55.007	53.661	50.460
14	55.341	54.837	50.898
15	54.798	54.810	51.289
16	55.767	55.480	51.708
17	53.678	54.219	51.959
18	55.372	55.026	52.266
19	55.782	55.555	52.595
20	54.648	54.920	52.827
21	54.889	54.898	53.034
22	54.669	54.738	53.205
23	54.554	54.609	53.345
24	55.472	55.213	53.532
25	55.837	55.650	53.744

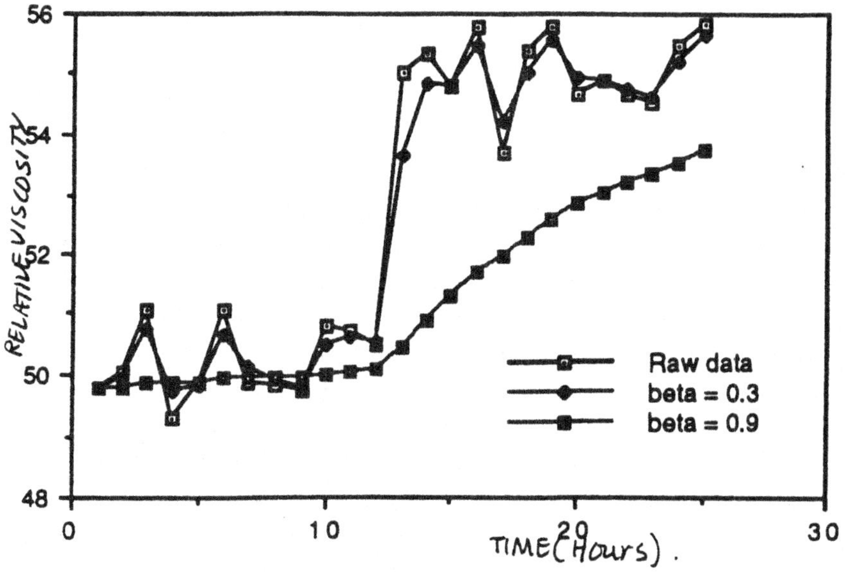

FIG S23.2 Raw and filtered viscosity data: Effect of abrupt shifts.

23.2 Represent each transfer function element by x_i; $i = 1, 2, 3, 4$; i.e. let

$$x_1(s) = \frac{12.8e^{-s}}{16.7s+1} u_1(s) \;; \quad x_2(s) = \frac{-18.9e^{-3s}}{21.0s+1} u_2(s)$$

$$x_3(s) = \frac{6.6e^{-7s}}{10.9s+1} u_1(s) \;; \quad x_4(s) = \frac{-19.4e^{-3s}}{14.4s+1} u_2(s)$$

with $y_1 = x_1 + x_2$; $y_2 = x_3 + x_4$

Then the most straightforward set of differential eqns to give rise to these transfer functions is:

$$\left.\begin{aligned} 16.7\frac{dx_1}{dt} + x_1 &= 12.8\, u_1(t-1) \\ 21.0\frac{dx_2}{dt} + x_2 &= -18.9\, u_2(t-3) \\ 10.9\frac{dx_3}{dt} + x_3 &= 6.6\, u_1(t-7) \\ 14.4\frac{dx_4}{dt} + x_4 &= -19.4\, u_2(t-3) \end{aligned}\right\} \quad (23.1)$$

together with

$$\left.\begin{aligned} y_1 &= x_1 + x_2 \\ y_2 &= x_3 + x_4 \end{aligned}\right\} \quad (23.2)$$

These equations constitute one possible realization of the transfer function matrix model.

The corresponding discrete-time model is easily obtained from here (either from TABLE 23.1, 2nd Entry [P. 834 main text] or from Eq 23.24, p. 837, with accomodation for the delay):

for the first equation, with $\Delta t = 1$ min, obtain

$$x_1(k+1) = e^{-1/16.7} x_1(k) + 12.8\left[1 - e^{-1/16.7}\right] u_1(k-1)$$

simplify to:

$$x_1(k+1) = 0.942 x_1(k) + 0.744 u_1(k-1) \qquad (23.3a)$$

similarly obtain for the remaining equations:

$$x_2(k+1) = 0.953 x_2(k) - 17.947 u_2(k-3) \qquad (23.3b)$$
$$x_3(k+1) = 0.912 x_3(k) + 5.688 u_1(k-7) \qquad (23.3c)$$
$$x_4(k+1) = 0.933 x_4(k) - 18.467 u_2(k-3) \qquad (23.3d)$$

along with

$$\left.\begin{aligned} y_1(k) &= x_1(k) + x_2(k) \\ y_2(k) &= x_3(k) + x_4(k) \end{aligned}\right\} \qquad (23.4)$$

23.3 The postulated discrete first-order model is:

$$y(k) = a y(k-1) + b u(k-1) \qquad (23.5)$$

Fitting this model to the supplied data, using linear regression, results in the parameter estimates:

$$\hat{a} = 0.6144 \qquad (23.6)$$
$$\hat{b} = 1.6979 \qquad (23.7)$$

FIG S23.3 shows the model predictions (solid line) and the actual data (circles) in the top half; the bottom half shows the residuals $(y - \hat{y})$. This figure shows a reasonable but not excellent fit.

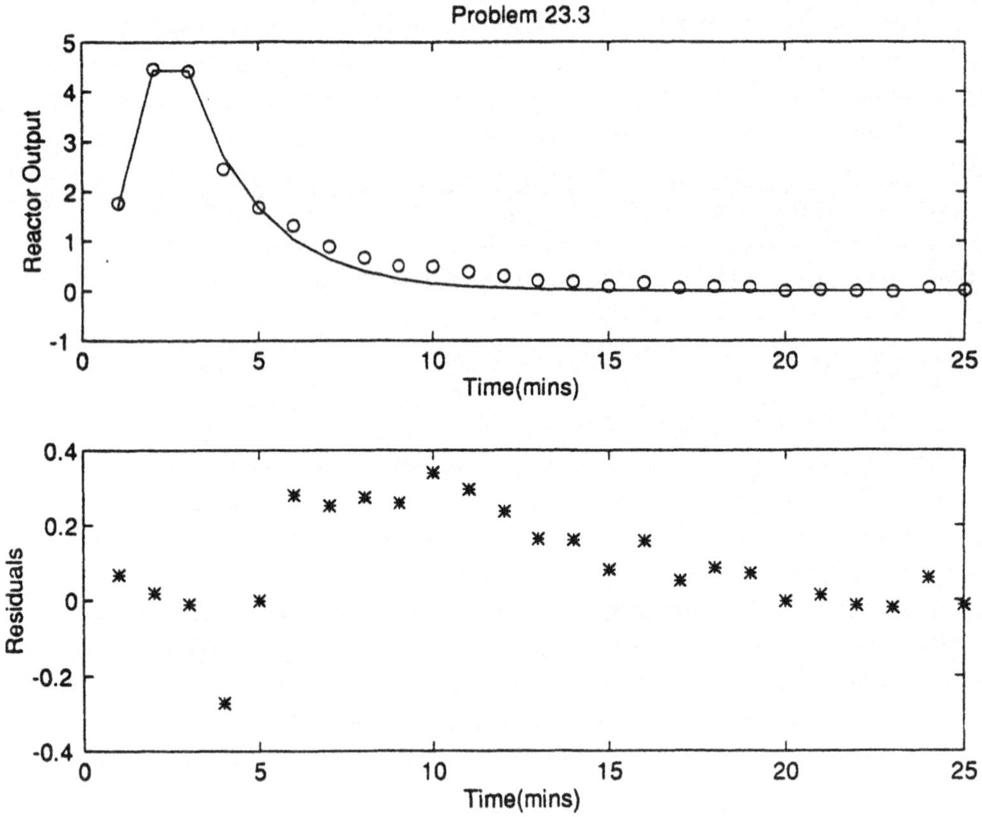

FIG S23.3 First-order model vs data for pilot plant polymer reactor pulse test identification.

23.4 Fitting the postulated second-order model to the data using linear regression produces the following parameter estimates:

$$\hat{a}_1 = 0.8367 \; ; \; \hat{a}_2 = -0.0997$$

$$\hat{b}_1 = 1.8096 \; ; \; \hat{b}_2 = -0.5564$$

FIG S23.4 compares the second-order model fit to the first-order model fit. The second-order model provides a better fit to the data, especially in the middle section.

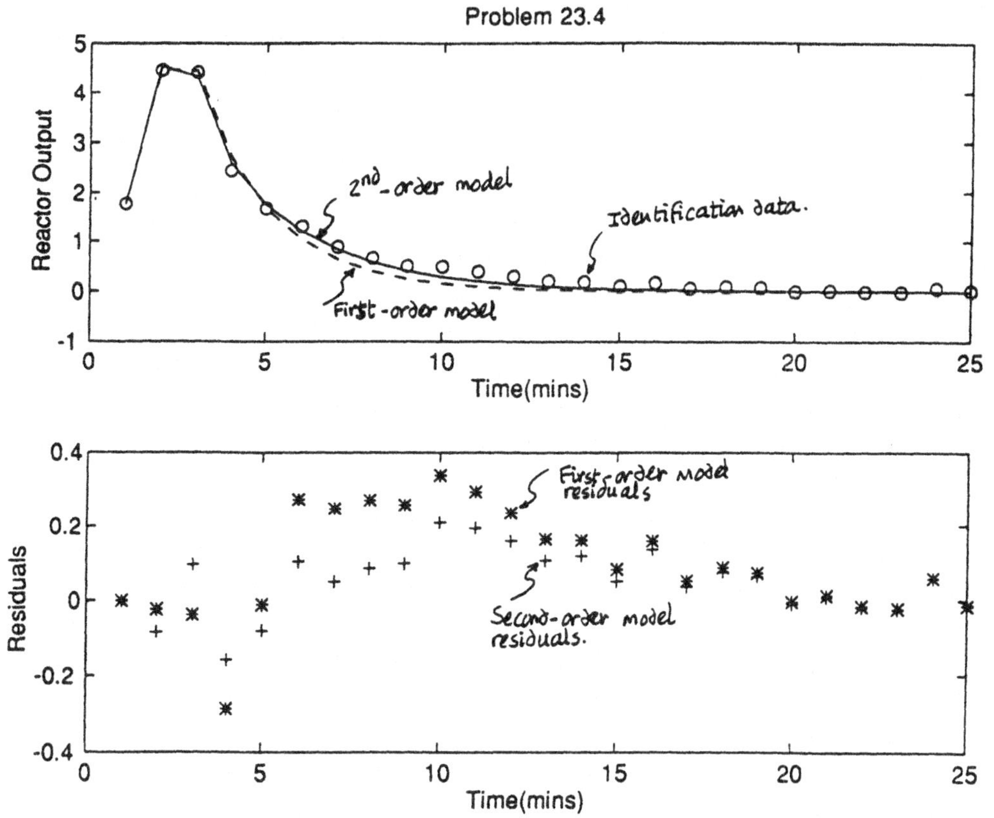

FIG S23.4 Comparison of first- and second-order model fit to polymer reactor identification data.

Chapter 24

24.1 Given $f(t) = \cos\omega t$, by definition,

$$f(z) = \sum_{k=0}^{\infty} \cos(k\omega\Delta t)\, z^{-k} \qquad (24.1)$$

or:

$$f(z) = \sum_{k=0}^{\infty} \tfrac{1}{2}\left\{\exp(j\omega k\Delta t) + \exp(-j\omega k\Delta t)\right\} z^{-k}$$

$$= \frac{1}{2}\sum_{k=0}^{\infty} \exp(j\omega k\Delta t)\, z^{-k} + \frac{1}{2}\sum_{k=0}^{\infty} \exp(-j\omega k\Delta t)\, z^{-k} \qquad (24.2)$$

$$= \frac{1}{2\left[1 - \exp(j\omega\Delta t) z^{-1}\right]} + \frac{1}{2\left[1 - \exp(-j\omega\Delta t) z^{-1}\right]}$$

$$= \frac{1}{2}\, \frac{2 - \{\exp(j\omega\Delta t) + \exp(-j\omega\Delta t)\} z^{-1}}{1 - \{\exp(j\omega\Delta t) + \exp(-j\omega\Delta t)\} z^{-1} + z^{-2}}$$

$$= \frac{1}{2}\, \frac{2 - 2\cos(\omega\Delta t)\, z^{-1}}{1 - 2\cos(\omega\Delta t) z^{-1} + z^{-2}}$$

or, finally,

$$f(z) = \frac{1 - z^{-1}\cos(\omega\Delta t).}{1 - 2z^{-1}\cos(\omega\Delta t) + z^{-2}} \qquad (24.3)$$

as required.

24.2 Use the procedure outlined on p. 872 of the main text:

$$\tilde{g}(t) = \mathcal{L}^{-1}\left\{\frac{g(s)}{s}\right\} = \mathcal{L}^{-1}\left\{\frac{K(\xi s+1)}{(\tau_1 s+1)(\tau_2 s+1)}\cdot\frac{1}{s}\right\}$$

so that $$\tilde{g}(t) = K\left[1 - \frac{(\tau_1-\xi)}{(\tau_1-\tau_2)}e^{-t/\tau_1} - \frac{(\tau_2-\xi)}{(\tau_2-\tau_1)}e^{-t/\tau_2}\right]$$

and therefore

$$\tilde{g}(k\Delta t) = K\left[1 - \left(\frac{\tau_1-\xi}{\tau_1-\tau_2}\right)e^{-k\Delta t/\tau_1} - \left(\frac{\tau_2-\xi}{\tau_2-\tau_1}\right)e^{-k\Delta t/\tau_2}\right] \quad (24.4)$$

Upon taking z-transforms, obtain

$$\tilde{g}(z) = K\left[\frac{1}{1-z^{-1}} - \left(\frac{\tau_1-\xi}{\tau_1-\tau_2}\right)\frac{1}{1-e^{-\Delta t/\tau_1}z^{-1}} - \left(\frac{\tau_2-\xi}{\tau_2-\tau_1}\right)\cdot\frac{1}{1-e^{-\Delta t/\tau_2}z^{-1}}\right]$$

and now, $g(z) = (1-z^{-1})\tilde{g}(z)$ gives

$$g(z) = K\left[1 - \left(\frac{\tau_1-\xi}{\tau_1-\tau_2}\right)\frac{1-z^{-1}}{1-e^{-\Delta t/\tau_1}z^{-1}} - \left(\frac{\tau_2-\xi}{\tau_2-\tau_1}\right)\frac{1-z^{-1}}{1-e^{-\Delta t/\tau_2}z^{-1}}\right] \quad (24.5)$$

Now define $\phi_1 = e^{-\Delta t/\tau_1}$; $\phi_2 = 1-e^{-\Delta t/\tau_2}$, and (24.5) above becomes

$$g(z) = K\left[1 - \left(\frac{\tau_1-\xi}{\tau_1-\tau_2}\right)\frac{1-z^{-1}}{1-\phi_1 z^{-1}} - \left(\frac{\tau_2-\xi}{\tau_2-\tau_1}\right)\frac{1-z^{-1}}{1-\phi_2 z^{-1}}\right] \quad [24.6)$$

Let $\gamma_1 = \dfrac{\tau_1 - \xi}{\tau_1 - \tau_2}$ and $\gamma_2 = \dfrac{\tau_2 - \xi}{\tau_2 - \tau_1}$ (24.7)

and (24.6) becomes

$$g(z) = K\left[1 - \frac{\gamma_1(1-z^{-1})}{(1-\phi_1 z^{-1})} - \frac{\gamma_2(1-z^{-1})}{1-\phi_2 z^{-1}}\right] \quad (24.8)$$

or $g(z) = N(z)/D(z)$ where

$$N(z) = K\Big\{1-(\phi_1+\phi_2)z^{-1} + \phi_1\phi_2 z^{-2} - \gamma_1\left[1-(1+\phi_2)z^{-1}+\phi_2 z^{-2}\right] - \gamma_2\left[1-(1+\phi_1)z^{-1}+\phi_1 z^{-2}\right]\Big\} \quad (24.9)$$

and $D(z) = 1 - (\phi_1 + \phi_2)z^{-1} + \phi_1\phi_2 z^{-1}$ (24.10)

$N(z)$ in (24.9) simplifies to:

$$N(z) = K\Big\{(1-\gamma_1-\gamma_2) + (\gamma_1+\gamma_2+\gamma_1\phi_2+\gamma_2\phi_1-\phi_1-\phi_2)z^{-1} + (\phi_1\phi_2 - \gamma_1\phi_2 - \gamma_2\phi_1)z^{-2}\Big\} \quad (24.11)$$

Now recall from (24.7) that

$$\gamma_1 + \gamma_2 = 1$$

Thus: $1-\gamma_1-\gamma_2 = 0$; and $\gamma_2 - 1 = -\gamma_1$; and $\gamma_1 - 1 = -\gamma_2$

With these identities, (24.11) simplifies further to:

$$N(z) = (1 - \gamma_1\phi_1 - \gamma_2\phi_2)z^{-1} + (\phi_1\phi_2 - \gamma_1\phi_2 - \gamma_2\phi_1)z^{-2}$$

Now, by defining $a_1 = -(\phi_1 + \phi_2)$; $a_2 = \phi_1\phi_2$

$D(z)$ becomes $D(z) = 1 + a_1 z^{-1} + a_2 z^{-2}$

and by defining

$$b_1 = K\left[1 - \gamma_1 \phi_1 - \gamma_2 \phi_2\right] \quad (24.12)$$

$$b_2 = K\left[\phi_1 \phi_2 - \gamma_1 \phi_2 - \gamma_2 \phi_1\right] \quad (24.13)$$

$g(z)$ now becomes

$$g(z) = \frac{b_1 z^{-1} + b_2 z^{-2}}{1 + a_1 z^{-1} + a_2 z^{-2}}$$

as required. Note that (P24.4d) and (P24.4e) are obtained by introducing (24.7) above for γ_1 & γ_2 in (24.12) and (24.13).

24.3 (a) From the given transfer function, use the results of Problem 24.2 to obtain, for $\Delta t = 1$ min,

$$\phi_1 = 0.607, \quad \phi_2 = 0.819 \quad \text{so that}$$

$$a_1 = -1.426 \quad ; \quad b_1 = -0.345$$
$$a_2 = 0.497 \quad ; \quad b_2 = 0.487$$

and finally obtain the required $g(z)$ as:

$$g(z) = \frac{-0.345 z^{-1} + 0.487 z^{-2}}{1 - 1.426 z^{-1} + 0.497 z^{-2}} \quad (24.14)$$

(b) For $\Delta t = 7$ mins,

$$\phi_1 = 0.0302 \; ; \quad \phi_2 = 0.247$$

and now

$$a_1 = -0.2772 \; ; \quad b_1 = 0.783$$
$$a_2 = 0.00746 \; , \quad b_2 = 0.677$$

and

$$g(z) = \frac{0.783 z^{-1} + 0.677 z^{-2}}{1 - 0.2772 z^{-1} + 0.00746 z^{-2}} \qquad (24.15)$$

(c) For $\Delta t = 1$, and the $g(z)$ in (24.14), obtain

$$(1 - 1.426 z^{-1} + 0.497 z^{-2}) y(z) = (-0.345 z^{-1} + 0.487 z^{-2}) u(z)$$

from where inverse z-transformation gives

$$y(k) - 1.426 y(k-1) + 0.497 y(k-2) = -0.345 u(k-1) + 0.487 u(k-2)$$

which rearranges to give the required difference equation model:

$$y(k) = 1.426 y(k-1) - 0.497 y(k-2) - 0.345 u(k-1) + 0.487 u(k-2) \qquad (24.16)$$

Similarly, for $\Delta t = 7$ and the $g(z)$ given in (24.15) obtain the required difference equation model as:

$$y(k) = 0.2772 y(k-1) - 0.00746 y(k-2) + 0.783 u(k-1) + 0.677 u(k-2) \qquad (24.17).$$

A plot of the unit step responses is shown in FIG S24.1

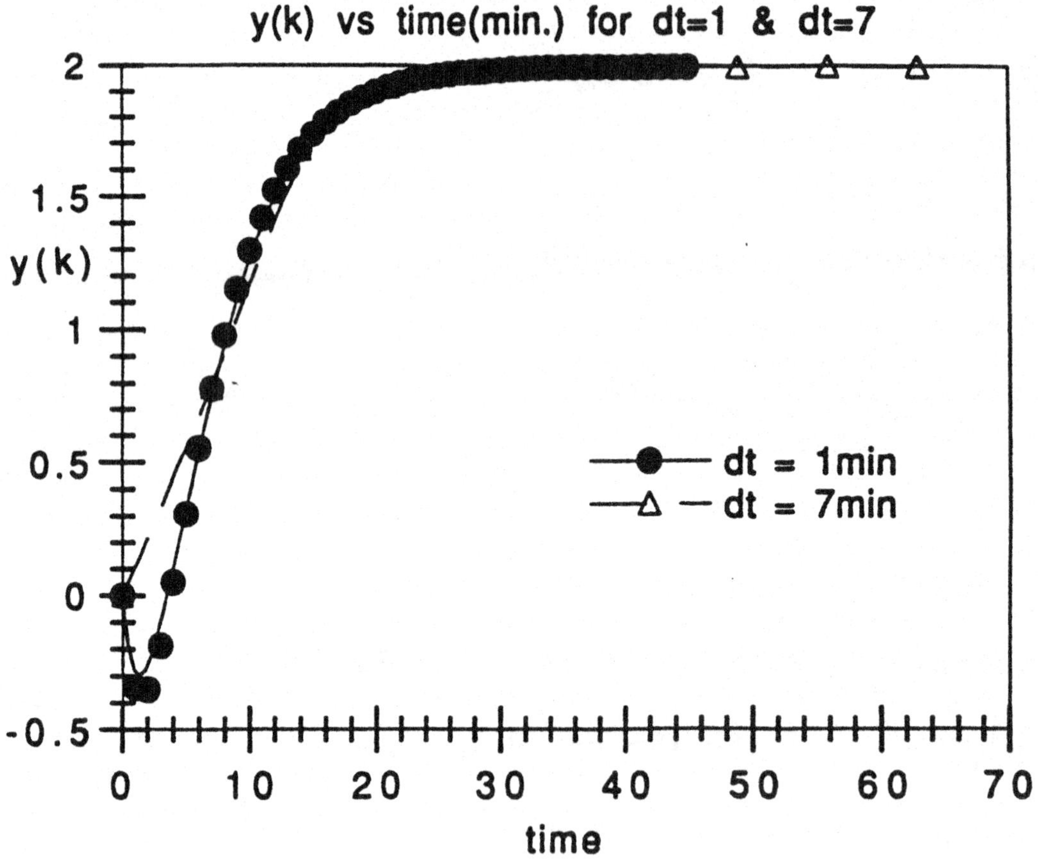

FIG S24.1 Unit step response of inverse response system: effect of sampling time, Δt.

Observe that with $\Delta t = 1$ min, the expected inverse response behavior is captured. By increasing the sample time to 7 mins, the initial inverse response is missed entirely and the apparent system behavior does not indicate the inherent inverse response.

24.4 From the given $g(s)$, obtain

$$\tilde{g}(t) = \mathcal{L}^{-1}\left\{\frac{g(s)}{s}\right\} = \mathcal{L}^{-1}\left\{\frac{-0.55}{s(5s-1)(2s+1)}\right\} \qquad (24.18)$$

or

$$\tilde{g}(t) = 0.55 - \frac{1.1}{7}e^{-t/2} - \frac{2.75}{7}e^{+t/5}$$

$$\tilde{g}(k\Delta t) = 0.55 - \frac{1.1}{7}e^{-k\Delta t/2} - \frac{2.75}{7}e^{+k\Delta t/5}$$

From where z-transformation yields:

$$\tilde{g}(z) = \frac{0.55}{1-z^{-1}} - \frac{1.1}{7(1-e^{-\Delta t/2}z^{-1})} - \frac{2.75}{7(1-e^{\Delta t/5}z^{-1})}$$

For $\Delta t = 0.5$, we now have

$$g(z) = (1-z^{-1})\tilde{g}(z)$$

$$= 0.55 - \frac{0.157(1-z^{-1})}{(1-e^{-0.25}z^{-1})} - \frac{0.393(1-z^{-1})}{(1-e^{0.1}z^{-1})}$$

or

$$g(z) = \frac{-0.0065z^{-1} - 0.0062z^{-1}}{1 - 1.884z^{-1} + 0.861z^{-2}}. \qquad (24.19)$$

24.5 $$\tilde{g}(t) = \mathcal{L}^{-1}\left\{\frac{g(s)}{s}\right\} = \mathcal{L}^{-1}\left\{\frac{0.9(0.3s-1)}{s^2(2.5s+1)}\right\}$$

$$= \mathcal{L}^{-1}\left\{0.9\left[-\frac{1}{s^2} + \frac{2.8}{s} - \frac{7}{(2.5s+1)}\right]\right\}$$

or $$\tilde{g}(t) = 0.9\left[-t + 2.8 - 7(1-e^{-t/2.5})\right] \qquad (24.20)$$

From which we obtain

$$\tilde{g}(z) = 0.9\left[\frac{\Delta t z^{-1}}{(1-z^{-1})^2} + \frac{2.8}{(1-z^{-1})} - \frac{7}{(1-0.961z^{-1})}\right]$$

and finally

$$g(z) = (1-z^{-1})\tilde{g}(z)$$

or

$$g(z) = \frac{0.9\left[9.8 - 19.591z^{-1} + 9.787z^{-2}\right]}{(1-z^{-1})(1-0.961z^{-1})} \qquad (24.21)$$

24.6 (a) Upon introducing the given physical parameters, obtain the following Laplace domain transfer function matrices:

$$\underline{G}(s) = \begin{bmatrix} \dfrac{0.308}{163.6s+1} & \dfrac{0.308}{163.6s+1} \\[2ex] \dfrac{0.481}{81.8s+1} & \dfrac{-0.113}{81.8s+1} \end{bmatrix} \qquad (24.22)$$

and (using $T_{d_s} = 35°C$)

$$\underline{G}_d(s) = \begin{bmatrix} \dfrac{0.308}{163.6s+1} & 0 \\[2ex] \dfrac{0.072}{81.8s+1} & \dfrac{0.106}{81.8s+1} \end{bmatrix} \qquad (24.23)$$

And now recall that for a first order transfer function

$$g(s) = \frac{K}{\tau s+1}$$

the corresponding pulse transfer function incorporating a ZOH element is given by

$$g(z) = \frac{K[1-\phi]z^{-1}}{(1-\phi z^{-1})} \quad \text{with } \phi = e^{-\Delta t/\tau}$$

we obtain immediately the following pulse transfer function matrices for the mixing process: ($\Delta t = 10$ secs)

$$\underline{G}(z) = \begin{bmatrix} \dfrac{0.018z^{-1}}{1-0.941z^{-1}} & \dfrac{0.018z^{-1}}{1-0.941z^{-1}} \\ \\ \dfrac{0.055z^{-1}}{1-0.885z^{-1}} & \dfrac{-0.013z^{-1}}{1-0.885z^{-1}} \end{bmatrix} \quad (24\text{-}24)$$

and

$$\underline{G}_d(z) = \begin{bmatrix} \dfrac{0.018z^{-1}}{1-0.941z^{-1}} & 0 \\ \\ \dfrac{0.008z^{-1}}{1-0.885z^{-1}} & \dfrac{0.012z^{-1}}{1-0.885z^{-1}} \end{bmatrix} \quad (24\text{-}25)$$

(b) The effect of such a delay is to cause the g_{21} and g_{22} elements of $\underline{G}$ (those associated with y_2, tank temperature) to acquire an additional 6 sample-time-delay: i.e.

g_{21} becomes $\dfrac{0.055z^{-7}}{1-0.885z^{-1}}$

g_{22} becomes $\dfrac{-0.013z^{-7}}{1-0.885z^{-1}}$

also, if the disturbance flow rate is subject to this delay then $g_{d_{21}}$ becomes $\dfrac{0.008z^{-2}}{1-0.885z^{-1}}$, the only $\underline{G}_d$ element to be so affected.

Chapter 25

25.1 For the given system model, use procedure illustrated on p.886 main text to obtain the required step response (for a yet unspecified ϕ) as:

$$y(k) = 2.0(1 - \phi^k) \quad (25.1)$$

(a) For $\phi = 0.5$,

$$y(k) = 2.0[1 - (0.5)^k] \quad (25.2)$$

(b) for $\phi = 0.8$

$$y(k) = 2.0[1 - (0.8)^k] \quad (25.3)$$

(c) for $\phi = -0.5$

$$y(k) = 2.0[1 - (-0.5)^k] \quad (25.4)$$

(d) for $\phi = 1.5$

$$y(k) = 2.0[1 - (1.5)^k] \quad (25.5)$$

The plots for $\phi = 0.5$, 0.8 and -0.5 are shown in FIG S25.1(a); for $\phi = 1.5$, the plot is shown in FIG S25.1(b).

The response for $\phi = 0.5$ is faster than for $\phi = 0.8$; for the negative value of ϕ, (but still with $|\phi| < 1$) the response alternates between positive and negative deviations from the final settling value of 2.0. For $\phi = 1.5$ (i.e. $\phi > 1$), the response is entirely unstable, whereas for the others, for which $|\phi| < 1$, the responses are all stable.

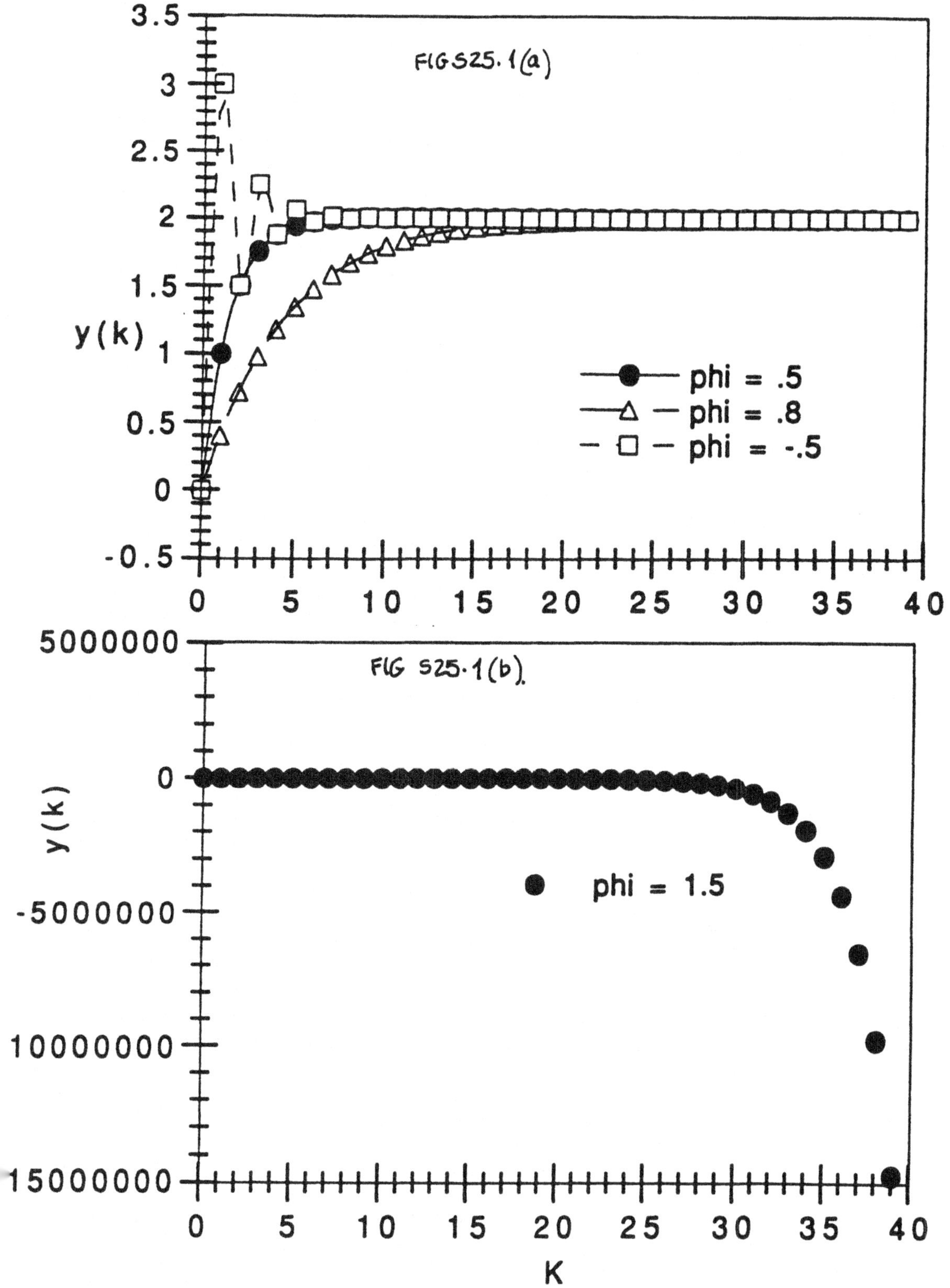

FIG S25.1 Unit step responses: First-order discrete-time system with various values of ϕ.

25.2 1. Given $g(z) = \dfrac{0.5z^{-1}}{1-0.8z^{-1}}$

Process steady state gain, $K = \lim_{z\to 1} g(z) = 2.5$

Poles: Single pole located at $z = 0.8$;

Zeros: None

Difference Equation model:

$$y(k) = 0.8y(k-1) + 0.5u(k-1)$$

Unit step response:

$$y(k) = 2.5\left[1 - (0.8)^k\right]$$

Process Type: Discrete first-order with ZOH.

2. Given $g(z) = 0.5z^{-1}$

$K = 0.5$; No pole; No zero

Diff. Eqn Model:

$$y(k) = 0.5u(k-1)$$

Unit step response:

$$y(k) = 0.5 \qquad k \geq 1$$

Process Type: Discrete PURE GAIN with ZOH.

3. Given $g(z) = \dfrac{0.5z^{-1}}{1-z^{-1}}$

Process gain: Does not exist

Poles: One at $z = 1$; No zeros

Diff. Equation model

$$y(k) = y(k-1) + 0.5u(k-1)$$

Unit step response

$$y(k) = 0.5k$$

Process type: PURE CAPACITY with ZOH.

4. Given $g(z) = \dfrac{4.5z^{-1} - 4z^{-2}}{1 - 0.8z^{-1}}$

Rearrange to $g(z) = \dfrac{(4.5 - 4.0z^{-1})z^{-1}}{1 - 0.8z^{-1}}$

Process Gain: $K = \lim_{z \to 1} g(z) = 2.5$

Poles: One, at $z = 0.8$

Zeros: One, at $z = 8/9$

Difference Eq^{n} model:

$$y(k) = 0.8y(k-1) + 4.5u(k-1) - 4u(k-2) \quad (25.6)$$

Unit step response: (solve Eq (25.6) and rearrange)

$$y(k) = 4.0(0.8)^{k-1} + 2.5\left[1 - (0.8)^{k}\right] \quad (25.7)$$

Process Type: LEAD/LAG system with ZOH.

5. Given $g(z) = \dfrac{6.0z^{-1} - 0.375z^{-2}}{1 + z^{-1} + 1.25z^{-2}}$

Process SS Gain: Does not exist: System open loop unstable!

Poles: Two, at $z = -\frac{1}{2} \pm j$ (Note $\|z\| = 1.118$; outside unit circle!)

Zeros: One, at $z = 0.0625$

Difference Eq^{n} model:

$$y(k) = -y(k-1) + 6u(k-1) - 0.375u(k-2) \quad (25.8)$$

Unit step response: Use Eqs (25.16) and (25.138) of main text; obtain

$$y(k) = \sum_{i=1}^{k} C(1.118)^{i-1} \cos\left[1.107(i-1) + \phi\right] \quad (25.9)$$

which will be "unstable".

Process Type: OPEN-LOOP UNSTABLE 2ND-ORDER with ZOH.

6. Given $g(z) = \dfrac{(-0.431 + 0.6093 z^{-1}) z^{-1}}{1 - 1.4252 z^{-1} + 0.4965 z^{-2}}$

Process SS Gain: $K = 2.50$
Poles: Two, at $z = 0.6063$; $z = 0.8189$
Zeros: One, at $z = 1.414$ (Note: zero outside unit circle).

Difference Eqn Model:

$$y(k) = 1.4252\, y(k-1) - 0.4965\, y(k-2) - 0.431\, u(k-1) + 0.6093\, u(k-2).$$

Unit step response:

$$y(k) = 2.5 + 4.16(0.6063)^k - 6.66(0.8189)^k$$

Process Type: SECOND ORDER SYSTEM, INVERSE RESPONSE, with ZOH.

7. Given $g(z) = \dfrac{\left[(0.0995 + 0.0789 z^{-1}) z^{-1}\right] z^{-6}}{1 - 4.252 z^{-1} + 0.4965 z^{-2}}$

Process SS Gain: $K = 2.50$
Poles: Two, at $z = 0.6063$; $z = 0.8189$
Zeros: One at $z = 0.7930$; 6 sample-time delays.

Difference Eqn Model:

$$y(k) = 1.4252\, y(k-1) - 0.4965\, y(k-2) + 0.0995\, u(k-7) + 0.0789\, u(k-8)$$

Unit step response

$$y(k) = \begin{cases} 0 & k < 6 \\ 2.50 + 33.5(0.6063)^{(k-6)} - 13.8(0.8189)^{(k-6)} & k \geq 7 \end{cases}$$

Process Type: SECOND-ORDER with ZOH, and DELAY of 6 samples.

8\. Given $g(z) = \dfrac{0.5z^{-1}}{1-1.2z^{-1}}$

Process SS Gain: Does not exist; System open loop unstable.
Pole: One at $z = 1.2$; outside the unit circle; system therefore open loop unstable.
Zero: None
Difference Eqn Model.

$$y(k) = 1.2y(k-1) + 0.5u(k-1)$$

Unit step response

$$y(k) = -2.5\left[1 - (1.2)^k\right]$$

Process Type: FIRST ORDER, OPEN LOOP UNSTABLE, with ZOH.

25.3 (a) For a discrete-time system to exhibit inverse response, its pulse transfer function must have a zero outside of the unit disk. Thus, for $g(z)$ as given in (P25.4) consolidation yields:

$$g(z) = \frac{b_1 z^{-1}(1-p_2 z^{-1}) - b_2 z^{-1}(1-p_1 z^{-1})}{(1-p_1 z^{-1})(1-p_2 z^{-1})}$$

and the zero is obtained by solving for z in the numerator; i.e. solving

$$z^{-1}\left[(b_1 - b_2) - (b_1 p_2 - b_2 p_1)z^{-1}\right] = 0$$

with the result:

$$z = \frac{b_1 p_2 - b_2 p_1}{b_1 - b_2}$$

and the required condition for inverse response is:

$$\frac{b_1 p_2 - b_2 p_1}{b_1 - b_2} > 1 \qquad (25.10)$$

(b) The relationships between (P25.2) and (P25.4) are:

$$p_1 = e^{-\Delta t/\tau_1} \quad ; \quad p_2 = e^{-\Delta t/\tau_2}$$

$$b_1 = K_1[1 - p_1] \quad ; \quad b_2 = K_2[1 - p_2]$$

Now use the result

$$\lim_{\Delta t \to 0} e^{-\Delta t/\tau} = 1 - \Delta t/\tau$$

so that

$$\lim_{\Delta t \to 0} p_1 = 1 - \Delta t/\tau_1 \; ; \quad \lim_{\Delta t \to 0} p_2 = 1 - \Delta t/\tau_2$$

$$\lim_{\Delta t \to 0} b_1 = K_1 \Delta t/\tau_1 \; ; \quad \lim_{\Delta t \to 0} b_2 = K_2 \Delta t/\tau_2$$

Thus: $$\lim_{\Delta t \to 0} \left\{ \frac{b_1 p_2 - b_2 p_1}{b_1 - b_2} \right\} = \frac{\cancel{\Delta t}\left[\frac{K_1}{\tau_1} - \frac{K_2}{\tau_2} + \left(\frac{K_2 - K_1}{\tau_1 \tau_2}\right)\Delta t\right]}{\left(\frac{K_1}{\tau_1} - \frac{K_2}{\tau_2}\right)\cancel{\Delta t}}$$

$$= 1 + \frac{\left(\frac{K_2 - K_1}{\tau_1 \tau_2}\right)\Delta t}{\frac{K_1}{\tau_1} - \frac{K_2}{\tau_2}} . \qquad (25.11)$$

Observe that $1 + \alpha > 1 \Rightarrow \alpha > 0$. Thus, from (25.11) above, the condition for inverse response obtained in (25.10) becomes, in the limit as $\Delta t \to 0$,

$$\left(\frac{K_2 - K_1}{\tau_1 \tau_2}\right)\Delta t \Big/ \left(\frac{K_1}{\tau_1} - \frac{K_2}{\tau_2}\right) \quad > 0 \qquad (25.12)$$

Now, we are given that $K_1 > K_2$; and since Δt is positive as is τ_1 and also τ_2, then the numerator in Eqn (25.12) above is <u>negative</u>, and the condition for inverse response will be satisfied if, and only if, the denominator is also negative, i.e.

$$\frac{K_1}{\tau_1} - \frac{K_2}{\tau_2} < 0$$

$$\text{or} \quad \frac{K_1}{\tau_1} < \frac{K_2}{\tau_2}$$

as stated in (P25.3), as required.

25.4

$$\underline{y}(k) = \underline{\phi}\,\underline{y}(k-1) + \underline{B}\,\underset{\sim}{u}(k-1)$$

for $\underset{\sim}{u}(0) = \underset{\sim}{1}$, obtain

$$\underline{y}(1) = \cancelto{0}{\underline{\phi}\,\underline{y}(0)} + \underset{\sim}{B}\cdot\underset{\sim}{1} \qquad \text{for } k = 1$$

For $k = 2$, obtain (since $\underset{\sim}{u}(1) = \underset{\sim}{1}$)

$$\underline{y}(2) = \underline{\phi}\,\underline{y}(1) + \underset{\sim}{B}\cdot\underline{1}$$

$$\text{or} \quad \underline{y}(2) = \underline{\phi}\,\underset{\sim}{B}\,\underset{\sim}{1} + \underset{\sim}{B}\cdot\underline{1}$$

and

$$\underline{y}(3) = \underline{\phi}\,\underline{y}(2) + \underline{B}\cdot\underline{1}$$

$$= \underline{\phi}^2\underline{B}\cdot\underline{1} + \underline{\phi}\,\underline{B}\cdot\underline{1} + \underline{B}\cdot\underline{1}$$

observe therefore that, in general,

$$\underset{\sim}{y}(k) = \underline{\phi}^{k-1}\underline{B}\cdot\underline{1} + \underline{\phi}^{k-2}\underline{B}\cdot\underline{1} + \cdots + \underline{\phi}\,\underline{B}\cdot\underline{1} + \underset{\sim}{B}\cdot\underset{\sim}{1} \qquad (25.13)$$

or $$\underline{y}(k) = [\underline{\Phi}^{k-1} + \underline{\Phi}^{k-2} + \cdots + \underline{\Phi} + \underline{I}]\,\underline{B}\cdot\underline{1} \tag{25.14}$$

Now let $$\underline{S}(k) = \underline{\Phi}^{k-1} + \underline{\Phi}^{k-2} + \cdots + \underline{\Phi} + \underline{I} \tag{25.15}$$

then
$$\underline{\Phi}\,\underline{S}(k) = \underline{\Phi}^{k} + \underline{\Phi}^{k-1} + \cdots + \underline{\Phi}^{2} + \underline{\Phi}$$

and by subtraction obtain
$$(\underline{I}-\underline{\Phi})\underline{S}(k) = \underline{I} - \underline{\Phi}^{k}$$

so that $$\underline{S}(k) = (\underline{I}-\underline{\Phi})^{-1}(\underline{I}-\underline{\Phi}^{k}) \tag{25.16}$$

Thus, the solution for $\underline{y}(k)$ is:
$$\underline{y}(k) = (\underline{I}-\underline{\Phi})^{-1}(\underline{I}-\underline{\Phi}^{k})\,\underline{B}\cdot\underline{1}$$

as reqd.

25.5 CASE I: No ZOH element.
In this case,

$$y(z) = g_2 g_{1NH}(z)\, u(z)$$

where
$$g_2 g_{1NH}(z) = \mathcal{Z}\left\{\left(g_2(s) g_1(s)\right)^*\right\}$$
$$= \mathcal{Z}\left\{\left[\frac{5}{s(3s+1)}\right]^*\right\} \tag{25.17}$$

From Table C.2 in Appendix C (entry 7a), obtain (using $\Delta t = 0.5$)
$$g_2 g_{1NH}(z) = \frac{0.768 z^{-1}}{(1-z^{-1})(1-0.846 z^{-1})} \tag{25.18}$$

CASE II: With the ZOH element.

Here,

$$g(z) = (1-z^{-1})\mathcal{Z}\left\{\left[\frac{g_1(s)\,g_2(s)}{s}\right]^*\right\}$$

$$= (1-z^{-1})\mathcal{Z}\left\{\left[\frac{5}{s^2(3s+1)}\right]^*\right\} \qquad (25.19)$$

Now, either by explicitly evaluating the RHS of eqn (25-19) or by recognizing that $g(z)$ is the pulse transfer function for a system with $g(s) = \frac{5}{s(3s+1)}$ and incorporating a ZOH, [in the latter case, one can then use Table C.3 entry 5 directly], obtain

$$g(z) = \frac{5\,[0.038z^{-1} + 0.039z^{-2}]}{(1-z^{-1})(1-0.846z^{-1})} \qquad (25.20)$$

25.6 Solution not provided.

25.7 Solution not provided.

25.8 Solution not provided.

25.9 (a) From Example 25.8 (main text), for pure proportional control (digital) of a first-order system,

$$\Delta t_{critical} = -\tau \ln\left(\frac{KK_c - 1}{KK_c + 1}\right) \qquad (25.21)$$

Observe that since (from (P25.6)) $KK_c = \tau/\tau_r$, then for the indicated design choice $\tau/\tau_r = 3$, $KK_c = 3$ and (25.21) above becomes

$$\Delta t_{critical} = -\tau \ln(1/2)$$
$$= 0.693\tau \qquad (25.22)$$

with the following implication: for a choice of $\Delta t > 0.693\tau$, the closed loop system will be unstable. Thus, the experienced engineer's recommendation that Δt should not exceed $0.7\Delta t$ is thus quite correct.

(b) For $\Delta t = \tau$, in this case,

$$\phi = e^{-\Delta t/\tau} = 0.3679$$

so that the critical value for KK_c (same as τ/τ_r), from Eqn (25.207) is:

$$(KK_c)_{critical} = \frac{1+\phi}{1-\phi} = 2.164 \qquad (25.23)$$

The implication is that the maximum allowable value for τ/τ_r under the given conditions is 2.164 if the closed loop system is to remain stable.

25.10 (a) Under continuous time proportional-only feedback control, the closed-loop characteristic equation

$$1 + gg_c = 0$$

becomes, in this case:

$$1 + \frac{1.5K_c}{(12s+1)(5s+1)} = 0$$

or

$$60s^2 + 17s + (1 + 1.5K_c) = 0 \qquad (25.24)$$

Since Eq (25.24) is a quatratic equation whose coefficients are all positive for all values of $K_c > 0$, its roots will all lie in the LHP for all values of $K_c > 0$; hence the closed-loop system will always be stable (for all values of $K_c > 0$).

(b). Under digital proportional feedback control (incorporating a ZOH), the closed-loop transfer function in this case will be:

$$\frac{g_f(z) K_c}{1 + g_\ell(z) K_c}$$

with a characteristic equation (see Eq (25.191))

$$1 + g_\ell(z) K_c = 0 \qquad (25.25)$$

Here, $g_f(z) = g_\ell(z) = g(z)$ for the given 2nd order system (with a ZOH) because there are no other elements in the loop. For the given $g(s)$,

$$g(s) = \frac{1.5}{(12s+1)(5s+1)}$$

obtain in the usual fashion (with a ZOH incorporated) the required $g(z)$ as

$$g(z) = \frac{0.0117z^{-1} + 0.01z^{-2}}{1 - 1.739z^{-1} + 0.753z^{-2}} \qquad (25.26)$$

In this case, the characteristic equation becomes, after some rearrangement,

$$z^2 + (0.0117K_c - 1.739)z + (0.753 + 0.01K_c) = 0$$

Employing the Jury test to establish the stability conditions, we have the following:

- CONDITION 1:

$$|a_2| < a_0$$

$$\Rightarrow \quad 0.753 + 0.01K_c < 1 \quad \Rightarrow \quad \boxed{K_c < 24.7} \qquad (25.27)$$

- CONDITION 2:

$$P(1) = 1 - 1.739 + 0.0117K_c + 0.753 + 0.01K_c$$

$$= 0.014 + 0.0217K_c > 0 \qquad (25.28)$$

(satisfied for all values of $K_c > 0$).

- CONDITION 3

$n = 2$ (even) thus

$$P(-1) = 3.492 - 0.0017K_c > 0$$

$$\Rightarrow \quad K_c < 2054.11 \qquad (25.29)$$

(superceded by condition 1)

thus for closed-loop stability, $K_c < 24.7$ is the required range.

(c). For $\Delta t = 5$ mins,

$$g(z) = \frac{0.20z^{-1} + 0.124z^{-2}}{1 - 1.027z^{-1} + 0.243z^{-2}}$$

and, under digital proportional control, the characteristic equation is:

$$1 - 1.027z^{-1} + 0.243z^{-2} + (0.2K_c)z^{-1} + (0.124K_c)z^{-2} = 0$$

which rearranges to:

$$z^2 + (0.2K_c - 1.027)z + (0.243 + 0.124K_c) = 0$$

From where the Jury test gives rise to the following conditions for closed-loop stability:

- CONDITION 1

$$0.243 + 0.124\,K_c < 1 \Rightarrow \boxed{K_c < 6.105} \qquad (25\text{-}30)$$

- CONDITION 2

$P(1) = 0.216 + 0.324\,K_c > 0$; satisfied for all $K_c > 0$;

- CONDITION 3

$n = 2$ (even), so that

$P(-1) = 2.27 - 0.076\,K_c > 0 \Rightarrow K_c < 29.868$

superceded by condition 1.

Thus for closed-loop stability, K_c must satisfy Eqn (25-30) i.e. $K_c < 6.105$.

Compared with the stability range for $\Delta t = 1$ ($K_c < 24.7$) observe that as Δt increases, the range of K_c values for which the closed-loop system remains stable narrows down considerably.

25.11 In continuous time, the characteristic equation for the closed loop system is:

$$1 - \frac{0.55K_c}{(5s-1)(2s+1)} = 0$$

or

$$10s^2 + 3s - (1 + 0.55 K_c) = 0$$

and with $K_c = -4$, it becomes

$$10s^2 + 3s + 1.2 = 0$$

a quadratic equation with all its roots in the LHP thus the closed loop system is indeed stable, in continuous time, with $K_c = -4.0$

Under digital control, the transfer function

$$g(s) = \frac{-0.55}{(5s-1)(2s+1)}$$

with a ZOH, and $\Delta t = 0.5$ mins gives rise to a pulse transfer function $g(z)$ given by

$$g(z) = \frac{-0.0065 z^{-1} - 0.0062 z^{-2}}{1 - 1.884 z^{-1} + 0.861 z^{-2}}$$

(see Problem 24.4). And now, the characteristic equation is:

$$1 - 1.884 z^{-1} + 0.861 z^{-2} - 0.0065 K_c z^{-1} - 0.0062 K_c z^{-2} = 0$$

so that for $K_c = -4$, we need to investigate the roots of

$$z^2 - 1.858 z + 0.8858 = 0 \qquad (25.31)$$

to establish closed-loop stability.

These roots are easily obtained as

$$z_{1,2} = 0.929 \pm 0.151 j \qquad (25.32)$$

and since $\| z_{1,2} \| = 0.941 < 1$, we conclude that

the closed-loop system is indeed stable if the same proportional controller (with $K_c = -4.0$) is employed, with $\Delta t = 0.5$ mins.

25.12 The characteristic equation whose roots are to be investigated is:

$$P(z) = 1 + \frac{2.0\left(0.15z^{-1} - 0.29z^{-2} + 0.32z^{-3}\right)}{1 - 1.6z^{-1} + 0.5z^{-2} - 0.4z^{-3}} = 0$$

which rearranges to:

$$P(z) = z^3 - 1.3z^2 - 0.08z + 0.24 = 0 \qquad (25.33)$$

(a) Using the bilinear transformation requires substituting

$$z = \frac{w+1}{w-1}$$

into (25.33) to obtain

$$\pi(w) = \frac{(w+1)^3}{(w-1)^3} - 1.3\frac{(w+1)^2}{(w-1)^2} - 0.08\frac{(w+1)}{(w-1)} + 0.24 = 0$$

or

$$\pi(w) = (w+1)^3 - 1.3(w+1)^2(w-1) - 0.08(w+1)(w-1)^2 + 0.24(w-1)^3 = 0$$

where, upon carrying out the required algebraic manipulations, the equation simplifies to

$$\pi(w) = -0.14w^3 + 1.06w^2 + 5.1w + 1.98 = 0 \qquad (25.34)$$

To apply Routh's test to this equation (see p. 489 of the main text), recast in the "standard form":

$$\pi(w) = 0.14w^3 - 1.06w^2 - 5.1w - 1.98 = 0 \qquad (25.35)$$

and now since 3 of the coefficients are negative, we need not proceed further; $\pi(w)$ has <u>at least</u> one root in the RHP, implying that the closed-loop system under investigation is UNSTABLE.

(b) Applying Jury's test directly to Eqn (25.33) above gives rise to the following results:

- CONDITION 1

$$|a_3| \overset{?}{<} a_0 \text{ ; in this case}$$

$$\boxed{0.24 < 1}$$

- CONDITION 2.

$$P(1) \overset{?}{>} 0$$

In this case $P(1) = 1 - 1.3 - 0.08 + 0.24 = -0.14$

Thus condition 2 is <u>NOT</u> satisfied!

- CONDITION 3

$n = 3$ (odd), $P(-1) \overset{?}{<} 0$

In this case $P(-1) = -1 - 1.3 + 0.08 + 0.24 = -1.98 < 0$

Since condition 2 is <u>not</u> satisfied, again, we need not proceed further; $P(z)$ has at least one root outside the unit disk, implying that the closed-loop system under investigation is UNSTABLE.

Observe that the Jury test involves less effort to apply.

25.13 The closed-loop system's characteristic equation is:

$$1 + g_\ell(z) g_c(z) = 0$$

In this case, since there are no other elements in the loop apart from the process (whose $g(s)$ is as given) we now have that, with $\Delta t = 1$,

$$g_\ell(z) = g(z) = \frac{K(1-\phi)z^{-1}}{1-\phi z^{-1}}, \quad \phi = e^{-1/6.7}$$

or

$$g_\ell(z) = \frac{0.092 z^{-1}}{1 - 0.861 z^{-1}}$$

Given

$$g_c(z) = \frac{6.7}{0.66\lambda}\left[1 + \frac{1}{6.7(1-z^{-1})}\right]$$

The characteristic equation becomes, upon simplification:

$$P(z) = (6.7\lambda)z^2 + (1.07 - 12.469\lambda)z + (5.769\lambda - 0.931) = 0 \qquad (25.36)$$

Applying Jury's test to (25.36) produces the following conditions to be satisfied for closed-loop stability:

- CONDITION 1.

$$|5.769\lambda - 0.931| < 6.7\lambda \qquad (25.37)$$

- CONDITION 2. $P(1) > 0$

$0.139 > 0$ (Always satisfied)

- CONDITION 3.

$$24.938\lambda - 2 > 0 \Rightarrow \boxed{\lambda > 0.08} \qquad (25.38)$$

And now care must be exercised in converting cond^{n} 1 into an explicit condition for λ. This may be done as follows:

(i) If $a_2 = 5.769\lambda - 0.931$ is negative, in which case

$$\lambda < 0.1614$$

stability will require that

$$-5.769\lambda + 0.931 < 6.7\lambda$$

or

$$\lambda > 0.075$$

Thus, a_2 negative results in the joint requirement

$$\boxed{0.075 < \lambda < 0.1614} \qquad (25.39a)$$

(ii) If a_2 is positive, in which case $\lambda > 0.1614$, stability will require

$$5.769\lambda - 0.931 < 6.7\lambda$$

or $\lambda > -1$

and therefore a_2 positive results in the requirement

$$\boxed{\lambda > 0.1614} \qquad (25.39b)$$

Thus, the explicit conditions implied in Eq. (25.37) above are obtained from the union of the two sets in Eq (25.39a, b); i.e.

$$\boxed{\lambda > 0.075} \qquad (25.39c)$$

Taking this now in conjunction with condition 3 gives the required result: For closed loop stability, λ must satisfy the condition

$$\boxed{\lambda > 0.08}$$

25.14 (a) The required closed-loop characteristic equation is:

$$1 + \frac{b z^{-(m+1)} K_c}{1 - \phi z^{-1}} = 0 \qquad (25.40)$$

or:

$$1 - \phi z^{-1} + b K_c z^{-(m+1)} = 0$$

or

$$\boxed{z^{m+1} - \phi z^{m} + b K_c = 0} \qquad (25.41)$$

(b) For $m = 0$, $b = 0.09$, $\phi = 0.86$, Eqn (25.41) becomes

$$z - 0.86 + 0.09 K_c = 0$$

From where stability requires

$$|-0.86 + 0.09 K_c| < 1 \qquad (25.42)$$

Combining the requirements when $a_1 = -0.86 + 0.09 K_c$ is negative ($-1.556 < K_c < 9.556$) with the requirements when a_1 is positive ($9.556 < K_c < 20.667$) (and the fact that $K_c > 0$), obtain the explicit requirement implied in Eq (25.42) above as:

$$\boxed{0 < K_c < 20.667} \qquad (25.43)$$

(c) For $m = 1$, the characteristic equation becomes

$$z^2 - 0.86 z + 0.09 K_c = 0 \qquad (25.44)$$

From where the Jury test gives rise to the following conditions for stability:

• CONDITION 1

$$0.09 K_c < 1$$

$$\Rightarrow \boxed{K_c < 11.1} \qquad (25.45)$$

• CONDITION 2 $P(1) > 0$

$0.14 + 0.09 K_c > 0$ (Always satisfied for all $K_c > 0$)

• CONDITION 3

$P(-1) > 0$ ($n = 2$, even)

$1.86 + 0.09 K_c > 0$ Always satisfied for all the K_c.

Thus for closed-loop stability when $m = 1$, it is required that

$$\boxed{K_c < 11.1}$$

Compared with eqn (25.43) above (no delay) not that the stability margin for K_c has been cut in half as a result of the presence of a unit delay.

For $m = 2$, on the other hand, the characteristic equation becomes

$$z^3 - 0.86 z^2 + 0.09 K_c = 0$$

The Jury test gives rise to the following new conditions for closed-loop stability:

• CONDITION 1:

$$0.09 K_c < 1 \quad \Rightarrow \quad K_c < 11.1$$

• CONDITION 2: $P(1) > 0$

$0.14 + 0.09 K_c > 0$ (Always satisfied for all $K_c > 0$)

• CONDITION 3:

$n = 3$, $P(-1) < 0$

$$-1 + 0.86 + 0.09 K_c < 0$$

or

$$0.09 K_c < 0.14$$

$$\Rightarrow \quad K_c < 1.555 \qquad (25.46)$$

Taken together, closed-loop stability now requires

$$\boxed{K_c < 1.555}$$

And now the presence of a 2-unit delay is seen to cut the stability margin from $K_c < 20.667$ (with no delay) to $K_c < 1.555$.
Thus the presence of delays is seen to narrow down the range of K_c values required for closed-loop stability; For a unit delay, the range is essentially cut in half; for two units of delay, the range is reduced by an order of magnitude!

25.15 Solution not provided.

Chapter 26

26.1 (a) The consolidated transfer function for the process is given by. $g(s) = g_v(s)\, g_i(s)\, g_f(s)$, or:

$$g(s) = \frac{150e^{-0.2s}}{(0.5s+1)(0.01s+1)^2} \quad (^{\circ}F/psig) \qquad (26.1)$$

When combined with a ZOH element, the resulting discrete process pulse transfer function may be obtained using the standard procedure presented in Chapter 25. For $\Delta t = 0.1$ sec, the result is:

$$g(z) = (1-z^{-1})\left\{\frac{150}{1-z^{-1}} - \frac{156.2}{1-0.8187z^{-1}} + \frac{30.6e^{-10}z^{-1}}{(1-e^{-10}z^{-1})^2} + \frac{6.185}{(1-e^{-10}z^{-1})}\right\} z^{-2}$$

which may be simplified to give:

$$g(z) = \frac{(-0.015z^3 + 22.12z^2 + 5.060z + 9.074\times 10^{-4})\, z^{-2}}{(z-0.8187)(z-e^{-10})^2} \qquad (26.2)$$

For the controller, $u(z) = g_c(z)\, \varepsilon(z)$ with

$$g_c(z) = \frac{0.01 - 0.008z^{-1}}{1 - 0.75z^{-1} - 0.25z^{-2}} \qquad (26.3)$$

(b) With $g(z)$ as in Eq (26.2) and $g_c(z)$ as in (26.3) the block diagram for this closed loop system is as shown below:

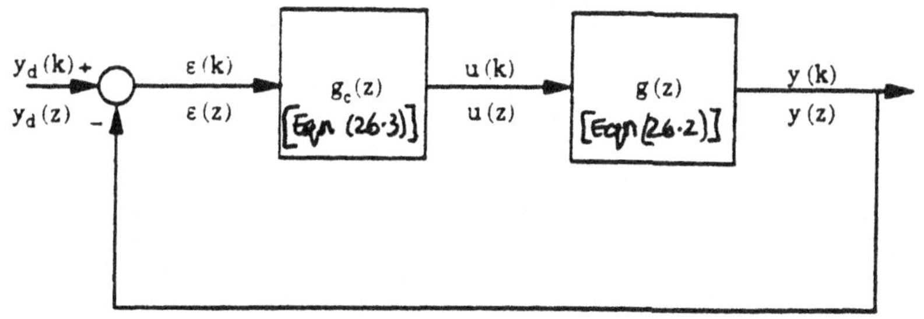

FIG S26.1 Block diagram for industrial gas furnace control system

The closed loop transfer function is given by

$$g_{CL}(z) = \frac{g(z)\,g_c(z)}{1 + g(z)\,g_c(z)}$$

and the characteristic equation is:

$$1 + g(z)\,g_c(z) = 0$$

which, upon introducing Eqns (26.2) and (26.3) becomes

$$p(z) = z^4(z - 0.8187)(z^2 - 0.75z - 0.25) + 0.001(10z - 8)(-0.015z^3 + 22.12z^2 + 5.062z + 1.138\times10^{-3}) = 0 \quad (26.4)$$

Expanding out to:

$$p(z) = z^7 - 1.5689z^6 + 0.3642z^5 + 0.2046z^4 + 0.2213z^3 - 0.1263z^2 - 0.04049z - 9.101\times10^{-6} = 0 \quad (26.5).$$

Apply the Jury test to Eqn (26.5) to determine closed-loop system stability.

(i) $|a_n| \overset{?}{<} a_0$

In this case

$9.101 \times 10^{-6} < 1$ $\Rightarrow$ (i) satisfied.

(ii) $p(1) \overset{?}{>} 0$:

Here $p(1) = 0.0543 > 0$ $\Rightarrow$ (ii) satisfied

(iii) $p(-1) \overset{?}{<} 0$, $n = 7$, odd.

Here $p(-1) = -3.04 < 0$ $\Rightarrow$ (iii) satisfied

Since all three screening criteria are satisfied, proceed to the full Jury test: (see procedure on p 939 main text)

JURY ARRAY (TABLE):

a:	1	-1.5689	0.3642	0.2046	0.2213	-0.1263	-0.04049	-9.101×10^{-6}
	-9.101×10^{-6}	-0.04049	-0.1263	0.2213	0.2046	0.3642	-1.5689	1
b:	1	-1.5689	0.3642	0.2046	0.2213	-0.1263	-0.0405	
	-0.0405	-0.1263	0.2213	0.2046	0.3642	-1.5689	1	
c:	0.9984	-1.574	0.3732	0.2129	0.2361	-0.1898		
	-0.1898	0.2361	0.2129	0.3732	-1.574	0.9984		
d:	0.9607	-1.527	0.4130	0.2834	-0.06315			
	-0.06315	0.2834	0.4130	-1.527	0.9607			
e:	0.9189	-1.449	0.4228	0.1758				
	0.1758	0.4228	-1.449	0.9189				
f:	0.8135	-1.406	0.6433					
	0.6433	-1.406	0.8135					

And now observe from this table that:

$1 > 0.0405$ for the b row; $0.9984 > 0.1898$ for c;

$0.9607 > 0.06315$ for d; $0.9189 > 0.1758$ for e;

and $0.8135 > 0.6433$ for f;

indicating that no root of the characteristic equation (26.5) lies outside the unit circle. We therefore conclude that the closed-loop system is stable.

26.2 (a) From the supplied process transfer function, deduce the process characteristic parameters as:

$$K = 0.55; \quad \alpha = 8, \quad \tau = 7.5$$

so that the Cohen-Coon rules in Table 15.3 give the following controller parameters:

$$K_c = \frac{1}{K}\left(\frac{\tau}{\alpha}\right)\left[0.9 + \frac{1}{12}\left(\frac{\alpha}{\tau}\right)\right] = 1.686 \qquad (26.6)$$

and

$$\tau_I = \alpha\left[\frac{30 + 3(\alpha/\tau)}{9 + 20(\alpha/\tau)}\right] = 8.76 \qquad (26.7)$$

The resulting PI controller is therefore given by

$$g_c(s) = 1.686\left(1 + \frac{1}{8.76\,s}\right) \qquad (26.8)$$

Discretize this as in Eq (26.49) of the main text to obtain

$$g_c(z) = 1.686\left(1 + \frac{1}{4.38(1-z^{-1})}\right) \qquad (26.9)$$

for $\Delta t = 2$.

Implementing this on the continuous process for a unit step change in the basis weight set-point gives the response shown in FIG S26.2 below.

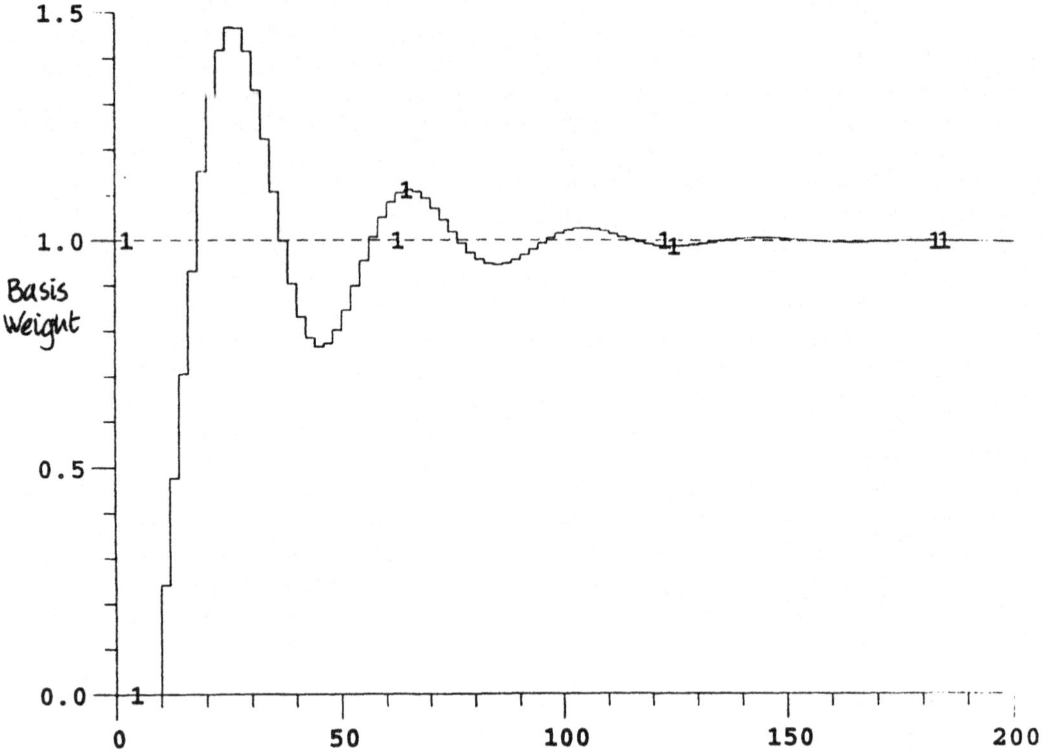

FIG S26.2 Basis weight response for paper machine under discrete PI control: $\Delta t = 2$

(b) when Δt is increased to 4, the discrete controller becomes

$$g_c(z) = 1.686\left(1 + \frac{1}{2.19(1-z^{-1})}\right) \qquad (26.10)$$

and under the given circumstances, this controller now produces the response shown in FIG S26.3. Observe that the <u>overshoot</u> is now more pronounced, and that it is taking longer for the system to settle than in FIG S26.2

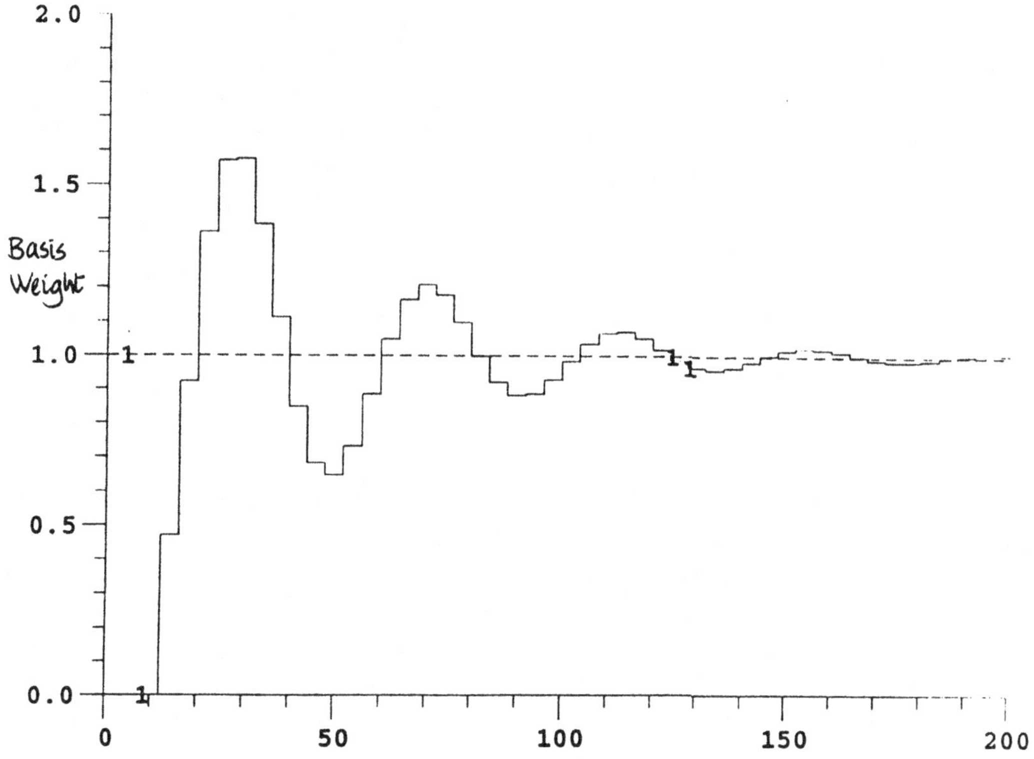

FIG S26.3 Basis weight response for paper machine under discrete PI control: $\Delta t = 4$

(c) If we now use $K = 0.55$, $\tau = 7.5$
but $\alpha' = 8 + \Delta t/2 = 10$,

obtain the "modified" Cohen-Coon controller parameters as.

$$K_c = 1.379, \quad \tau_I = 9.533$$

so that now

$$g_c(z) = 1.379\left[1 + \frac{4}{9.533(1-z^{-1})}\right] \qquad (26.11)$$

If we now implement this controller (with $\Delta t = 4$) we obtain the response shown in FIG S26.4, where we observe immediately that there is a significant advantage in using this strategy of adding $\Delta t/2$ to the process delay before designing the controller. Observe that the process responds faster, with less overshoot, and also settles faster. In fact, this response is better than the one shown in part (a), FIG S26.2 with $\Delta t = 2$.

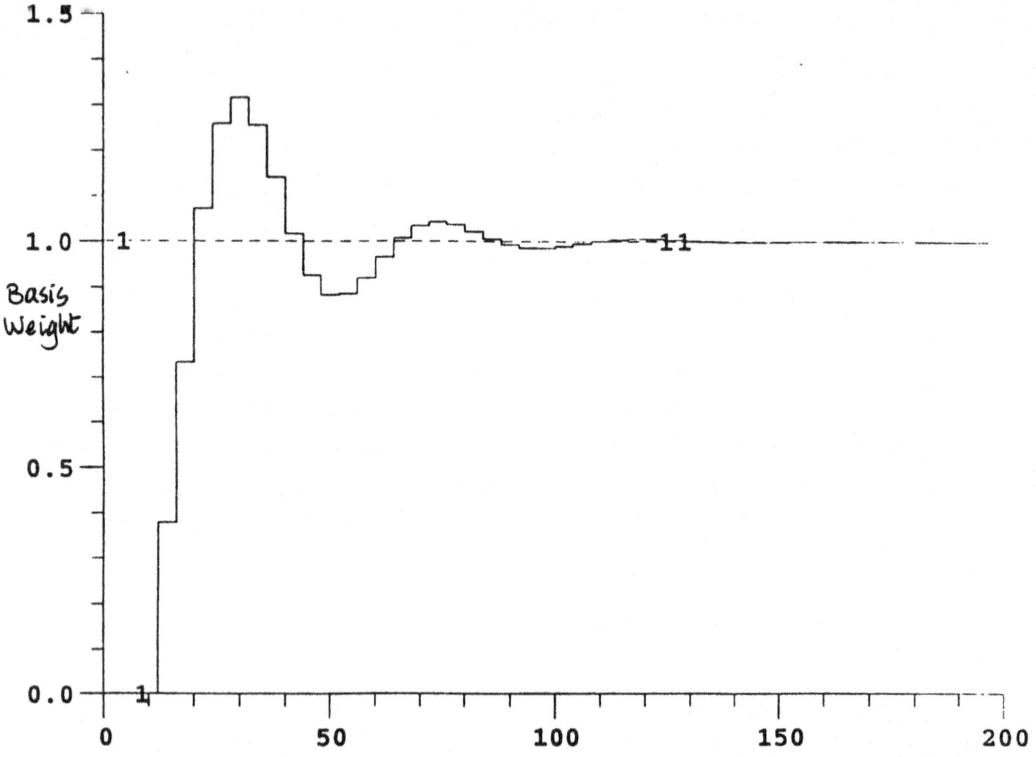

FIG S26.4 Basis weight response for paper machine "Modified" discrete PI control: $\Delta t = 4$, $\alpha' = \alpha + \Delta t/2$

26.3 The "continuous" feedforward controller transfer function is:

$$g_{ff}(s) = \frac{0.261(6.7s+1)e^{-9.5s}}{(6.2s+1)} \qquad (26.12)$$

and for $\Delta t = 0.5$ min, the delay term in the discrete transfer function is z^{-19}; the other aspect of $g_{ff}(z)$ are determined as follows.

(a) A forward difference approximation is obtained as:

$$g_{ff}(z) = g_{ff}(s)\Big|_{s=\left(\frac{1-z^{-1}}{z^{-1}\Delta t}\right)}$$

$$= \frac{0.21\left(6.7-6.2z^{-1}\right)}{\left(6.2-5.7z^{-1}\right)} z^{-19}$$

or

$$g_{ff}(z) = \frac{\left(0.227-0.21z^{-1}\right)}{\left(1-0.919z^{-1}\right)} z^{-19} \qquad (26.13)$$

(b) For the backward difference approximation, use

$$s \approx \frac{1-z^{-1}}{\Delta t}$$

and obtain

$$g_{ff}(z) = \frac{0.21\left(7.2-6.7z^{-1}\right)}{\left(6.7-6.2z^{-1}\right)} z^{-19}$$

or

$$g_{ff}(z) = \left(\frac{0.226-0.21z^{-1}}{1-0.925z^{-1}}\right) z^{-19} \qquad (26.14)$$

(c) For the Tutsin approximation, use $s \approx \frac{2}{\Delta t}\left(\frac{1-z^{-1}}{1+z^{-1}}\right)$ and obtain

$$g_{ff}(z) = \frac{0.21\,(13.9 - 12.9z^{-1})}{(12.9 - 11.9z^{-1})}\, z^{-19}$$

or

$$g_{ff}(z) = \left(\frac{0.226 - 0.21z^{-1}}{1 - 0.922z^{-1}}\right) z^{-19} \qquad (26.15)$$

Note how very similar all three expressions are; note in particular how the backward difference approximation produces a $g_{ff}(z)$ that is almost indistinguishable from that produced by the more "complicated" Tustin approx.

26.4 (a) The equation for the direct synthesis controller obtained in Example 26.2 (Eq. 26.53) is.

$$\Delta u(k) = \left(\frac{\tau + \Delta t}{K\tau_r}\right)\epsilon(k) - \left(\frac{\tau}{K\tau_r}\right)\epsilon(k-1) \qquad (26.16)$$

The general expression for the discrete PID controller (Eq (26.39) main text) is:

$$\Delta u(k) = K_c\left[\left(1 + \frac{\Delta t}{\tau_I} + \frac{\tau_D}{\Delta t}\right)\epsilon(k) - \left(\frac{2\tau_D}{\Delta t} + 1\right)\epsilon(k-1) + \frac{\tau_D}{\Delta t}\epsilon(k-2)\right]$$

Comparing these two equations yields the following results:

<u>Coefficients of $\epsilon(k-2)$</u>

$$\frac{\tau_D}{\Delta t} = 0 \quad \Rightarrow \quad \tau_D = 0 \qquad (26.17)$$

Coeff. of $\epsilon(k-1)$

$$K_c\left(\frac{2\tau_D}{\Delta t}+1\right) = \frac{\tau}{K\tau_r}$$

and using (26.17) reduces this to

$$\boxed{K_c = \frac{\tau}{K\tau_r}} \qquad (26.18)$$

Coeff of $\epsilon(k)$

$$K_c\left(1+\frac{\Delta t}{\tau_I}+\frac{\tau_D}{\Delta t}\right) = \frac{\tau+\Delta t}{K\tau_r}$$

Introduce (26.17) and (26.18) to obtain

$$\frac{\tau}{K\tau_r}\left(1+\frac{\Delta t}{\tau_I}\right) = \frac{\tau+\Delta t}{\tau_r}$$

simplify to obtain:

$$\boxed{\tau_I = \tau} \qquad (26.19)$$

Observe therefore from (26.17), (26.18) and (26.19) that the controller parameter values for K_c and τ_I are entirely independent of Δt.

It is desirable to have the controller parameters depend on Δt because closed-loop system behavior, especially closed-loop stability, is influenced significantly by changes in Δt. In particular, as Δt increases, the controller parameter values required for closed-loop stability become more conservative. It will be desirable to have a controller design technique that produces as results controller parameters that depend on Δt; in this case, the controller parameters can "automatically" adjust to changes in Δt, thereby maintaining the original design criteria even when the sampling time is altered.

(b) Introduce the approximation

$$s \approx \frac{2}{\Delta t}\left(\frac{1-z^{-1}}{1+z^{-1}}\right)$$

in the expression

$$g_c(s) = \frac{1}{\tau_r s} \cdot \left(\frac{\tau s + 1}{K}\right).$$

to obtain

$$g_c(z) \approx \frac{1\ \Delta t(1+z^{-1})}{\tau_r\ 2(1-z^{-1})} \cdot \frac{2\tau(1-z^{-1}) + \Delta t(1+z^{-1})}{K \Delta t(1+z^{-1})}$$

which simplifies to

$$g_c(z) = \frac{2\tau + \Delta t}{2K\tau_r}\left(\frac{1 - b_1 z^{-1}}{1 - z^{-1}}\right) \qquad (26.20a)$$

$$\text{where } b_1 = \left(\frac{2\tau - \Delta t}{2\tau + \Delta t}\right) \qquad (26.20b)$$

The time domain realization of this controller is:

$$u(k) = u(k-1) + \left(\frac{2\tau + \Delta t}{2K\tau_r}\right)\epsilon(k) - \left(\frac{2\tau - \Delta t}{2K\tau_r}\right)\epsilon(k-1) \qquad (26.21)$$

Comparing this with the general PID controller equation yields the following results:

<u>Coeff. of $\epsilon(k-2)$</u>

$$\frac{\tau_D}{\Delta t} = 0 \quad \Rightarrow \quad \tau_D = 0$$

<u>Coeff. of $\epsilon(k-1)$.</u>

$$K_c\left(\frac{2\tau_D}{\Delta t} + 1\right) = \frac{2\tau - \Delta t}{2K\tau_r}$$

$$\Rightarrow \quad \boxed{K_c = \frac{\tau - \Delta t/2}{K\tau_r}} \qquad (26.22)$$

and <u>coeff of $\epsilon(k)$</u>

$$K_c\left(1 + \frac{\Delta t}{\tau_I}\right) = \frac{2\tau + \Delta t}{2K\tau_r}$$

use (26.22) above and simplify to obtain

$$\boxed{\tau_I = \tau - \Delta t/2.} \qquad (26.23)$$

Observe now that these controller parameters depend on Δt. Also observe that the dependence on Δt is such that as $\Delta t \uparrow$, $K_c \downarrow$, which is in perfect keeping with our knowledge of how K_c should be modified in the face of increased sample time. Note also that this controller parameterization limits Δt to no more than 2τ — Again a very reasonable restriction. If one is compelled to sample a process with a Δt greater than twice the time constant, the controller design should be changed.

(c) Following the recommendation, obtain the required controller by setting $D = 1$ in Eq (26.59) main text:

$$u(k) = \left(\frac{\tau_r}{\tau_r + \Delta t}\right) u(k-1) + \left(\frac{\Delta t}{\tau_r + \Delta t}\right) u(k-1) + \left[\frac{\tau + \Delta t}{K(\tau_r + \Delta t)}\right] \epsilon(k) - \left[\frac{\tau}{K(\tau_r + \Delta t)}\right] \epsilon(k-1)$$

which rearranges to give

$$u(k) = u(k-1) + \frac{\tau + \Delta t}{K(\tau_r + \Delta t)} \epsilon(k) - \frac{\tau}{K(\tau_r + \Delta t)} \epsilon(k-1) \qquad (26.24)$$

recognizable as a PI controller in velocity form. Comparing this controller with the Direct Synthesis controller obtained in Eq (26.53) main text, observe that the two controllers are nearly identical but for the singular exception that the denominator of the coefficients of $\epsilon(k)$ and $\epsilon(k-1)$ has been altered from $K\tau_r$ in the Direct Synthesis controller, to $K(\tau_r + \Delta t)$ in the "Modified Smith predictor" controller. This difference is particularly important because as $\Delta t \uparrow$, the coefficient of $\epsilon(k)$ in the Direct Synthesis controller goes up indefinitely while the coeff. of $\epsilon(k-1)$ is entirely unaffected. In Eq (26.24) above, however, as $\Delta t \uparrow$, the coefficient of $\epsilon(k)$ has a finite limiting value of $1/K$; and the coeff. of $\epsilon(k-1)$ goes to zero.

From another perspective, recall that, in terms of the standard discrete PI controller, the Direct synthesis controller in Eq (26.53) main text has K_c and τ_I parameters that are independent of Δt. This modified controller on the other hand can be presented in the standard PI form with

$$K_c = \frac{\tau}{K(\tau_r + \Delta t)}$$

and $\tau_I = \tau$

so that $K_c \rightarrow 0$ as $\Delta t \rightarrow \infty$.

26.5 Solution not provided. (see Prob. 26.2)

26.6 Problem is open-ended; solution not provided.

26.7 Solution not provided.

26.8 Solution not provided.

26.9 (a) The discrete process pulse transfer function is obtained from the given $g(s)$ in Eq (P26.5) as.

$$g_m(z) = \frac{0.55(0.234)\, z^{-5}}{(1 - 0.766\, z^{-1})} \qquad (26.25)$$

using $\Delta t = 2$. And now, for the given $\tau_r = 3.0$ and $\gamma = 8.0$, use Eq (26.74) main text to derive the Dahlin controller:

$$g_c(z) = \frac{1 - 0.766 z^{-1}}{0.1287 \cancel{z^{-5}}} \left\{ \frac{(1 - e^{-2/3})\, \cancel{z^{-5}}}{1 - e^{-2/3} z^{-1} - (1 - e^{-2/3})\, z^{-5}} \right\}$$

and simplify to give:

$$g_c(z) = \frac{3.781(1-0.766z^{-1})}{1-0.513z^{-1}-0.487z^{-5}} \qquad (26.26)$$

From here, since the controller operates according to

$$u(z) = g_c(z)\,\epsilon(z)$$

obtain an explicit time-domain expression for the controller given in Eq (26.26) above as:

$$u(k) = 0.513u(k-1) + 0.487u(k-5) + 3.781\epsilon(k) - 2.896\epsilon(k-1) \qquad (26.27)$$

Upon implementing this controller on the "true" process as given in Eq (P26.11), obtain the response shown in FIG S26.5 in the solid line. Note the "bumpiness" of the response

(b). With $\tau_r = 6.0$, again use Eq (26.74) main text and simplify to obtain the new Dahlin controller:

$$g_c(z) = \frac{2.203(1-0.766z^{-1})}{1-0.717z^{-1}-0.283z^{-5}} \qquad (26.28)$$

whose explicit time-domain realization is given by.

$$u(k) = 0.717u(k-1) + 0.283u(k-5) + 2.203\epsilon(k) - 1.687\epsilon(k-1). \qquad (26.29)$$

when implemented on the "true process" the response is shown in the dashed line in FIG S26.5. This response is obviously a little "slower" than with $\tau_r = 3$, but observe the smoothness. The "bumpiness" evident with the more aggressive controller is a direct effect of plant/model mismatch. (The Dahlin controller is based on "cancelling" process poles, and replacing them with desired ones. Plant/model mismatch makes exact cancellation impossible, leaving residual "bumpiness" in the response). The more conservative controller minimizes the observable effect of plant/model mismatch: i.e. it is more robust.

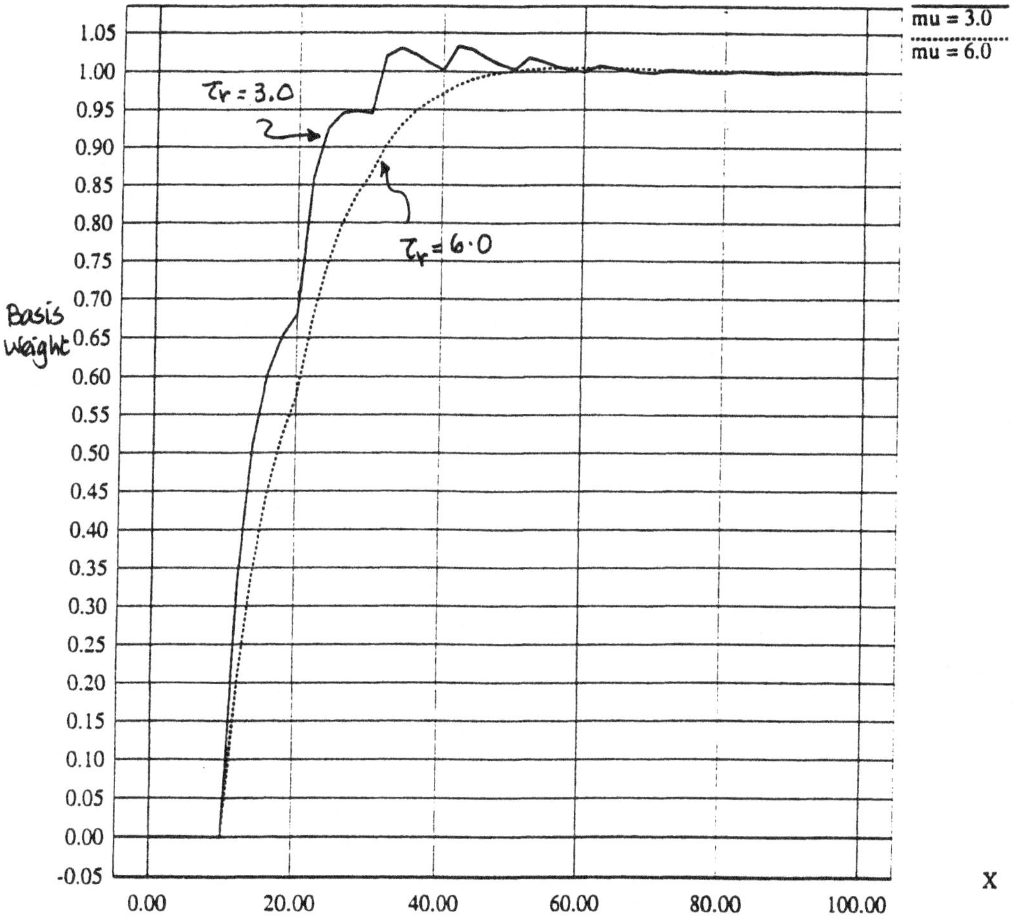

FIG S26.5 Basis Weight response using Dahlin controllers. Effect of plant/model mismatch and Dahlin tuning parameter.

26.10 Problem is open-ended; Solution not provided. Ideal for a 'Take-home' exam.

26.11 (a) The required pulse transfer function is obtained using the standard procedure (or using Table C.3, Appendix C, Entry # 4) as:

$$g(z) = \frac{-0.121 z^{-1}}{1 - 1.22 z^{-1}} \qquad (26.30)$$

(with $\Delta t = 1$). The corresponding difference equation is

$$y(k) = 1.22 y(k-1) - 0.121 u(k-1) \qquad (26.31)$$

(b) The given controller equation,

$$u(k) = u(k-1) - 1.2 K \epsilon(k) + K \epsilon(k-1)$$

may be rewritten as:

$$\Delta u(k) = -1.2 K \epsilon(k) + K \epsilon(k-1) \qquad (26.32)$$

Advance Eq (26.31) in time by one step to obtain

$$y(k+1) = 1.22 y(k) - 0.121 u(k) \qquad (26.33)$$

and subtract (26.31) from it to obtain

$$\Delta y(k+1) = 1.22 \Delta y(k) - 0.121 \Delta u(k) \qquad (26.34)$$

where $\Delta y(k+1) = y(k+1) - y(k)$; $\Delta u(k) = u(k) - u(k-1)$.
Now introduce the controller equation (26.32) into (26.34) and obtain:

$$\Delta y(k+1) = 1.22\,\Delta y(k) + 0.1452\,K\,\epsilon(k) - 0.121\,K\,\epsilon(k-1)$$

or

$$y(k+1) = 2.22\,y(k) - 1.22\,y(k-1) + 0.1452\,K\epsilon(k) - 0.121\,K\epsilon(k-1)$$

Now introduce $\epsilon(k) = y_d(k) - y(k)$, assuming $y_d(k) = y_d$ for all k (i.e. y_d is constant), and obtain

$$y(k+1) = (2.22 - 0.1452\,K)\,y(k) + (0.121\,K - 1.22)\,y(k-1) + 0.0242\,K\,y_d$$

as the required closed loop difference equation relating y to y_d. This may also be "stepped-back" one step in discrete time to give

$$y(k) = (2.22 - 0.1452\,K)\,y(k-1) + (0.121\,K - 1.22)\,y(k-2) + 0.0242\,K\,y_d \qquad (26.35)$$

The characteristic equation from which the stability of this closed loop system is to be determined is easily obtained from here as:

$$z^2 - (2.22 - 0.1452K) - (0.121K - 1.22) = 0$$

or

$$P(z) = z^2 + (0.1452K - 2.22) + (1.22 - 0.121K) = 0 \qquad (26.36)$$

We may now employ Jury's test on Eq. (26.36). The conditions to be satisfied for stability are as follows:

• COND^N 1

$$|1.22 - 0.121K| < 1 \qquad (26.37)$$

• COND^N 2

$$P(1) > 0$$

$$\Rightarrow \quad 0.242K > 0 \quad \Rightarrow \quad \boxed{K > 0} \qquad (26.38)$$

• COND^N 3

$P(-1) > 0$, or

$$4.44 - 0.2662K > 0 \Rightarrow \boxed{K < 16.679} \qquad (26.39)$$

Observe that conditions 2 and 3 are explicit, but not condition 1. For the explicit conditions implied in Eq(26.37) proceed as follows:

Because of the absolute value sign, we have two possibilities:

Either (a) $a_2 = 1.22 - 0.121K$ is positive, or

(b) $a_2 = 1.22 - 0.121K$ is negative

(a). If a_2 is positive, in which case

$$1.22 - 0.121K > 0$$

$$\Rightarrow \quad K < 10.08 \qquad (26.40a)$$

then Eq (26.37) implies that

$$1.22 - 0.121K < 1$$

or that $K > 1.82$ (26.40b)

Under these circumstances, combining (26.40a,b) yields the first half of the condition: i.e.

$$\boxed{1.82 < K < 10.08} \qquad (26.41)$$

(b) if a_2 is <u>negative</u>, in which case

$$1.22 - 0.121K < 0$$

$$\Rightarrow \quad K > 10.08 \qquad (26.42a)$$

then Eq (26.37) implies that

$$-1.22 + 0.121K < 1$$

or that $$K < 18.3 \qquad (26.42b)$$

Under these circumstances, combining (26.42a, b) yields the last half of the condition: i.e.

$$\boxed{10.08 < K < 18.3} \qquad (26.43)$$

Thus, condition 1 explicitly requires the <u>union</u> of the two sets indicated in Equs (26.41) and (26.43); i.e.

$$\boxed{1.82 < K < 18.3} \qquad (26.44).$$

The Jury test will be <u>violated</u> if any one of the 3 conditions in (26.44), (26.39), (26.38) is violated. Thus the <u>intersection</u> of these 3 sets is the set that guarantees that none of the 3 conditions will be violated. Thus, for stability, K must lie in the set:

$$\boxed{1.82 < K < 16.679} \qquad (26.45)$$

This is the range of acceptable values for K if the system is to remain stable.

As a check, investigate P(z) in Eq(26.36) for K = 1.8

(just outside the lower limit), $K = 17$, (just outside the upper limit).

For $K = 1.8$,

$$P(z) = z^2 - 1.95864z + 1.0022 = 0 \qquad (26.46)$$

and condition 1 is not satisfied $(1.0022 \nless 1)$;

and for $K = 17$,

$$P(z) = z^2 + 0.2484 - 0.837 = 0 \qquad (26.47)$$

and while conditions 1 and 2 are satisfied, condition 3 is not $[P(-1) = -0.0854 \ngtr 0]$.

Also, check with $K = 1.8$, and $K = 16$, just inside the range & confirm that the conditions are met.

26.12 Solution not provided.

CPSIA information can be obtained
at www.ICGtesting.com
Printed in the USA
BVOW04s1914261017
498757BV00007B/148/P

9 780195 119374